Monographs in Theoretical Computer Science

An EATCS Series

Monographs in Theoretical Computer Science

An EATCS Series

Editors: W. Brauer J. Hromkovič G. Rozenberg A. Salomaa

On behalf of the European Association for Theoretical Computer Science (EATCS)

Advisory Board:

G. Ausiello M. Broy C.S. Calude A. Condon D. Harel J. Hartmanis T. Henzinger T. Leighton M. Nivat C. Papadimitriou D. Scott

Ernst-Erich Doberkat

Stochastic Coalgebraic Logic

Springer

Author
Prof. Dr. Ernst-Erich Doberkat
Lehrstuhl für Software-Technologie
Fakultät für Informatik
Technische Universität Dortmund
Germany
ernst-erich.doberkat@udo.edu

Series Editors
Prof. Dr. Wilfried Brauer
Institut für Informatik der TUM
Boltzmannstr. 3
85748 Garching, Germany
brauer@informatik.tu-muenchen.de

Prof. Dr. Grzegorz Rozenberg
Leiden Institute of Advanced
Computer Science
University of Leiden
Niels Bohrweg 1
2333 CA Leiden, The Netherlands
rozenber@liacs.nl

Prof. Dr. Juraj Hromkovič
ETH Zentrum
Department of Computer Science
Swiss Federal Institute of Technology
8092 Zürich, Switzerland
juraj.hromkovic@inf.ethz.ch

Prof. Dr. Arto Salomaa
Turku Centre of Computer Science
Lemminkäisenkatu 14 A
20520 Turku, Finland
asalomaa@utu.fi

ISSN 1431-2654
ISBN 978-3-642-26147-3 e-ISBN 978-3-642-02995-0
DOI 10.1007/978-3-642-02995-0
Springer Heidelberg Dordrecht London New York

ACM Computing Classification (1998): F.4.1, G.3

Cover design: KünkelLopka GmbH, Heidelberg

Printed on acid-free paper

Springer is part of Springer Science+Business Media (www.springer.com)

… und für Lotta.

Preface

Motivation

Modal logics are usually interpreted through Kripke models, branching logics find their interpretation through models which deal with infinite paths. These seemingly structurally different interpretations can be unified by considering coalgebras which model the underlying worlds suitably; the predicates through which the formulas are represented in their interpretation are modelled using natural transformations between functors, to which the functor that underlies the coalgebra contributes. The basic functor is usually based on the power set functor. Adopting this general approach, we see that a fairly general and uniform way of interpreting modal logics and their step twins arises through coalgebras and the generalization of predicates into suitable natural transformations. We will show in this treatise that coalgebras based on the subprobability functor are amenable to these ideas as well.

Having arrived at this general approach of interpreting a rather broad family of logics, the question of comparing different models relative to a given logic presents itself. So we ask (and answer) the question about the conditions under which the well investigated relationship of logical equivalence, bisimilarity, and behavioral equivalence holds in this generalized, uniform scenario.

This is one of two driving topics in this book. The other one notes that stochastic interpretations — such as the ones indicated above — rest on stochastic relations, which in turn are the Kleisli morphisms for the Giry monad. The morphisms considered so far are based on measurable maps which are the morphisms of the base category. But at least bisimilarity and behavioral equivalence are formulated through morphisms, viz., through the existence of a span or a cospan, respectively. These formulations apply verbatim to Kleisli morphisms as well. Generalizing logical equivalence to distributional equivalence lifts the entire stage to the level of the Kleisli category, and, again, the problem of the relationship of the various behavioral descriptions presents itself. Since morphisms and congruences are very closely related, a study of these issues needs to be accompanied by a careful investigation of congruences on the Kleisli category.

This is what you will find in this book.

Overview

A brief survey of the contents of the individual chapters is in order.

Chapter 1

Stochastic relations form the mathematical basis for a probabilistic interpretation of coalgebraic logics. They provide also a foundation for Markov transition systems. These relations in turn are based on transition probabilities, and because we do not confine ourselves to the probabilistic case but rather accept models in which mass vanishes, we base the theory on transition subprobabilities. Capturing the properties of these relations and applying them to the logics under consideration requires a substantial amount of Measure Theory when models are required that are based on spaces containing more than a finite or a countable number of elements. Thus we study measures on general measurable spaces, and, since these spaces are usually too general, we make a topological assumption and consider Polish spaces as well as spaces that arise from their images under measurable maps, i.e., analytic spaces. Analytic spaces find usually the right balance of generality (for representing the application) and manageability (for the model). They provide the stage on which our play is going to be performed.

Chapter 1 is devoted to giving an overview of Borel sets and measures on analytic spaces, as far as this theory is needed; the purpose of this chapter is to provide a self-contained introduction to these topics. In contrast to Chapter 1 in the present author's book on stochastic relation [20], however, not all proofs are given, and whenever necessary we provide a pointer to that book's *Gentle Tutorial to All Things Considered*, where an attempt has been made to collect useful topics from the somewhat scattered literature. Nevertheless, the reader finds all necessary information on the topological and measure-theoretic base of what is going to be discussed in the latter chapters.

Because we will deal extensively with equivalence relations that are countably generated, and with the lifting of these relations to the space of all subprobabilities over the base space, we devote two sections to their study and consider this an investment which pays off largely in the later chapters.

Oh, yes: we assume a very basic familiarity with the theory of categories; this is assembled as well in this chapter (in fact, the reader is welcomed with a bunch of definitions from categories). Coalgebras together with the fundamental notions of behavioral equivalence and bisimilarity are discussed here as well; although in most applications to follow a detailed definition is given, the coalgebraic definitions provided here give a frame of reference. More advanced concepts of categories like monads are introduced whenever we need them; in this case, monads are not discussed before Section 3.2.

Chapter 2

The probabilistic interpretation of some fairly well understood logics is discussed in this chapter. We first consider bisimilar stochastic relations and give a criterion for bisimilarity; this discussion is fairly basic also for the more coalgebraically oriented logics that follow later, so we discuss this concept carefully and with a view towards generalizations. It is applied to modal logics and to continuous time stochastic logics. These logics have been studied already [20]. We deal with them nevertheless in the present treatise in the context of general coalgebraic logic. The reason for this is that we feel that it is difficult to appreciate the step from modal logics to coalgebraic logics without an understanding of these approaches for well understood logics. It becomes only then visible where the interesting and critical points are. This applies in particular to the interplay of the functor and predicate liftings, the latter being the constructs replacing modal operators. The similarities of and the differences between interpretations of modal logics and continuous time stochastic logics are emphasized, and we encounter these topics again when discussing coalgebraic logics.

After all these interpretations the patient reader is given a vacation from analytic spaces: we show that Kripke models for a simple Hennessy-Milner logic are logical equivalent if and only if they are behavioral equivalent without imposing any topological assumptions. This is just an application of the techniques developed so far, but care has to be exercised, since working with just these topological assumptions is fairly tempting. A comparison of the general measurable setting against the analytic scenario shows that some phenomena which coincide in the analytic case display a fairly wide gap in the general measurable case, albeit the basic technical ideas are the same.

Given that Chapters 1 and 2 provide the fundamental tools, the discussion now splits into two branches, both of which are generalizations of interpretations for modal logics. The first one generalizes the state space, the second one generalizes the logic proper.

Chapter 3

A Kripke model discusses validity on the basis of states: it defines the validity of formulas for individual states. But this is sometimes not adequate. Consider a large economy: here the behavior of an individual customer is only interesting insofar as it influences larger entities, the coalitions. Hence the question arises how to deal with these larger entities adequately, and we propose comparing the behavior of subprobability distributions rather than states proper — two distributions behave in the same way with respect to the logic iff they assign the same probability to all formulas' extensions. This gives rise to new ways of comparing the behavior of Kripke models, viz., on the distributional level rather than on the level of individual states.

The discussion in the previous chapter has shown that these questions are answered through an investigation of morphisms. Hence we need custom-tailored morphisms for this purpose. They are provided through the Kleisli category of the associated monad (the Giry monad), and we develop a theory of morphisms in this particular Kleisli category. Actually, we do a bit more than we need to, and discuss factoring for these morphisms as well, but this does not hurt (on the contrary, it provides some insight into the inner workings of these morphisms).

An investigation of the different ways for comparing the behavior of Kripke models is in order, and we show how the different ways relate to each other. An important tool in these discussions is provided by ergodic morphisms; we borrow the notion of ergodicity from the theory of dynamical systems for the purpose of modelling a very close relationship of Kleisli morphisms to the logic via the equivalence relations which they induce.

Chapter 4

Modal logics are characterized by their modal operators. Take for example the logic that has the diamond as a modal operator in addition to the usual Boolean operations. Then an interpreting Kripke model $\mathcal{R} = (S, R)$ with state space S and relation $R \subseteq S \times S$ has $\mathcal{R}, s \models \Diamond\phi$ iff we can find a successor t to s that satisfies ϕ, hence iff $R(s) \cap [\![\phi]\!] \neq \emptyset$. Reformulating the latter condition says $s \in \bigl(R^{-1} \circ \lambda_S\bigr)([\![\phi]\!])$, where $\lambda_S(A) := \{B \subseteq S \mid B \cap A \neq \emptyset\}$. Then λ is a natural transformation for a power set functor, and $\mathcal{R}$ is perceived as a coalgebra for this functor with relation R as its dynamics.

Hence a modal operator is viewed as a natural transformation, and the transition structure in the Kripke model gives rise to a coalgebra. The required natural transformations will have some additional properties; they are called predicate liftings. This is basically the theme which will be developed for stochastic coalgebraic logics in this Chapter.

To this end we will look at two structurally different scenarios by developing the theory separately for left and for right coalgebras. A left coalgebra for functor $\mathfrak{F}$ is a coalgebra for the functor $\mathfrak{S} \circ \mathfrak{F}$. These coalgebras correspond to A. Sokolova's generative systems; in contrast, a right coalgebra for functor $\mathfrak{F}$ is a coalgebra for the functor $\mathfrak{F} \circ \mathfrak{S}$, corresponding to reactive systems in Sokolova's taxonomy. It is shown that modal logics can be modelled through left coalgebras (this applies to, e.g., continuous time stochastic logics), and that Markov transition systems are special cases of right coalgebras. We investigate the corresponding logics for both kinds of coalgebras.

For left coalgebras the subprobability functor is the dominating one, and it turns out that the techniques for dealing with modal logics developed so far from stochastic relations may be applied, albeit with a grain of salt. For right coalgebras, however, the dominant functor is $\mathfrak{F}$, and here some new techniques have to be found, in particular the selection of morphisms through suitable selection theorems from the theory of Borel sets becomes important.

Interestingly, the resulting characterization of logical equivalence, bisimilarity, and behavioral equivalence, which is formulated on a global level (i.e., for entire coalgebras) can be made to work for a local characterization of these properties (i.e., for individual states). This yields also a new characterization of bisimilar states in a Markov transition system.

The graph below indicates the dependencies among the sections.

Acknowledgements

Christoph Schubert convinced me to tackle the problems at hand with at least a small load of categories under the belt. I want to thank him for more than a few discussions and suggestions. The Deutsche Forschungsgemeinschaft (DFG) supported the investigations into the algebraic properties of stochastic relations for four years (*Algebraische Eigenschaften stochastischer Relationen*, grant DO-263/8-1/2). This is gratefully acknowledged.

The author is also grateful to Alberto Casagrande, Alexander Kurz, Carla Piazza, Chunlai Zhou, Dieter Pumplün, Eugenio Omodeo, Giovanna d'Agostino, Ingo Battenfeld, Johannes Köster, Klaus Keimel, Larry Moss, Lutz Schröder, Peter Buchholz, Peter Padawitz, Vincent Danos, and Yixiang Chen for discussions related in one way or other to the topics discussed here; the anonymous reviewer also made some insightful comments. Ingo Battenfeld and in particular Christoph Schubert offered numerous suggestions for improving the representation.

Special help was provided by Alla Stankjawitschene, Helmut Henning gave some hints when I needed to write my own indexing programs, and, again, Stefan Dissmann's organizational skills made many things easier for me. Petra

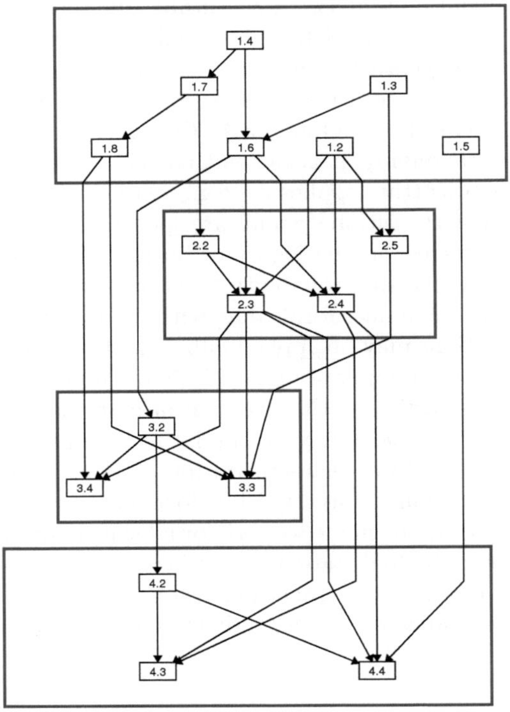

Mutzel provided the drawing of the dependency graph. The cooperation with Ronan Nugent from Springer-Verlag was a pleasure. The author is grateful for all this help.

Bochum and Dortmund, June 2009 Ernst-Erich Doberkat

Contents

Chapter 1
Borel Sets, Measures and All That

1.1 Introduction

Stochastic coalgebraic logic works in a general probabilistic setting, in contrast to some more traditionally oriented approaches that use finite or countably infinite state spaces. Since the universe of discourse is no longer finite, one needs to look at properties of general measurable spaces; these spaces have a rich structure which needs to be exploited, so we will deal with measurable sets and maps, and with σ-algebras and their generators. Here the π-λ-Theorem is of particular prominence. But sometimes a measurable structure is not rich enough for some properties, so we will specialize this structure to the Borel sets of a topological space. We will quickly see that it is the class of Polish spaces which attracts most attention, together with the class of analytic spaces, i.e., the continuous images of Polish spaces. We give the basic properties of these spaces in Sections 1.3 and 1.4. In some situations we will need the possibility of selecting objects in a measurable way; this leads to investigating measurable selectors in Section 1.5. Then we talk about measures, specifically about subprobability measures. Together with Borel sets, they yield the most important tools in our investigations. The space of all subprobability measures is introduced in Section 1.6 together with some important properties (which will be pursued further when investigating the categorical structure in Chapter 3), and it is shown that the subprobabilities over a Polish space form themselves a Polish space with the weak-$*$-σ-algebra, i.e., the σ-algebra that is induced on it by the bounded continuous functions. We briefly mention the Prohorov metric and the Huntingdon metric, two metrics that have been used for metrizing this topology. When discussing modal or coalgebraic logics, we will want to know whether or not worlds can be separated through a formula, which means that there exists a formula that is satisfied by one state and not by the other. This leads to equivalence relations that have the particularly pleasant property of being countably generated (we call these equivalence relations smooth).

E.-E. Doberkat, *Stochastic Coalgebraic Logic*, Monographs in Theoretical Computer Science. An EATCS Series, DOI 10.1007/978-3-642-02995-0_1,

These relations have some properties that make them interesting from our point of view, in particular they make sure that factoring does not leave the realm of analytic spaces, so we study them. This happens in Section 1.7. When one deals with smooth relations, the invariant Borel sets determine the factor structure; using this σ-algebra, one can define a derived equivalence relation on the set of all subprobability measures, which turns out to be smooth again. The interrelationship between smooth relations on the base space and on the subprobabilities is studied in Section 1.8. The class of near-grounded relations is studied a bit closer (because they are interesting when factoring through weak morphisms), Souslin's Theorem helps us to establish some helpful isomorphisms.

The results on measurable, Polish, and analytic spaces are nearly all folklore in Measure Theory; a collection of results in this area that did turn out to be helpful in the study of Markov transition systems through stochastic relations is contained in [20, Chapter 1]. We borrow freely from there and point the reader to the proofs given there or in the original literature, of which a short survey is given in the Bibliographic Notes in Section 2.5. From time to time we do give a proof, mainly in order to convey the flavor of the argumentation. Since the results of Section 1.8 did not yet appear in book form, we provide the complete proofs here.

Before we start a discussion of measure-theoretic constructions, we have a brief glimpse at categories and functors. They will be used occasionally in this book. We assume a basic knowledge of categories and refer the reader for further information to [57, 11, 12, 4, 68].

1.2 Categories and Functors

Let $\mathbf{C}$ be a category, then the hom set $\mathbf{C}(a, b)$ contains all $\mathbf{C}$-morphisms $a \to b$. Functors $\mathbf{C} \to \mathbf{D}$ are covariant, unless otherwise stated. A *natural transformation* $\eta : \mathfrak{F} \stackrel{\bullet}{\to} \mathfrak{G}$ for the functors $\mathfrak{F}, \mathfrak{G} : \mathbf{C} \to \mathbf{D}$ is a family of $\mathbf{D}$-morphisms $\eta_a : \mathfrak{F}(a) \to \mathfrak{G}(a)$ indexed by the objects a of $\mathbf{C}$ such that for every $\mathbf{C}$-morphism $f : a \to b$ this diagram commutes:

$$\begin{array}{ccc} \mathfrak{F}(a) & \xrightarrow{\eta_a} & \mathfrak{G}(a) \\ {\scriptstyle \mathfrak{F}(f)}\downarrow & & \downarrow{\scriptstyle \mathfrak{G}(f)} \\ \mathfrak{F}(b) & \xrightarrow[\eta_b]{} & \mathfrak{G}(b) \end{array}$$

Recall products and coproducts in a category. For a family $(x_j)_{j\in J}$ of objects in a category $\mathbf{C}$ the object p together with morphisms $\pi_j : p \to x_j$ is the *product* of $(x_j)_{j\in J}$, denoted by $\prod_{j\in J} x_j$, iff, given an arbitrary object q in $\mathbf{C}$ and morphisms $\zeta_j : q \to x_j$, there exists a unique morphism $\varpi : q \to p$

with $\zeta_j = \pi_j \circ \varpi$ for all $j \in J$. Dually, the object c together with morphisms $i_j : x_j \to c$ is the *coproduct* (or the *sum*) of $(x_j)_{j \in J}$, denoted by $\coprod_{j \in J} x_j$, iff, given an arbitrary object d of $\mathbf{C}$ and morphisms $k_j : x_j \to d$, there exists a unique morphism $\ell : c \to d$ with $k_j = \ell \circ i_j$ for all $j \in J$.

Pullbacks

Given a pair $a \xrightarrow{f} c \xleftarrow{g} b$ of morphisms in $\mathbf{C}$, an object p together with morphisms $a \xleftarrow{k} p \xrightarrow{\ell} b$ is called a *semi-pullback* iff this diagram commutes:

$$\begin{array}{ccc} p & \xrightarrow{\ell} & b \\ {\scriptstyle k}\downarrow & & \downarrow{\scriptstyle g} \\ a & \xrightarrow[f]{} & c \end{array}$$

A semi-pullback p, k, ℓ is called a *weak pullback* iff, given an object q and morphisms $a \xleftarrow{t} q \xrightarrow{s} b$ with $f \circ t = g \circ s$, there exists a morphism $w : q \to p$ such that $t = k \circ w$ and $s = \ell \circ w$. It is called a *pullback* iff the morphism w is uniquely determined.

Coalgebras

Let $\mathfrak{F} : \mathbf{C} \to \mathbf{C}$ be an endofunctor on the category $\mathbf{C}$; then (a, γ) is called an $\mathfrak{F}$-*coalgebra* iff a is an object in $\mathbf{C}$, and if $\gamma : a \to \mathfrak{F}(a)$ is a $\mathbf{C}$-morphism. The morphism γ is sometimes called the *dynamics* or the *transition law* of the coalgebra. A $\mathbf{C}$-morphism $f : a \to a'$ is a *coalgebra morphism* $(a, \gamma) \to (a', \gamma')$ iff $\gamma' \circ f = \mathfrak{F}(f) \circ \gamma$, so that this diagram commutes:

$$\begin{array}{ccc} a & \xrightarrow{f} & a' \\ {\scriptstyle \gamma}\downarrow & & \downarrow{\scriptstyle \gamma'} \\ \mathfrak{F}(a) & \xrightarrow[\mathfrak{F}(f)]{} & \mathfrak{F}(a') \end{array}$$

1.2.1 Bisimilarity and Behavioral Equivalence

The coalgebras (a, γ) and (a', γ') are called *bisimilar* iff there exists a span of morphisms, i.e., there exists a coalgebra (a'', γ'') and morphisms

$$(a,\gamma) \xleftarrow{f} (a'',\gamma'') \xrightarrow{f'} (a',\gamma')\ .$$

The coalgebra (a'',γ'') in the middle is sometimes called *mediating*. Similarly, call the coalgebras (a,γ) and (a',γ') *behaviorally equivalent* iff there exists a cospan of morphisms, i.e., there exists a coalgebra (a'',γ'') and morphisms

$$(a,\gamma) \xrightarrow{f} (a'',\gamma'') \xleftarrow{f'} (a',\gamma')\ .$$

The significance of these names will become clear in due course.

1.2.2 Yoneda's Lemma

The *hom functor* $\mathbf{C}(a,-) : \mathbf{C} \to \mathbf{Set}$ is covariant with a contravariant step twin $\mathbf{C}(-,b) : \mathbf{C} \to \mathbf{Set}$. Each morphism $f : y \to x$ defines a natural transformation $Y(f) : \mathbf{C}(x,-) \xrightarrow{\bullet} \mathbf{C}(y,-)$ upon setting $Y(f)_a : \mathbf{C}(x,a) \ni g \mapsto g \circ f \in \mathbf{C}(y,a)$. The famous Yoneda Lemma [4, Corollary 4.5.4] states

Lemma 1.2.1. *Every natural transformation $\mathbf{C}(x,-) \xrightarrow{\bullet} \mathbf{C}(y,-)$ is given by the composition with a unique morphism $x \to y$. The natural transformation is an isomorphism iff the corresponding morphism is an isomorphism.* □

1.3 σ-Algebras and Their Generators

A measurable space $(M,\mathcal{M})$ consists of a set M with a σ-algebra $\mathcal{M}$, which is a Boolean algebra of subsets of M that is closed under countable unions (hence countable intersections or countable disjoint unions). The elements of M are usually referred to as *measurable sets.* If $\mathcal{M}_0$ is a family of subsets of M, then

$$\sigma(\mathcal{M}_0) = \bigcap\{\mathcal{M} \mid \mathcal{M} \text{ is a } \sigma\text{-algebra on } M \text{ with } \mathcal{M}_0 \subseteq \mathcal{M}\}$$

is the smallest σ-algebra on M which contains $\mathcal{M}_0$; $\mathcal{M}_0$ is called a *generator* for $\sigma(\mathcal{M}_0)$. This construction works since the power set $\mathcal{P}(M)$ is a σ-algebra on M. Take for example as a generator $\mathcal{I}$ all open intervals in the real numbers $\mathbb{R}$; then $\sigma(\mathcal{I}) =: \mathcal{B}(\mathbb{R})$ is the σ-algebra of real *Borel sets*. We will encounter the Borel sets again in Section 1.4.

π-λ-Theorem

An important tool is the π-λ-Theorem, which makes it sometimes simpler to identify the σ-algebra generated from some family of sets. It hides a proof principle that is applied quite often in this area. The principle is akin to mathematical induction: one shows that a property holds for a generator $\mathcal{P}$ of a σ-algebra which is closed under finite intersections, and then one shows that the sets for which a property holds is closed under complements and countable disjoint unions. Then one may conclude that all elements of $\sigma(\mathcal{P})$ enjoy this property. We discuss the proof in detail because it conveys the flavor of the argumentation.

Theorem 1.3.1. *(π-λ-**Theorem**) Let $\mathcal{P}$ be a family of subsets of a set X that is closed under finite intersections (a* π-class*). Then $\sigma(\mathcal{P})$ is the smallest λ-class containing $\mathcal{P}$, where a family $\mathcal{L}$ of subsets of X is called a* λ-class *iff it is closed under complements and countable disjoint unions.*

Proof 1. Let $\mathcal{L}$ be the smallest λ-class containing $\mathcal{P}$; then we show that $\mathcal{L}$ is a σ-algebra.

2. We show first that it is an algebra. Being a λ-class, $\mathcal{L}$ is closed under complementation. Let $A \subseteq X$, then $\mathcal{L}_A := \{B \subseteq X \mid A \cap B \in \mathcal{L}\}$ is a λ-class again: if $A \cap B \in \mathcal{L}$, then

$$A \cap (X \setminus B) = A \setminus B = X \setminus ((A \cap B) \cup (X \setminus A)),$$

which is in $\mathcal{L}$, since $(A \cap B) \cap X \setminus A = \emptyset$, and since $\mathcal{L}$ is closed under disjoint unions.

If $A \in \mathcal{P}$, then $\mathcal{P} \subseteq \mathcal{L}_A$, because $\mathcal{P}$ is closed under intersections. Because $\mathcal{L}_A$ is a λ-system, this implies $\mathcal{L} \subseteq \mathcal{L}_A$ for all $A \in \mathcal{P}$. Now take $B \in \mathcal{L}$, then the preceding argument shows that $\mathcal{P} \subseteq \mathcal{L}_B$, and again we may conclude that $\mathcal{L} \subseteq \mathcal{L}_B$. Thus we have shown that $A \cap B \in \mathcal{L}$, provided $A, B \in \mathcal{L}$, so that $\mathcal{L}$ is closed under finite intersections. Thus $\mathcal{L}$ is a Boolean algebra.

3. $\mathcal{L}$ is a σ-algebra as well. It is enough to show that $\mathcal{L}$ is closed under countable unions. But since

$$\bigcup_{n \in \mathbb{N}} A_n = \bigcup_{n \in \mathbb{N}} \left(A_n \setminus \bigcup_{i=1}^{n-1} A_i \right),$$

this follows immediately. □

Measurable Maps

If $(N, \mathcal{N})$ is a measurable space, then a map $f : M \to N$ is called $\mathcal{M}$-$\mathcal{N}$-*measurable* iff the inverse image under f of each set in $\mathcal{N}$ is a member of $\mathcal{M}$, hence iff $f^{-1}[G] \in \mathcal{M}$ holds for all $G \in \mathcal{N}$.

Checking measurability is made easier by the observation that it suffices for the inverse images of a generator to be measurable sets.

Lemma 1.3.2. *Let* $(M, \mathcal{M})$ *and* $(N, \mathcal{N})$ *be measurable spaces, and assume that* $\mathcal{N} = \sigma(\mathcal{N}_0)$ *is generated by a family* $\mathcal{N}_0$ *of subsets of* N*. Then* $f : M \to N$ *is* $\mathcal{M}$-$\mathcal{N}$*-measurable iff* $f^{-1}[G] \in \mathcal{M}$ *holds for all* $G \in \mathcal{N}_0$*.*

Proof Clearly, if f is $\mathcal{M}$-$\mathcal{N}$-measurable, then $f^{-1}[G] \in \mathcal{M}$ holds for all $G \in \mathcal{N}_0$.

Conversely, suppose $f^{-1}[G] \in \mathcal{M}$ holds for all $G \in \mathcal{N}_0$. The set $\mathcal{G}$ for which the assertion is true,

$$\mathcal{G} := \{G \in \mathcal{N} \mid f^{-1}[G] \in \mathcal{M}\}$$

is a σ-algebra, so that $\sigma(\mathcal{G}) = \mathcal{G}$ holds. By assumption, $\mathcal{N}_0 \subseteq \mathcal{G}$, so that

$$\mathcal{N} = \sigma(\mathcal{N}_0) \subseteq \sigma(\mathcal{G}) = \mathcal{G} \subseteq \mathcal{N}$$

is inferred. Thus all elements of $\mathcal{N}$ have their inverse image in $\mathcal{M}$. □

An example is furnished by a real-valued function $f : M \to \mathbb{R}$ on M, which is $\mathcal{M}$-$\mathcal{B}(\mathbb{R})$-measurable iff $\{m \in M \mid f(m) \ltimes t\} \in \mathcal{M}$ holds for each $t \in \mathbb{R}$; the relation $\ltimes$ may be taken from $<, \leq, \geq, >$. This observation will be used frequently.

If $(M, \mathcal{M})$ is a measurable space and $f : M \to N$ is a map, then

$$\mathcal{N} := \{D \subseteq N \mid f^{-1}[D] \in \mathcal{M}\}$$

is the largest σ-algebra $\mathcal{N}_0$ on N that renders f $\mathcal{M}$-$\mathcal{N}_0$-measurable ($\mathcal{N}$ is the *final* σ-algebra w.r.t. f and $\mathcal{M}$). In fact, because the inverse set operator f^{-1} is compatible with the Boolean operations, it is immediate that $\mathcal{N}$ is closed under the operations for a σ-algebra, and a moment's reflection shows that this is also the largest σ-algebra with this property.

This is an intrinsic, universal characterization of the final σ-algebra for a single map.

Lemma 1.3.3. *Let* $(M, \mathcal{M})$ *be a measurable space and* $f : M \to N$ *be a map. The following conditions are equivalent:*

a. The σ*-algebra* $\mathcal{N}$ *on* N *is final with respect to* f*.*
b. If $(P, \mathcal{P})$ *is a measurable space, and* $g : N \to P$ *is a map, then the* $\mathcal{M}$-$\mathcal{P}$*-measurability of* $g \circ f$ *implies the* $\mathcal{N}$-$\mathcal{P}$*-measurability of* g*.* □

Thus, in the case of a final σ-algebra we may infer measurability of g from measurability of $g \circ f$, given the commutative diagram

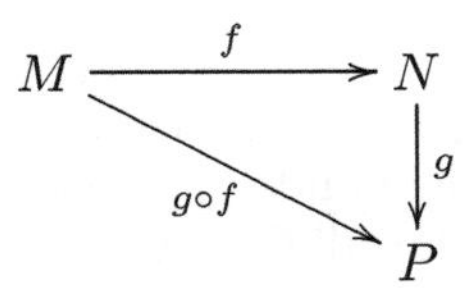

We will use the final σ-algebra mainly for factoring through an equivalence relation. In fact, let α be an equivalence relation on a set X, where $(X, \mathcal{M})$ is a measurable space. Then the factor map

$$\eta_\alpha : \begin{cases} X & \to X/\alpha \\ x & \mapsto [x]_\alpha \end{cases}$$

that maps each element to its class can be made measurable by taking the final σ-algebra $\mathcal{M}/\alpha$ with respect to η_α and $\mathcal{M}$ as the σ-algebra on X/α.

Symmetrically, if $g : P \to M$ is a map, then

$$g^{-1}[\mathcal{M}] := \{g^{-1}[E] \mid E \in \mathcal{M}\}$$

is the smallest σ-algebra $\mathcal{P}_0$ on P that renders g $\mathcal{P}_0$-$\mathcal{M}$ measurable (accordingly, $g^{-1}[\mathcal{M}]$ is called *initial* w.r.t. g and $\mathcal{M}$). Similarly, $g^{-1}[\mathcal{M}]$ is a σ-algebra, and it is fairly clear that this is the smallest one with the desired property. In particular, the inclusion $i_Q : Q \to M$ becomes measurable for a subset $Q \subseteq M$ when Q is endowed with the σ-algebra $\{Q \cap B \mid B \in \mathcal{M}\}$. It is called the *trace of* $\mathcal{M}$ *on* Q and is denoted sometimes — in a slight abuse of notation — by $\mathcal{M} \cap Q$.

Dually to Lemma 1.3.3, the initial σ-algebra is characterized.

Lemma 1.3.4. *Let $(N, \mathcal{N})$ be a measurable space and $f : M \to N$ be a map. The following conditions are equivalent:*

a. The σ-algebra $\mathcal{M}$ on M is initial with respect to f.
b. If $(P, \mathcal{P})$ is a measurable space, and $g : P \to M$ is a map, then the $\mathcal{P}$-$\mathcal{N}$-measurability of $f \circ g$ implies the $\mathcal{P}$-$\mathcal{M}$-measurability of g. □

Again dually, we infer for the initial σ-algebra the measurability of g from the measurability of $g \circ f$ in this commutative diagram:

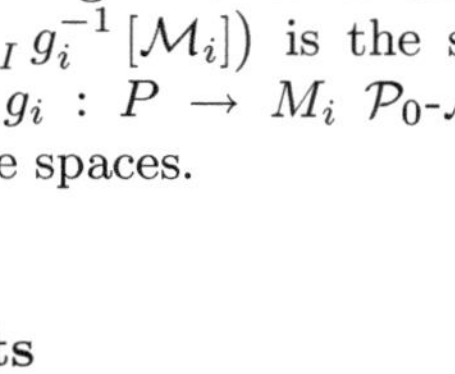

Initial and final σ-algebras generalize in an obvious way to families of maps. For example, $\sigma\left(\bigcup_{i \in I} g_i^{-1}[\mathcal{M}_i]\right)$ is the smallest σ-algebra $\mathcal{P}_0$ on P which makes all the maps $g_i : P \to M_i$ $\mathcal{P}_0$-$\mathcal{M}_i$-measurable for a family $\left((M_i, \mathcal{M}_i)\right)_{i \in I}$ of measurable spaces.

Products and Coproducts

Let $\left((M_i, \mathcal{M}_i)\right)_{i \in I}$ be a family of measurable spaces. The product-σ-algebra $\bigotimes_{i \in I} \mathcal{M}_i$ denotes the initial σ-algebra on $\prod_{i \in I} M_i$ for the projections

$$\pi_j : \langle m_i \mid i \in I \rangle \mapsto m_j.$$

It is not difficult to see that $\bigotimes_{i \in I} \mathcal{M}_i = \sigma(\mathcal{Z})$ with

$$\mathcal{Z} := \{\prod_{i \in I} E_i \mid \forall i \in I : E_i \in \mathcal{M}_i, E_i = M_i \text{ for almost all indices}\}$$

as the collection of *cylinder sets* (use Theorem 1.3.1 and the observation that $\mathcal{Z}$ is closed under intersection).

For $I = \{1, 2\}$, the σ-algebra $\mathcal{M}_1 \otimes \mathcal{M}_2$ is generated from the set of *measurable rectangles*

$$\{E_1 \times E_2 \mid E_1 \in \mathcal{M}_1, E_2 \in \mathcal{M}_2\}.$$

Dually, the sum $(M_1 + M_2, \mathcal{M}_1 + \mathcal{M}_2)$ of the measurable spaces $(M_1, \mathcal{M}_1)$ and $(M_2, \mathcal{M}_2)$ is defined through the final σ-algebra on the sum $M_1 + M_2$ for the injections $M_i \to M_1 + M_2$. This is the special case of the coproduct $(M, \mathcal{M}) := \coprod_{i \in I}(M_i, \mathcal{M}_i)$, where the σ-algebra $\mathcal{M}$ is initial with respect to the injections.

Real-valued Functions

We need occasionally the representation of sets through indicator functions. Define for $A \subseteq N$ the *indicator function*

$$\chi_A(x) := \begin{cases} 1, & \text{if } x \in A, \\ 0, & \text{if } x \notin A. \end{cases}$$

Clearly, if $\mathcal{N}$ is a σ-algebra on N, then $A \in \mathcal{N}$ iff χ_A is a $\mathcal{N}$-$\mathcal{B}(\mathbb{R})$-measurable function. This is so since we have for the inverse image of an interval under χ_A

$$\chi_A^{-1}[[0, q]] = \begin{cases} \emptyset, & \text{if } q < 0, \\ N \setminus A, & \text{if } 0 \leq q < 1, \\ N, & \text{if } q \geq 1. \end{cases}$$

A measurable *step function*

$$f = \sum_{i=1}^{n} \alpha_i \cdot \chi_{A_i}$$

is a linear combination of indicator functions with $\alpha_i \in \mathbb{R}, A_i \in \mathcal{N}$. The following statement is most useful in measure theory [43, Chapter V], where it is used among others for the construction of the Lebesgue integral. It will come in quite handy in many situations when we have information about the behavior of a construction for measurable sets (i.e., for indicator functions), when the construction is linear, and when we can guarantee closedness under monotone convergence.

Proposition 1.3.5. *Denote for a measurable space* $(N, \mathcal{N})$ *by*

$$\mathcal{F}(N, \mathcal{N}) := \{f : N \to \mathbb{R} \mid f \text{ is } \mathcal{N} - \mathcal{B}(\mathbb{R}) \text{ measurable and bounded}\}$$

the linear space of all bounded measurable real functions on N. *Then*

a. For $f \in \mathcal{F}(N, \mathcal{N})$ *with* $f \geq 0$ *there exists an increasing sequence* $(f_n)_{n\in\mathbb{N}}$ *of step functions* $f_n \in \mathcal{F}(N, \mathcal{N})$ *with* $f(x) = \sup_{n\in\mathbb{N}} f_n(x)$ *for all* $x \in X$.
b. For $f \in \mathcal{F}(N, \mathcal{N})$ *there exists a sequence* $(f_n)_{n\in\mathbb{N}}$ *of step functions* $f_n \in \mathcal{F}(N, \mathcal{N})$ *with* $f(x) = \lim_{n\to\infty} f_n(x)$ *for all* $x \in X$. □

Convention

Measurability of real-valued functions always means measurability with respect to the Borel sets $\mathcal{B}(\mathbb{R})$ of the real numbers, unless otherwise stated.

1.4 Polish and Analytic Spaces

General measurable spaces are sometimes too general for supporting specific structures. We deal with Polish and analytic spaces which are general enough to support interesting applications but have specific properties which help establish vital properties. We remind the reader first of some basic facts and provide then some helpful tools for working with Polish spaces, and their more general cousins, analytic spaces.

Topological spaces

A *topology* $\mathcal{T}$ on a set X is a family of subsets that is closed under finite intersections and arbitrary unions, and that contains the empty set and the entire set X; the pair $(X, \mathcal{T})$ is called a *topological space*. The elements of $\mathcal{T}$ are called the *open sets*; their complements are called *closed sets*. The space $(X, \mathcal{T})$ is called a *Hausdorff space* iff two distinct points can be separated by disjoint open sets. Thus, given $x \neq y$, there exist disjoint open sets U, V with $x \in U, y \in V$.

A family $\mathcal{B}$ of open subsets of X is called a *base* for the topology $\mathcal{T}$ iff each element of $\mathcal{T}$ can be represented as the union of elements of $\mathcal{B}$. This is equivalent to saying that $\bigcup\{B \mid B \in \mathcal{B}\} = X$, and that we can find for each $x \in B_1 \cap B_2$ with $B_1, B_2 \in \mathcal{B}$ an element $B_3 \in \mathcal{B}$ with $x \in B_3 \subseteq B_1 \cap B_2$. A *subbase* $\mathcal{S}$ for $\mathcal{T}$ has the property that the set $\{\bigcap \mathcal{F} \mid \mathcal{F} \subseteq \mathcal{S} \text{ finite}\}$ of finite intersections of elements of $\mathcal{S}$ forms a base for $\mathcal{T}$.

Given another topological space $(Y, \mathcal{S})$, a map $f : X \to Y$ is called $\mathcal{T}$-$\mathcal{S}$-*continuous* iff the inverse image of an open set from Y is open in X again, i.e., iff $f^{-1}[\mathcal{S}] \subseteq \mathcal{T}$. The topological spaces $(X, \mathcal{T})$ and $(Y, \mathcal{S})$ are called *homeomorphic* iff there exists a $\mathcal{T}$-$\mathcal{S}$-continuous bijection $f : X \to Y$ the inverse of which is $\mathcal{S}$-$\mathcal{T}$-continuous.

Proceeding in analogy to measurable spaces, a topology $\mathcal{T}$ on a set X is called *initial* for a map $f : X \to Y$ with a topological space $(Y, \mathcal{S})$ iff $\mathcal{T}$ is the smallest topology $\mathcal{T}_0$ on X rendering f a $\mathcal{T}_0$-$\mathcal{S}$-continuous map. The characterization of initial resp. final topologies are similar to those for their measurable cousins (just replace *measurable* by *continuous*), see Lemma 1.3.3 resp. Lemma 1.3.4. For example, if $Y \subseteq X$ is a subset, then the topological subspace $(Y, \{Y \cap G \mid G \in \mathcal{T}\})$ is just the initial topology with respect to the inclusion map $i_Y : Y \to X$.

Dually, if $(X, \mathcal{T})$ is a topological space and $f : X \to Y$ is a map, then the *final topology* $\mathcal{S}$ on Y is the largest topology $\mathcal{S}_0$ on Y making f $\mathcal{T}$-$\mathcal{S}_0$-continuous. Both initial and final topologies generalize to families of spaces and maps.

The *topological product* $\prod_{i\in I}(X_i, \mathcal{T}_i)$ of the topological spaces $((X_i, \mathcal{T}_i))_{i\in I}$ is the Cartesian product $\prod_{i\in I} X_i$ endowed with the initial topology with respect to the projections, and the *topological sum* $\coprod_{i\in I}(X_i, \mathcal{T}_i)$ of the topological spaces $((X_i, \mathcal{T}_i))_{i\in I}$ is the direct sum $\coprod_{i\in I} X_i$ endowed with the final topology with respect to the injections.

Borel sets

Given a topological space $(X, \mathcal{T})$, a measurable structure comes for free: denote by $\mathcal{B}(X, \mathcal{T})$ the smallest σ-algebra on X that contains the open sets, so that $\mathcal{B}(X, \mathcal{T}) = \sigma(\mathcal{T})$. These sets are called the *Borel sets* of $(X, \mathcal{T})$; measurability of maps with respect to the Borel sets is referred to as *Borel measurability*.

An immediate consequence of Lemma 1.3.2 is that continuity implies Borel measurability.

Lemma 1.4.1. *Let $(X, \mathcal{T})$ and $(Y, \mathcal{S})$ be topological spaces. Then $f : X \to Y$ is $\mathcal{B}(X, \mathcal{T})$-$\mathcal{B}(Y, \mathcal{S})$ measurable, provided f is $\mathcal{T}$-$\mathcal{S}$-continuous.* □

1.4.1 Metric Spaces

Metric spaces are particularly important topological spaces.

Definition 1.4.2. *A* metric *d on a set X is a map $d : X \times X \to \mathbb{R}_+$ such that for all $x, y, z \in X$*

a. $d(x, y) = 0 \Leftrightarrow x = y$,

b. $d(x, y) = d(y, x)$ *(symmetry),*
c. $d(x, z) \leq d(x, y) + d(y, z)$ *(triangle inequality)*

holds. The pair (X, d) *is called a* metric space.

Example 1.4.3. Let $X := \mathbb{R}^2$ be the Euclidean plane, and define

a. $d_1(\langle x_1, x_2\rangle, \langle y_1, y_2\rangle) := |x_1 - y_1| + |x_2 - y_2|$,
b. $d_2(\langle x_1, x_2\rangle, \langle y_1, y_2\rangle) := \left(|x_1 - y_1|^2 + |x_2 - y_2|^2\right)^{1/2}$,
c. $d_\infty(\langle x_1, x_2\rangle, \langle y_1, y_2\rangle) := \max\{|x_1 - y_1|, |x_2 - y_2|\}$.

Then these are metrics on X for the usual Euclidean topology. Each set M can be made into a metric space (M, d) upon setting

$$d(x, y) := \begin{cases} 1, & \text{if } x \neq y \\ 0, & \text{if } x = y \end{cases}$$

(d is called the *discrete metric*). ⊣

Given a metric space (X, d), define for $x \in X, r > 0$ the *open ball around* x *with radius* r

$$B_{r,d}(x) := B_r(x) := \{x' \in X \mid d(x, x') < r\}.$$

Example 1.4.4. If (M, d) is discrete, then

$$B_r(x) = \begin{cases} \{x\}, & \text{if } r \leq 1 \\ M, & \text{if } r > 1. \end{cases}$$

⊣

Metric topology

Call $G \subseteq X$ *open* iff given $x \in G$ there exists $r > 0$ such that the open ball around x with radius r is entirely contained in G, so that $B_r(x) \subseteq G$. Open balls are open sets: let $G := B_r(x)$ be the open ball around x, and take $y \in B_r(x)$, so that $s := r - d(x, y) > 0$. We claim that $B_s(y) \subseteq B_r(x)$. In fact, if $z \in B_s(y)$, then

$$d(x, z) \leq d(x, y) + d(y, z) < d(x, y) + r - d(x, y) = r$$

by the triangle inequality.

The open sets for a metric space form a topology (the *metric topology*); the topology is said to be metrized through the metric. Conversely, a topological space is called metrizable iff its topology comes from a metric; we sometimes talk simply about *metrizable spaces*. A metrizable topology is Hausdorff:

given $x \neq y$, we know that $s := d(x,y) > 0$, so balls with radius $s/3$ are open sets that separate these points. Continuity now takes this — probably more familiar — form:

Lemma 1.4.5. *Let (X_1, d_1) and (X_2, d_2) be metric spaces, and denote by $\mathcal{T}_i$ the topology on X_i induced by d_i. The following conditions are equivalent for a map $f : X_1 \to X_2$.*

a. f is $\mathcal{T}_1$-$\mathcal{T}_2$-continuous.
b. Given $x_1 \in X_1$ and $\varepsilon > 0$, there exists $\delta > 0$ such that

$$d_1(x_1, x_1') < \delta \Rightarrow d_2(f(x_1), f(x_1')) < \varepsilon. \ \square$$

Let (X, d) be metric. Define for $x \in X$ and for $A \subseteq X$ the distance $d(x, A)$ of x to A through

$$d(x, A) := \inf_{y \in A} d(x, y)$$

(with $(d(x, \emptyset) := 0)$. Thus we know that a point x has a distance to $A \neq \emptyset$ of less than $r > 0$ iff we can find $y \in A$ with $d(x, y) < r$. Because

$$|d(x, A) - d(y, A)| \leq d(x, y),$$

we know that $x \mapsto d(x, A)$ is uniformly continuous. If $F \subseteq X$ is closed, then $x \in F$ iff $d(x, F) = 0$ (take a sequence $(x_n)_{n \in \mathbb{N}}$ of elements in F with $d(x_n, x) < 1/n$). F can be written as

$$F = \{x \in X \mid d(x, F) = 0\} = \bigcap_{n \in \mathbb{N}} \{x \in X \mid d(x, F) < \frac{1}{n}\},$$

the latter being a countable intersection of open sets. Thus a closed set in a metric space is a G_δ-set:

Definition 1.4.6. *The countable intersection of open sets in a topological space is called a G_δ-set; the countable union of closed sets is called an F_σ-set.*

Whenever feasible, we will omit the notation of a topology or a metric from a space.

Recall that a subset D in a topological space is said to be *dense* iff D meets each nonempty open set. A metric space has a countable dense subset iff its topology has a countable base. This is fairly obvious: take for a topology with a countable base $\mathcal{B}$ from each element of $\mathcal{B}$ an arbitrary element, then this will form a countable dense subset. Conversely, if D is the countable dense set, then all balls with rational radii, $\{B_r(x) \mid 0 < r \in \mathbb{Q}, d \in D\}$, will form a countable base. Spaces with a countable base can be embedded into the cube $[0, 1]^\infty := \prod_{n \in \mathbb{N}} [0, 1]$, the *Hilbert cube* as we will see in Theorem 1.4.14 (by the way, we define

$$X^{\infty} := \prod_{n \in \mathbb{N}} X = X^{\mathbb{N}}$$

for any set X).

Definition 1.4.7. *A metric space is called* separable *iff it has a countable dense subset, or, equivalently, iff its topology has a countable base.*

1.4.2 Polish Spaces: Elementary Properties

Neither general topological spaces nor metric spaces offer a structure rich enough for the study of the transition systems that we will enter. We need to restrict the class of topological spaces to a particularly interesting class of spaces that are traditionally called *Polish*.

Recall that a metric space (X, d) is called *complete* iff each Cauchy sequence has a limit.

Definition 1.4.8. *A* Polish space *X is a topological space the topology of which is metrizable through a complete metric, and which has a countable dense subset; the topology is also called* Polish. *A* Standard Borel space *$(X, \mathcal{A})$ is a measurable space for which there exists a Polish topology $\mathcal{T}$ on X such that $\mathcal{B}(X, \mathcal{T}) = \mathcal{A}$.*

We will talk about the Borel sets of a Standard Borel space; slightly abusing notation, a Standard Borel space $(X, \mathcal{A})$ will usually be denoted only through its base set X, and its σ-algebra as $\mathcal{B}(X)$.

Familiar spaces are Polish, as these examples show.

Example 1.4.9. Denote by $\mathbb{R}$ the reals with their usual topology, which is induced by the open intervals. Then this is a Polish space. ⊣

Example 1.4.10. The open unit interval $]0, 1[$ with the usual topology induced by the open intervals form a Polish space.

Take as a dense subset the rationals $\mathbb{Q} \cap]0, 1[$, and take as a complete metric that generates the topology

$$d(x, y) := \left| \ln \frac{x}{1-x} - \ln \frac{y}{1-y} \right|.$$

This metric is suitable since $x \mapsto \ln(x/(1-x))$ is a continuous bijection from $]0, 1[$ to $\mathbb{R}$, and the inverse $y \mapsto e^y/(1+e^y)$ is also a continuous bijection. ⊣

Structural properties

Polish spaces have a number of interesting and helpful properties: Closed subsets form Polish spaces in their own right, and they are closed under forming countable products as well as countable coproducts.

Lemma 1.4.11. *Let X be a Polish space, and assume that $F \subseteq X$ is closed; then the subspace F is Polish as well. If $(X_n)_{n\in\mathbb{N}}$ is a sequence of Polish spaces, then the product $\prod_{n\in\mathbb{N}} X_n$ and the coproduct $\coprod_{n\in\mathbb{N}} X_n$ are Polish spaces.* □

Example 1.4.12. The set $\mathbb{N}$ of natural numbers with the discrete topology is a Polish space on account of being the topological sum of its elements. Thus the set $\mathbb{N}^\infty$ of all infinite sequences is a Polish space. The sets

$$\Sigma_\alpha := \{\tau \in \mathbb{N}^\infty \mid \alpha \text{ is an initial piece of } \tau\}$$

for $\alpha \in \mathbb{N}^*$, the free monoid generated by $\mathbb{N}$, constitute a base for the product topology. ⊣

This last example is structurally interesting, since it will help in the construction of a prototypical Polish space, i.e., a Polish space which is homeomorphic to an arbitrary (uncountable) Polish space. Such a construction permits sometimes reducing the discussion of properties for general Polish spaces to an investigation of the corresponding properties of $\mathbb{N}^\infty$, the structure of the latter space being more easily accessible than that of a general space.

Subspaces of Polish spaces that are Polish in their own right can be completely characterized.

Theorem 1.4.13. *Let Y be a Polish space. Then the subspace $X \subseteq Y$ is a Polish space iff X is a G_δ-set.* □

Conversely, each Polish space can be represented as a G_δ-set in the *Hilbert cube* $[0, 1]^\infty$; this is the famous characterization of Polish spaces due to Alexandrov I.33.VI*[55].

Theorem 1.4.14. ***(Alexandrov)*** *Let X be a separable metric space. Then X is homeomorphic to a subspace of the Hilbert cube. If X is Polish, this subspace is a G_δ.* □

Alexandrov's Theorem presents a complete characterization of Polish spaces as the G_δ-sets of the Hilbert cube. This cube is a compact metric space; recall that a topological Hausdorff space X is *compact* iff each open cover of X contains a finite cover of X. These properties of compact spaces will be used from time to time.

Theorem 1.4.15. *Let X be a topological Hausdorff space. Then*

a. If Y is a compact subspace of X, then Y is closed.
b. If X is compact and Y is closed, then Y is compact.
c. The union of a finite number of compact sets is compact.
d. The continuous image of a compact set into a Hausdorff space is compact.
e. The product of compact spaces is compact (Tihonov's Theorem).

f. If X is metrizable, then X is compact iff every sequence has a convergent subsequence.

g. If X is metrizable, then X is compact iff it is complete and totally bounded (thus given $\epsilon > 0$, there exists a finite subset $\{x_1, \dots, x_n\} \subseteq X$ with $X \subseteq B_\epsilon(x_1) \cup \dots \cup B_\epsilon(x_n)$) (Bolzano-Weierstraß Theorem). □

The Bolzano-Weierstraß Theorem implies that compact metrizable spaces are Polish. It is inferred from Tihonov's Theorem that the Hilbert cube $[0,1]^\infty$ is compact, because the unit interval $[0,1]$ is compact, again by the Bolzano-Weierstraß Theorem. Commenting on Alexandrov's Theorem 1.4.14 again: it not only embeds a Polish space as a G_δ into a compact metric space, it also says that the closure of the image will be compact. This is so by Proposition 1.4.15, part *b*.

1.4.3 Manipulating Polish Topologies

We will show now that Borel maps between Polish spaces can be turned into continuous maps. Specifically, we will show that, given a measurable map between Polish spaces, we can find on the domain a finer Polish topology with the same Borel sets, which renders the map continuous. This will be established through a sequence of auxiliary statements, each of which will be of interest and of use in its own right. We sketch the construction.

A Polish space X with topology $\mathcal{T}$ is fixed for the time being. Recall that a set is *clopen* in a topological space iff it is both closed and open.

Lemma 1.4.16. *Let F be a closed set in X. Then there exists a Polish topology $\mathcal{T}'$ such that $\mathcal{T} \subseteq \mathcal{T}'$ (hence $\mathcal{T}'$ is finer than $\mathcal{T}$), F is clopen in $\mathcal{T}'$, and $\mathcal{B}(X, \mathcal{T}) = \mathcal{B}(X, \mathcal{T}')$.*

Proof Both F and $X \setminus F$ are Polish by Theorem 1.4.13, so the topological sum of these Polish spaces is Polish again by Lemma 1.4.11. The sum topology is the desired topology. □

Lemma 1.4.17. *Let $(\mathcal{T}_n)_{n\in\mathbb{N}}$ be a sequence of Polish topologies $\mathcal{T}_n$ with $\mathcal{T} \subseteq \mathcal{T}_n$.*

a. The topology $\mathcal{T}_\infty$ generated by $\bigcup_{n\in\mathbb{N}} \mathcal{T}_n$ is Polish.

b. If $\mathcal{T}_n \subseteq \mathcal{B}(X, \mathcal{T})$, then $\mathcal{B}(X, \mathcal{T}_\infty) = \mathcal{B}(X, \mathcal{T})$. □

As a consequence, we may add to a Polish topology a Borel set without destroying the property of the space to be Polish or changing the Borel sets. This is true as well for sequences of Borel sets, as we will see now.

Proposition 1.4.18. *If $(B_n)_{n\in\mathbb{N}}$ is a sequence of Borel sets in X, then there exists a Polish topology $\mathcal{T}_0$ on X such that $\mathcal{T}_0$ is finer than $\mathcal{T}$, $\mathcal{T}$ and $\mathcal{T}_0$ have the same Borel sets, and each B_n is clopen in $\mathcal{T}_0$.*

Proof 1. We show first that we may add just one Borel set to the topology without changing the Borel sets. In fact, call a Borel set $B \in \mathcal{B}(X, \mathcal{T})$ *neat* if there exists a Polish topology $\mathcal{T}_B$ that is finer than $\mathcal{T}$ such that $B \in \mathcal{T}_B$, and $\mathcal{B}(X, \mathcal{T}) = \mathcal{B}(X, \mathcal{T}_B)$.

$$\mathcal{H} := \{B \in \mathcal{B}(X, \mathcal{T}) \mid B \text{ is neat}\}.$$

Then $\mathcal{T} \subseteq \mathcal{H}$, and each closed set is a member of $\mathcal{H}$ by Lemma 1.4.16. Furthermore, $\mathcal{H}$ is closed under complements and under countable unions by Lemma 1.4.17. Thus we may now infer with the π-λ-Theorem 1.3.1 that $\mathcal{H} = \mathcal{B}(X, \mathcal{T})$, so that each Borel set is neat.

2. Now construct inductively Polish topologies $\mathcal{T}_n$ that are finer than $\mathcal{T}$ with $\mathcal{B}(\mathcal{T}) = \mathcal{B}(\mathcal{T}_n)$. Start with $\mathcal{T}_0 := \mathcal{T}$. Adding B_{n+1} to the Polish topology $\mathcal{T}_n$ according to the first part yields a finer Polish topology $\mathcal{T}_{n+1}$ with the same Borel sets. Thus the assertion follows from Lemma 1.4.17. □

This permits turning a Borel map into a continuous one, whenever the domain is Polish and the range is a second countable metric space.

Proposition 1.4.19. *Let $(Y, \mathcal{S})$ be a separable metric space with topology $\mathcal{S}$. If $f : X \to Y$ is a $\mathcal{B}(X, \mathcal{T})$-$\mathcal{B}(Y, \mathcal{S})$-Borel measurable map, then there exists a Polish topology $\mathcal{T}'$ on X with these properties:*

a. $\mathcal{T}'$ is finer than $\mathcal{T}$,
b. $\mathcal{T}$ and $\mathcal{T}'$ have the same Borel sets,
c. f is $\mathcal{T}'$-$\mathcal{S}$ continuous.

Proof [79, Corollary 3.2.6] or [20, Proposition 1.28]. □

This property is technically most useful, because it permits rendering measurable maps continuous, when they go into a second countable metric space (thus in particular into a Polish space). Consequently, one may capitalize on properties of continuous maps for Borel maps.

Having a countable dense subset for a metric space, we can use the corresponding base for a structurally interesting characterization of the Borel sets of a separable metric space. The next lemma says that the Borel sets are in this case countably generated.

Lemma 1.4.20. *Let Y be a separable metric space. Then*

$$\mathcal{B}(Y) = \sigma(\{B_r(d) \mid r > 0 \text{ rational}, d \in D\}),$$

where D is countable and dense. □

This representation implies that the Borel sets $\mathcal{B}(X)$ of the Polish space X are countably generated.

Borel sets in a metric space are characterized as the closure of the open (closed) sets under countable unions and countable intersections. This is a direct consequence of the fact that each closed set is a G_δ-set by Theorem 1.4.13.

Lemma 1.4.21. *The Borel sets in a metric space Y are the smallest collection of sets that contains the open (closed) sets and that are closed under countable unions and countable intersections.*

As a preparation for dealing with analytic sets, we mention that the Polish space X can be represented as the continuous image of the space $\mathbb{N}^\infty$.

Proposition 1.4.22. *For every Polish space X there exists a continuous map $f : \mathbb{N}^\infty \to X$ with $f[\mathbb{N}^\infty] = X$.* □

Proposition 1.4.22 permits sometimes establishing properties of a Polish space first for the space of infinite sequences, and then transferring the arguments to the Polish space under consideration. In studying the specific space $\mathbb{N}^\infty$ instead of the abstractly given space X, the former may permit some rather special constructions. This will be noticed in the investigation of some remarkable properties of analytic sets which we will study now.

1.4.4 Analytic Spaces

An *analytic set* B is the projection of a Borel subset of $X \times X$, where X is a Polish space; the complement of an analytic set is called a *co-analytic* set. One may wonder whether these projections are Borel sets, but we will show in a moment that there are strictly more analytic sets than Borel sets whenever the underlying Polish space is uncountable. Thus analytic sets are a proper extension to Borel sets. On the other hand, analytic sets arise fairly naturally from factoring Polish spaces through equivalence relations that are generated from a countable collection of Borel sets, see Proposition 1.7.5. Consequently, it is sometimes adequate to consider analytic sets rather than their Borel cousins, and sometimes one does not even have this choice.

This is a first characterization of analytic sets (using π_X for the projection to X).

Proposition 1.4.23. *Let X be a Polish space. Then the following statements are equivalent for $A \subseteq X$:*

a. A is analytic.
b. There exists a Polish space Y and a Borel set $B \subseteq X \times Y$ with $A = \pi_X[B]$.
c. There exists a continuous map $f : \mathbb{N}^\infty \to X$ with $f[\mathbb{N}^\infty] = A$.
d. $A = \pi_X[C]$ for a closed subset $C \subseteq X \times \mathbb{N}^\infty$. □

As an immediate consequence we obtain that a Borel set is analytic. Just for the record:

Corollary 1.4.24. *Each Borel set in a Polish space is analytic.* □

Not All Analytic Sets Are Borel

The converse does not hold, as we will show now. This statement is not only of interest in its own right. Historically, it initiated *Descriptive Set Theory* i.e, the study of analytic and co-analytic sets as a separate discipline in set theory.

Proposition 1.4.25. *Let X be an uncountable Polish space. Then there exists an analytic set that is not Borel.* □

Analytic sets have closure properties that are similar to those of Borel sets, but not quite the same: they are closed under countable unions and intersections, and under the inverse image of Borel maps. They are closed under the direct image of Borel maps as well, but suspiciously missing is the closure under complementation (which will give rise to Souslin's Theorem).

Proposition 1.4.26. *Analytic sets in a Polish space X are closed under countable unions and countable intersections. If Y is another Polish space, with analytic sets $A \subseteq X$ and $B \subseteq Y$, and $f : X \to Y$ is a Borel map, then $f[A] \subseteq Y$ is analytic in Y, and $f^{-1}[B]$ is analytic in X.* □

Roughly, an *analytic space* is Borel isomorphic to an analytic set in a Polish space; to be more specific:

Definition 1.4.27. *A measurable space $(M, \mathcal{M})$ is called an* analytic space *iff there exists a Polish space X and an analytic set A in X such that the measurable spaces $(M, \mathcal{M})$ and $(A, \mathcal{B}(X) \cap A)$ are Borel isomorphic. The elements of $\mathcal{M}$ are then called the* Borel sets of M. *$\mathcal{M}$ is denoted by $\mathcal{B}(M)$.*

Abusing notation, $\mathcal{B}(X) \cap A$ refers to the σ-algebra $\{B \cap A \mid B \in \mathcal{B}(X)\}$ which may be thought of as the σ-algebra of traces left by the Borel sets of X on the set A, see page 7.

We will usually omit the σ-algebra from the notation of an analytic space.

Souslin's Theorem

Analytic spaces share many favorable properties with analytic sets, and with Polish spaces, but they are a wee bit more general: whereas an analytic set lives in a Polish space, an analytic space requires only a Polish space to sit in the background somewhere and to be Borel isomorphic to it. This makes life considerably easier, since we are not always obliged to present a Polish space directly when dealing with properties of analytic spaces. An immediate consequence is that the image of an analytic space under a Borel map into a Polish space is analytic again.

The representation of an analytic set through a continuous map on $\mathbb{N}^\infty$ has the remarkable consequence that we can separate two disjoint analytic sets by disjoint Borel sets (Lusin's Theorem). This in turn implies a pretty characterization of Borel sets due to Souslin, which says that an analytic set is Borel iff it is co-analytic as well. We start with Lusin's Theorem.

Proposition 1.4.28. *Given disjoint analytic sets A and B in a Polish space X, there exist disjoint Borel sets E and F with $A \subseteq E$ and $B \subseteq F$.* □

As a consequence, we obtain Souslin's Theorem. It is usually formulated for Polish spaces. But since each Polish space is the continuous image of $\mathbb{N}^\infty$, the proof can be made to carry over to analytic spaces as well.

Theorem 1.4.29. ***(Souslin)*** *Let A be an analytic set in an analytic space Y. If $Y \setminus A$ is analytic, then A is a Borel set.* □

Souslin's Theorem is important when one wants to show that a set is a Borel set that is given, for example, through the image of another Borel set. A typical scenario for its use is establishing, given a Borel set B and a Borel map $f : X \to Y$, that both $A = f[B]$ and $Y \setminus A = f[X \setminus B]$ hold. Then one infers from Proposition 1.4.26 that both A and $Y \setminus A$ are analytic, and from Souslin's Theorem that A is a Borel set.

Take a Borel measurable bijection between two Polish spaces. It is not a priori clear whether or not this map is an isomorphism. Souslin's Theorem gives a helpful hand here as well. We will need this property in a moment for a characterization of countably generated sub-σ-algebras of Borel sets, but it appears to be interesting in its own right.

Before we state it, we give a very simple Lemma which is occasionally of use.

Lemma 1.4.30. *Let $f : M \to N$ be a map; call $A \subseteq M$* $\mathsf{ker}(f)$-invariant *(or simply f-invariant) iff $a \in A$ and $f(a) = f(a')$ together imply $a' \in A$. Then $f[A_1 \cap A_2] = f[A_1] \cap f[A_2]$, whenever A_1 and A_2 are f-invariant; also $f^{-1}[f[A]] = A$ for f-invariant $A \subseteq M$.* □

This is a characterization of the Borel sets of an analytic space which is the image of another one under a surjective Borel map as final.

Lemma 1.4.31. *Let X and Y be analytic spaces with $f : X \to Y$ Borel and onto. Then $\mathcal{B}(Y) = \{Q \subseteq Y \mid f^{-1}[Q] \in \mathcal{B}(X)\}$.*

Proof It is clear that $f^{-1}[Q] \in \mathcal{B}(X)$ for each $Q \in \mathcal{B}(Y)$. Conversely, assume that $Q \subseteq Y$ with $f^{-1}[Q] \in \mathcal{B}(X)$, then $Q = f\left[f^{-1}[Q]\right]$, since f is onto. Hence Q is analytic. On the other hand, $Y \setminus Q = f\left[f^{-1}[Y \setminus Q]\right]$ as well, so that $Y \setminus Q$ is analytic, too. From Souslin's Theorem 1.4.29 we obtain that $Q \in \mathcal{B}(Y)$. □

The image of an invariant Borel set under a measurable map is usually not a Borel set, but if the set is invariant, and if the map is onto, this holds.

Corollary 1.4.32. *Given X, Y, and f as above. If $A \subseteq X$ is an f-invariant Borel set, then $f[A]$ is a Borel set in Y.*

Proof By Lemma 1.4.30, $f^{-1}[f[A]] = A \in \mathcal{B}(X)$, so that the conclusion follows from Lemma 1.4.31. □

We obtain as an immediate consequence of Lemma 1.4.31 this characterization of Borel isomorphisms.

Proposition 1.4.33. *Let X and Y be analytic spaces and $f : X \to Y$ be a bijection that is Borel measurable. Then f is a Borel isomorphism.* □

Separable Measurable Spaces

Call a measurable space $(M, \mathcal{M})$ *separable* iff the σ-algebra $\mathcal{M}$ has a countable set $(A_n)_{n\in\mathbb{N}}$ of generators which separates points, i.e., given $x, x' \in M$ with $x \neq x'$ there exists A_n which contains exactly one of them. A Polish space is separable as a measurable space; so is an analytic space, as we will show now.

Separable measurable spaces are characterized through subsets of Polish spaces.

Lemma 1.4.34. *The measurable space $(M, \mathcal{M})$ is separable iff there exists a Polish space X and a subset $P \subseteq X$ such that the measurable spaces $(M, \mathcal{M})$ and $(P, \mathcal{B}(X) \cap P)$ are Borel isomorphic.* □

We obtain as a consequence that analytic spaces are separable.

Corollary 1.4.35. *An analytic space is a separable measurable space.* □

A second consequence is that separable measurable spaces are derived from separable metric spaces in a rather straightforward way.

Lemma 1.4.36. *For a separable measurable space $(X, \mathcal{A})$ there exists a separable metric topology $\mathcal{T}$ on X such that $\mathcal{B}(X, \mathcal{T}) = \mathcal{A}$.* □

By the way, this innocent looking statement has some remarkable consequences.

Corollary 1.4.37. *Let $(M, \mathcal{M})$ be a separable measurable space. Then*

a. *The diagonal is measurable in the product, i.e.,*

$$\Delta_M := \{\langle t, t\rangle \mid t \in M\} \in \mathcal{M} \otimes \mathcal{M}.$$

b. *If $f_i : X_i \to M$ is $\mathcal{A}_i$ - $\mathcal{M}$-measurable, where $(X_i, \mathcal{A}_i)$ is a measurable space ($i = 1, 2$), then*

$$f_1^{-1}[\mathcal{M}] \otimes f_2^{-1}[\mathcal{M}] = (f_1 \times f_2)^{-1}[\mathcal{M} \otimes \mathcal{M}]. \;□$$

With a similar argument, the measurability of the graph of a measurable map is established.

Lemma 1.4.38. *Let $(M, \mathcal{M})$ be a measurable space, and $f : M \to Z$ be a $\mathcal{M}$-$\mathcal{N}$-measurable map, where $(Z, \mathcal{N})$ is a separable measurable space. The graph of f,*

$$\mathsf{graph}(f) := \{\langle m, f(m)\rangle \mid m \in M\},$$

is a member if $\mathcal{M} \otimes \mathcal{N}$. □

The kernel of a Borel map is of interest when studying certain equivalence relations. As a consequence of the characterization above, we obtain measurability of the kernel whenever the range is a separable measurable space.

Corollary 1.4.39. *Let $f : M \to N$ be an $\mathcal{M}$-$\mathcal{N}$-measurable map, where $(M, \mathcal{M})$ and $(N, \mathcal{N})$ are measurable spaces, the latter being separable. Then the* kernel *of f*

$$\mathsf{ker}\,(f) := \{\langle m_1, m_2\rangle \mid f(m_1) = f(m_2)\}$$

is a member of $\mathcal{M} \otimes \mathcal{M}$.

Proof The map $f \times f : \langle m_1, m_2\rangle \mapsto \langle f(m_1), f(m_2)\rangle$ is $\mathcal{M} \otimes \mathcal{M}$–$\mathcal{N} \otimes \mathcal{N}$-measurable, and the diagonal Δ_N is a member of $\mathcal{N} \otimes \mathcal{N}$. Since $\mathsf{ker}\,(f) = (f \times f)^{-1}\left[\Delta_Z\right]$, the assertion follows. □

A second immediate consequence gives a characterization of Standard Borel spaces in terms of measurable maps that have a measurable inverse.

Proposition 1.4.40. *Let $(X, \mathcal{M})$ and $(Y, \mathcal{N})$ be measurable spaces, and assume that $(X, \mathcal{M})$ is separable. Let $f : X \to Y$ be $\mathcal{M}$-$\mathcal{N}$-measurable and onto. If there exists a $\mathcal{N}$-$\mathcal{M}$-measurable map $g : Y \to X$ such that $f \circ g = id_Y$, then*

a. $g\left[Y\right] \subseteq X$ is measurable,
b. g is a Borel isomorphism from Y to $g\left[Y\right]$,
c. if X is a Standard Borel space, so is Y.

Proof Let $F \subseteq X$ be the set of all fixed points of $g \circ f$, then clearly $F = g\left[Y\right]$. Since $(X, \mathcal{M})$ is separable, we know from Corollary 1.4.37 that the diagonal $\Delta_X \in \mathcal{M} \otimes \mathcal{M}$, and $F = r^{-1}\left[\Delta_X\right]$ with $r(x) := \langle x, (g \circ f)(x)\rangle$, which is measurable. Thus F is a measurable subset of X. Since $G : Y \to X$ and the restriction $f|_F : F \to Y$ of f to F are inverse and measurable, g is a Borel isomorphism. If X is Standard Borel, Y is Borel isomorphic to a Borel set, and so is Standard Borel itself. □

The Blackwell-Mackey Theorem

Returning to analytic spaces, have a brief look at countably generated sub-σ-algebras of an analytic space. This will help us to establish that the factor space for a particularly interesting and important class of equivalence relations is an analytic space. The proof shows a typical application of the structural properties established so far.

Proposition 1.4.41. *Let X be an analytic space, and a$\mathcal{B}_0$ a countably generated sub-σ-algebra of $\mathcal{B}(X)$ that separates points. Then $\mathcal{B}_0 = \mathcal{B}(X)$.*

Proof 1. $(X, \mathcal{B}_0)$ is a separable measurable space, so there exists a Polish space P and a subset $Y \subseteq P$ of P such that $(X, \mathcal{B}_0)$ is Borel isomorphic to $(P, \mathcal{B}(P) \cap Y)$ by Lemma 1.4.34. Let f be this isomorphism, then $\mathcal{B}_0 = f^{-1}[\mathcal{B}(P) \cap Y]$.

2. f is a Borel map from $(X, \mathcal{B}(X))$ to $(Y, \mathcal{B}(P) \cap Y)$; thus Y is an analytic set with $\mathcal{B}(Y) = \mathcal{B}(X) \cap P$ by Proposition 1.4.26. By Proposition 1.4.33, f is an isomorphism, hence $\mathcal{B}(X) = f^{-1}[\mathcal{B}(P) \cap Y]$. But this establishes the assertion. □

This gives an interesting characterization of measurable spaces to be analytic, provided they have a separating sequence of sets; to be specific:

Lemma 1.4.42. *Let X be analytic, and $f : X \to Y$ $\mathcal{B}(X)$-$\mathcal{N}$-measurable and onto for a separable measurable space $(Y, \mathcal{N})$. Then $(Y, \mathcal{N})$ is analytic.*

Proof 1. Let $(B_n)_{n \in \mathbb{N}}$ be the sequence of sets that separates points. Take an arbitrary set $N \in \mathcal{N}$ and define the σ-algebra $\mathcal{B}_0 := \sigma(\{B_n \mid n \in \mathbb{N}\} \cup \{N\})$. Then $(Y, \mathcal{B}_0)$ is a separable measurable space. So by Lemma 1.4.34 we can find a Polish space P with $Y \subseteq P$ and $\mathcal{B}_0$ is the trace of $\mathcal{B}(P)$ on Y. Proposition 1.4.26 tells us that $Y = f[X]$ is analytic with $\mathcal{B}_0 = \mathcal{B}(Y) = \sigma(\{B_n \mid n \in \mathbb{N}\})$, the latter equality being implied by Proposition 1.4.41. Because N was arbitrary, and because $B_n \in \mathcal{N}$, this yields $\mathcal{N} = \sigma(\{B_n \mid n \in \mathbb{N}\}) = \mathcal{B}(Y)$. □

We will use Lemma 1.4.42 for example for demonstrating that the factor space of an analytic space for a smooth equivalence relation is analytic again. This class of relations will be defined and briefly characterized in Section 1.7.

The Blackwell-Mackey-Theorem analyzes those Borel sets that are unions of $\mathcal{A}$-atoms for a sub-σ-algebra $\mathcal{A} \subseteq \mathcal{B}(X)$. Recall that a set $W \in \mathcal{A}$ is an *$\mathcal{A}$-atom* (or simply an atom) iff for each $V \in \mathcal{A}$ with $V \subseteq W$ either $V = \emptyset$ or $V = W$ holds. If $\mathcal{A}$ is countably generated by, say, $(A_n)_{n \in \mathbb{N}}$, then it is not difficult to see that an atom in $\mathcal{A}$ can be represented as

$$\bigcap_{i \in T} A_i \cap \bigcap_{i \in \mathbb{N} \setminus T} (X \setminus A_i)$$

for a suitable subset $T \subseteq \mathbb{N}$; see Lemma 1.7.11.

Theorem 1.4.43. ***(Blackwell-Mackey)*** *Let X be an analytic space and $\mathcal{A} \subseteq \mathcal{B}(X)$ be a countably generated sub-σ-algebra of the Borel sets of X. If $B \subseteq X$ is a Borel set that is a union of atoms of $\mathcal{A}$, then $B \in \mathcal{A}$.* □

1.5 Measurable Selectors

Assume that X and Z are sets. Consider a set-valued map $R : X \to \mathcal{P}(Z)$, or equivalently, a relation $R \subseteq X \times Z$. We will not distinguish too narrowly between relations and set-valued maps, so that for a relation R the set $R(x)$ will be defined as well. We define for R and a set $G \subseteq Z$ the *weak inverse* as

$$\exists R(G) := \{x \in X \mid R(x) \cap G \neq \emptyset\}$$

and the *strong inverse* as

$$\forall R(G) := \{x \in X \mid R(x) \subseteq G\}.$$

Example 1.5.1. Both weak and strong inverse of a relation invoke an analogy to modal logic. The formulas of simple modal logic are given through

$$\phi ::= \top \mid p \mid \neg\phi \mid \phi_1 \wedge \phi_2 \mid \Diamond\phi$$

with p an atomic sentence. Assume that $\mathcal{M} = (S, R, V)$ is a Kripke model for the basic modal logic, S being the state space, $R \subseteq S \times S$ the transition relation, and V the valuation for the atomic sentences, see Section 2.3. We define the semantics of formulas inductively, starting with $[\![p]\!]_{\mathcal{M}} := V(p)$. The meaning of negation and conjunction are given as usual, and the semantics for $\Diamond\phi$ is defined through

$$\mathcal{M}, s \models \Diamond\phi \Leftrightarrow \exists s' \in S : \langle s, s' \rangle \in R \wedge \mathcal{M}, s' \models \phi.$$

Thus

$$\begin{aligned} s \in [\![\Diamond\phi]\!]_{\mathcal{M}} &\Leftrightarrow \exists s' : s' \in R(s) \wedge s' \in [\![\phi]\!]_{\mathcal{M}} \\ &\Leftrightarrow R(s) \cap [\![\phi]\!]_{\mathcal{M}} \neq \emptyset \\ &\Leftrightarrow s \in \exists R([\![\phi]\!]_{\mathcal{M}}). \end{aligned}$$

Define as usual $\Box\phi$ through $\neg\Diamond\neg\phi$, then its semantics $[\![\Box\phi]\!]$ is shown similarly to be $\forall R([\![\phi]\!])$. ⊣

We will discuss briefly the relationship between measurable relations and those measurable maps that always select an element from the range of such a relation. For this, we endow X and Z with a measurable resp. a Polish structure. Assume that X is a measurable space, and that Z is Polish.

Let $R(x)$ always take closed and nonempty values. If the *weak inverse* $\exists R(G)$ is a measurable set, whenever $G \subseteq Z$ is open, then R is called a *weakly measurable relation on* $X \times Z$. Relation R is called a *measurable relation* iff the *strong inverse* $\forall R(F)$ is measurable whenever $F \subseteq Z$ is closed. Finally, R

is called *C-measurable* iff for any compact set $C \subseteq Z$ the weak inverse $\exists R(C)$ of C is a Borel set in X.

Castaing Representation

Weakly measurable relations can be represented through measurable selectors (sometimes called a *Castaing representation*). This representation implies in particular that a weakly measurable set-valued map has a measurable selector. Formally:

Definition 1.5.2. *Given the measurable space X and the Polish space Z, assume that $\emptyset \neq R(x) \subseteq Z$ always takes closed values.*

a. A measurable map $f : X \to Z$ is called a measurable selector *for R iff $f(x) \in R(x)$ holds for all $x \in X$.*
b. The sequence $(f_n)_{n \in \mathbb{N}}$ of measurable selectors $f_n : X \to Z$ is called a Castaing representation *for the relation $R \subseteq X \times Z$ iff $R(x)$ is the topological closure of $\{f_n(x) \mid n \in \mathbb{N}\}$ for all $x \in X$.*

Thus for relation R to have a Castaing representation it is a necessary condition that $R(x)$ be nonempty and closed; the Castaing representation is in this case a dense subset for each $x \in X$. We will use the following characterization.

Proposition 1.5.3. *Given the measurable space X and the Polish space Z, let $R \subseteq X \times Z$ be a relation such that $\emptyset \neq R(x) \subseteq Z$ is closed for every $x \in X$. Then the following conditions are equivalent:*

a. R is weakly measurable.
b. There exists a Castaing representation for R.
c. R is C-measurable. □

Standard Borel Covers

We relate analytic spaces to Standard Borel covers through measurable selectors. Given an analytic space A and a Standard Borel space P, call a surjective Borel map $e : P \to A$ a *Standard Borel cover* of A.

Proposition 1.5.4. *The following statements are equivalent for an analytic space A:*

a. A is Standard Borel;
b. A is projective with respect to Standard Borel covers*; that is, for any Standard Borel cover $e : P \to B$ of an analytic space B and any Borel map $f : A \to B$, there exists a Borel map $d : A \to P$ with $f = e \circ d$.*

Proof $a \Rightarrow b$: Choose a Polish topology on P which generates the σ-algebra such that e is continuous. We define a set-valued function $M : A \to \mathcal{P}(P)$ via $a \mapsto e^{-1}[\{f(a)\}]$. Each $M(a)$ is closed and nonempty by construction. Moreover, for any $C \subseteq P$ and $a \in A$, we have

$$\begin{aligned} a \in \exists M(C) &\Leftrightarrow \exists c \in C : c \in M(a) \\ &\Leftrightarrow \exists c \in C : e(c) = f(a) \\ &\Leftrightarrow f(a) \in e[C] \\ &\Leftrightarrow a \in f^{-1}[e[C]]. \end{aligned}$$

Now take any compact $C \subseteq P$. Then $e[C]$ is compact, hence closed, hence measurable, and thus $f^{-1}[e[C]] = \exists M(C)$ is measurable in A. Hence, by Proposition 1.5.3 there exists a measurable selector d for M, which satisfies $e \circ d = f$ by construction.

$b \Rightarrow a$: Assume A to be projective with respect to Standard Borel spaces and let $e : P \to A$ be a Standard Borel cover of A. We obtain a measurable function d such that

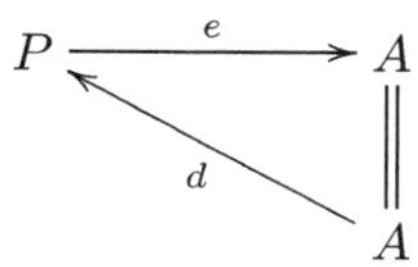

commutes. A is Standard Borel by Proposition 1.4.40. □

1.6 Probability Measures

Stochastic relations and Markov transition systems are based on subprobability measures. We will discuss these measures now and point out some salient features from which we will develop some algebraic properties.

Dealing first with properties of individual measures, the attention shifts soon to the set of all subprobabilities on a Polish space. This space is endowed with a topology, rendering it a Polish space again, and some properties of this topology will be investigated. We need in the sequel some particular constructions (like projective limits for the interpretation of logics operating on infinite sequences); these constructions will be provided here as well.

Subprobabilities

A *probability measure* on the measurable space $(N, \mathcal{N})$ is a monotone and σ-additive map $\mu : \mathcal{N} \to [0, 1]$ with $\mu(\emptyset) = 0$ and $\mu(N) = 1$. That μ is σ-additive means that

$$\mu(\bigcup_{i\in\mathbb{N}} D_i) = \sum_{i\in\mathbb{N}} \mu(D_i)$$

holds whenever $(D_n)_{n\in\mathbb{N}}$ is a countable family of mutually disjoint sets in $\mathcal{N}$. Denote by $\mathfrak{P}(N, \mathcal{N})$ the set of all probability measures on $(N, \mathcal{N})$. We will use *subprobability* as well: they are defined like probability measures with the exception that the entire space is assigned a mass which does not exceed unity; $\mathfrak{S}(N, \mathcal{N})$ is the set of all subprobability measures on $(N, \mathcal{N})$. If $N = \{1, \ldots, n\}$ is a discrete space, then $\mathfrak{S}(N)$ is identified with all positive convex combinations, i.e., with the set

$$\{\langle p_1, \ldots, p_n\rangle \mid p_i \geq 0 \text{ for } 1 \leq i \leq n, \sum_{i=1}^{n} p_i \leq 1\}.$$

A rather important tool is the well-known Monotone Convergence Theorem [43, Theorem 27.B], which yields the analogue to σ-additivity for the integral. Recall that the elements of $\mathcal{F}(M, \mathcal{M})$ are bounded.

Proposition 1.6.1. *Let $f \in \mathcal{F}(M, \mathcal{M})$ for the measurable space $(M, \mathcal{M})$ be a nonnegative and bounded measurable function with $f \geq 0$; assume that $0 \leq f_1 \leq f_2 \leq \ldots$ is a monotonically increasing sequence $(f_n)_{n\in\mathbb{N}} \subseteq \mathcal{F}(M, \mathcal{M})$ with $f = \sup_{n\in\mathbb{N}} f_n$, and let $\mu \in \mathfrak{S}(N, \mathcal{M})$ be a subprobability measure. Then*

$$\int_M f\, d\mu = \lim_{n\to\infty} \int_M f_n\, d\mu. \ \square$$

Choquet Representation

An easy and occasionally very practical first consequence of Proposition 1.6.1 is the representation of an integral with respect to an arbitrary measure through an integral on the real line, what is sometimes called the Choquet representation. This representation builds a bridge between the classical Riemann integral and Lebesgue integral, since it permits computing a Lebesgue integral for a nonnegative function f through the area

$$\{\langle x, t\rangle \in M \times \mathbb{R} \mid 0 \leq t < f(x)\}$$

under its graph.

Recall for the proof which follows that given $\mu_i \in \mathfrak{S}(N_i, \mathcal{N}_i)$ there exists a unique measure $\mu_1 \otimes \mu_2 \in \mathfrak{S}(N_1 \times N_2, \mathcal{N}_1 \otimes \mathcal{N}_2)$ such that for all $B_1 \in \mathcal{N}_1, B_2 \in \mathcal{N}_2$

$$\mu_1 \otimes \mu_2(B_1 \times B_2) = \mu_1(B_1) \cdot \mu_2(B_2).$$

The proof of this relationship between the area under the graph and the Lebesgue integral of a nonnegative function is presented in full in order to convey the flavor of the argument that will be encountered fairly often.

Proposition 1.6.2. *Let $f \in \mathcal{F}(M, \mathcal{M})$ for the measurable space $(M, \mathcal{M})$ be a nonnegative and bounded measurable function with $f \geq 0$, then*

$$\int_M f \, d\mu = \int_0^\infty \mu(\{x \in M \mid f(x) > t\}) \, dt.$$

Proof Define for $f \geq 0$ the set

$$C(f) := \{\langle x, t\rangle \in M \times \mathbb{R} \mid 0 \leq t < f(x)\},$$

we claim that $C(f) \in \mathcal{M} \otimes \mathcal{B}(\mathbb{R})$ and

$$\int_M f \, d\mu = (\mu \otimes \lambda)(C(f))$$

holds, where λ is Lebesgue measure, and $\mu \otimes \lambda$ is the product measure. Consider these cases:

1. If $f = \chi_A$ with $A \in \mathcal{M}$, then $C(f) = M \setminus A \times \{0\} \cup A \times [0, 1[\in \mathcal{M} \otimes \mathcal{B}(\mathbb{R})$ thus
$$\int_M \chi_A \, d\mu = \mu(A) = (\mu \otimes \lambda)(C(f)).$$
2. If f is represented as a step function with a finite number of mutually disjoint steps, say, $f = \sum_{i=1}^k r_i \cdot \chi_{A_i}$ with $r_i \geq 0$ and all $A_i \in \mathcal{M}$, then
$$C(f) = \left(M \setminus \bigcup_{i=1}^k A_i\right) \times \{0\} \cup \bigcup_{i=1}^k A_i \times [0, r_i[\in \mathcal{M} \otimes \mathcal{B}(\mathbb{R}),$$
and
$$\int_M f \, d\mu = \sum_{i=1}^k \int_{A_i} r_i \, d\mu = (\mu \otimes \lambda)(C(f)).$$
3. If f is represented as a monotone limit of step function $(f_n)_{n\in\mathbb{N}}$ with $f_n \geq 0$ according to Proposition 1.3.5, then $C(f) = \bigcup_{n\in\mathbb{N}} C(f_n)$, thus $C(f) \in \mathcal{M} \otimes \mathcal{B}(\mathbb{R})$, and, by Proposition 1.6.1,
$$\begin{aligned}\int_M f \, d\mu &= \lim_{n\to\infty} \int_M f_n \, d\mu \\ &= \lim_{n\to\infty} (\mu \otimes \lambda)(C(f_n)) \\ &= (\mu \otimes \lambda)(C(f)).\end{aligned}$$

Thus we have for $f \geq 0$ the representation

$$\int_M f \, d\mu = (\mu\otimes\lambda)(C(f)) = \int_0^\infty \mu(C(f)_t) \, dt = \int_0^\infty \mu(\{x \in M \mid f(x) > t\}) \, dt,$$

the latter equality being derived from Fubini's Theorem for product integration. □

An Extension Theorem

The existence of a semi-pullback in Polish spaces hinges on extending measures from sub-σ-algebras defined by measurable maps to the entire σ-algebra of Borel sets; to be specific:

Proposition 1.6.3. *Let X_1 and X_2 be Polish spaces, $(Z, \mathcal{C})$ a separable measurable space, and assume that*

$$\psi_i : X_i \to Z \ (i = 1, 2)$$

are measurable and surjective maps. Define

$$S := \{\langle x_1, x_2 \rangle \in X_1 \times X_2 \mid \psi_1(x_1) = \psi_2(x_2)\},$$

endow S with the trace $\mathcal{B}(S)$ of the product σ-algebra, and assume that subprobability measures $\mu_1 \in \mathfrak{S}(X_1), \mu_2 \in \mathfrak{S}(X_2), \theta \in \mathfrak{S}(S)$ are given such that

$$\forall E_i \in \psi_i^{-1}[\mathcal{C}] : \theta\big(\pi_i^{-1}[E_i]\big) = \mu_i(E_i) \ (i = 1, 2)$$

holds, where $\pi_1 : S \to X_1, \pi_2 : S \to X_2$ are the projections. Then there exists $\mu \in \mathfrak{S}(S)$ such that

$$\forall E_i \in \mathcal{B}(X_i) : \mu\big(\pi_i^{-1}[E_i]\big) = \mu_i(E_i) \ (i = 1, 2)$$

holds. □

The condition of the proposition says for $i = 1$ that $\theta\big(S \cap (E_1 \times X_2)\big) = \mu_1(E_1)$ holds for each Borel set $E_1 \in \mathcal{B}(X_1)$, so that θ is determined on the part of the strip $E_1 \times X_2$ that lies in the set S. This is of course not sufficient for extending the measure θ to all Borel sets of S with the constraint that the preassigned marginal distributions are satisfied by means of the usual extension theorems. The existence of the measure can, however, be ascertained through a construction involving some other tools from analysis; see [20] for details.

1.6.1 Weak Topology

We will define the topology of weak convergence on $\mathfrak{S}(X)$, and we will find metrics for it. If X is Polish, the weak topology will be shown to be Polish as well. The Borel sets for this topology will also be investigated. They will be a valuable tool for the investigations to follow even without the weak topology being present explicitly.

Let X be a metric space, then $\mathfrak{S}(X) = \mathfrak{S}(X, \mathcal{B}(X))$ is equipped with the *topology of weak convergence*. This is the smallest topology on $\mathfrak{S}(X)$ which makes the map $\mu \mapsto \int_X f\ d\mu$ continuous for each continuous and bounded $f : X \to \mathbb{R}$. Denote by $\mathcal{C}(X)$ the linear space of all these functions, and by $\rightharpoonup_w$ convergence in this topology.

A base for the weak topology is furnished through sets of the form

$$U(\mu_0, \varepsilon, f_1, \dots, f_n) := \{\mu \in \mathfrak{S}(X) \mid \left|\int_X f_i\ d\mu - \int_X f_i\ d\mu_0\right| < \varepsilon \text{ for } 1 \leq i \leq n\}$$

with $\varepsilon > 0, \mu_0 \in \mathfrak{S}(X), f_1, \dots, f_n \in \mathcal{C}(X)$. This topology is characterized through the famous Portmanteau Theorem:

Proposition 1.6.4. *The following conditions are equivalent for a sequence $(\mu_n)_{n\in\mathbb{N}}$ and a measure $\mu \in \mathfrak{S}(X)$ whenever X is a Polish space:*

a. $\mu_n \rightharpoonup_w \mu$.
b. $\int_X f\ d\mu_n \to \int_X f\ d\mu$ *for each bounded and continuous* $f : X \to \mathbb{R}$.
c. $\int_X f\ d\mu_n \to \int_X f\ d\mu$ *for each bounded and uniformly continuous* $f : X \to \mathbb{R}$.
d. $\liminf_{n\to\infty} \mu_n(G) \geq \mu(G)$ *for each open subset* $G \subseteq X$, *and* $\mu_n(X) \to \mu(X)$.
e. $\limsup_{n\to\infty} \mu_n(F) \leq \mu(F)$ *for each closed subset* $F \subseteq X$, *and* $\mu_n(X) \to \mu(X)$. □

It is clear that $\mu_n \rightharpoonup_w \mu$ does not imply that $\mu_n(A) \to \mu_0(A)$ holds for all Borel sets A (this would be a case for the topology of strong convergence, which is not that interesting — for us, that is). The following simple example illustrates the point. Recall that the *Dirac measure* δ_x is defined through $\delta_x(A) := \chi_A(x)$, so $\delta_x(A) = 1$ iff $x \in A$, and $\delta_x(A) = 0$ otherwise. Topologically, $x \mapsto \delta_x$ yields a homeomorphic embedding of X into $\mathfrak{S}(X)$.

Example 1.6.5. Take $X := [0, 1]$, let $\mu_n := \delta_{1/n}$, and consider the Borel set $A :=]0, 1]$. Since $\int_X f\ d\mu_n = f(1/n)$, we see that $\mu_n \rightharpoonup_w \delta_0$, but $\mu_n(A) = 1$, $\delta_0(A) = 0$. ⊣

Denote by $\mathsf{bd}(A)$ the *boundary* of a set $A \subseteq X$; thus

$$\mathsf{bd}(A) = \mathsf{cl}(A) \setminus \left(\bigcup\{G \mid G \subseteq A \text{ open}\}\right)$$

($\bigcup\{G \mid G \subseteq A \text{ open}\}$ is just the largest open set that is contained in A). Then the Portmanteau Theorem entails

Corollary 1.6.6. $\mu_n \rightharpoonup_w \mu$ *iff* $\lim_{n\to\infty} \mu_n(A) = \mu(A)$ *for each Borel set A with* $\mu(\mathsf{bd}(A)) = 0$. □

Hence in order to learn something about the weak limit of a sequence $(\mu_n)_{n\in\mathbb{N}}$ of measures, we could compute the setwise limit of $(\mu_n(A))_{n\in\mathbb{N}}$, but only for those sets A to which the boundary is assigned the value 0 by the limiting measure. Consequently, determining the limit probability through the limit of sequences of probabilities for sets is close to hopeless.

It is easy to find a dense subset in $\mathfrak{S}(X)$. As one might expect, the measures living on discrete subsets are dense.

Proposition 1.6.7. *Let X be a separable metric space. The set of*

$$\Big\{\sum_{k\in\mathbb{N}} r_k \cdot \delta_{x_k} \mid x_k \in X, r_k \geq 0, \sum_{k\in\mathbb{N}} r_k \leq 1\Big\}$$

discrete measures is dense in the topology of weak convergence. □

Corollary 1.6.8. *If X is a separable metric space, then $\mathfrak{S}(X)$ is a separable metric space in the topology of weak convergence.* □

1.6.2 Metrizability of the Weak Topology

We will discuss briefly two metrics that are known to give rise to the topology of weak convergence. The Prohorov metric is defined in terms of neighborhoods of measurable sets, while the Hutchinson metric makes use of integrals of Lipschitz maps.

Let (X, d) be a metric space, and define for $A \subseteq X, r > 0$ the r-neighborhood of A by

$$A^r := \{x \in X \mid d(x, A) < r\}.$$

Thus $\{x\}^r = B_{r,d}(x)$, so A^r builds a kind of a measured cloud around A.

Prohorov metric

Now define the *Prohorov metric* $\mathbf{d}_P$ on $\mathfrak{S}(X)$ through

$$\mathbf{d}_P(\mu_1, \mu_2) := \inf\{\varepsilon > 0 \mid \forall A \in \mathcal{B}(X) : \mu_1(A) \leq \mu_2(A^\varepsilon) + \varepsilon \text{ and } \mu_2(A) \leq \mu_1(A^\varepsilon) + \varepsilon\}.$$

Lemma 1.6.9. $(\mathfrak{S}(X), \mathbf{d}_P)$ *is a metric space.* □

Relating the Prohorov metric and the topology of weak convergence, we note first that metric convergence implies weak convergence. This is a simple application of the Portmanteau Theorem 1.6.4.

Lemma 1.6.10. *If $\mathbf{d}_P(\mu_n, \mu) \to 0$, then $\mu_n \rightharpoonup_w \mu$. Thus each set which is open in the metric topology is open in the weak topology.* □

The converse is considerably more involved, however. The following can be shown.

Theorem 1.6.11. *Let X be a separable metric space, then the Prohorov metric is a metric for the topology of weak convergence.* □

$\mathfrak{S}(X)$ is a Polish space provided X is one; thus applying the $\mathfrak{S}$-functor to a Polish space does not leave the realm of Polish spaces. We will show later that this functor is actually an endofunctor on the category of Polish spaces; for the time being, however, we will lay only the groundwork for this.

Proposition 1.6.12. *Let X be a Polish space. Then $\mathfrak{S}(X)$ is a Polish space in the topology of weak convergence.* □

The proof rests on the embedding of a Polish space into a compact metric space, solving the problem there, and transporting the solution back into the original environment. The solution for the compact case then depends on a classical tool like the Riesz Representation Theorem.

It can actually be shown that the Prohorov metric is complete. The proof of this is considerably more complicated. It rests also on the Riesz Theorem. But since we need merely the fact that $\mathfrak{S}(X)$ is Polish for Polish X, we do not want to delve into the complexities of that proof.

Hutchinson Metric

The Prohorov metric is sometimes a bit awkward to handle. In these cases, the Hutchinson metric may be an alternative. Although not needed in the sequel, we give its definition for completeness. Before defining it, recall that for a metric space (X, d) a map $f : X \to \mathbb{R}$ is called *Lipschitz* iff

$$\forall x, x' \in X : \left|f(x) - f(x')\right| \leq d(x, x').$$

Clearly, a Lipschitz map is uniformly continuous. Denote by $\mathbf{K}_1(X)$ the set of all Lipschitz maps f on X with $\sup_{x\in X} |f(x)| \leq 1$.

Define for $\mu, \mu' \in \mathfrak{S}(X)$ their *Hutchinson distance*

$$\mathfrak{d}_{\mathfrak{H}}(\mu, \mu') := \sup\Big\{\Big|\int_X f\ d\mu - \int_X f\ d\mu'\Big| \mid f \in \mathbf{K}_1(X)\Big\}.$$

This metric serves the same purposes as the Prohorov distance.

Proposition 1.6.13. *$\mathfrak{d}_{\mathfrak{H}}$ is a metric on $\mathfrak{S}(X)$ for the topology of weak convergence. The metric space $\big(\mathfrak{S}(X), \mathfrak{d}_{\mathfrak{H}}\big)$ is complete.* □

1.6.3 The Weak-σ-Algebra*

The σ-algebra of Borel sets for the topology of weak convergence is just the weak*-σ-algebra. The latter σ-algebra can be described for each measurable space.

Definition 1.6.14. *Let $(M, \mathcal{M})$ be a measurable space. The initial σ-algebra $\mathcal{M}^\bullet$ which makes all evaluation maps $\mu \mapsto \mu(E)$ for $E \in \mathcal{M}$ measurable is called the* weak*-σ-algebra.

Two remarks are in order. First, to show for a measurable space $(N, \mathcal{N})$ that a map $\Phi : N \to \mathfrak{S}(M, \mathcal{M})$ is $\mathcal{N}$-$\mathcal{M}^\bullet$-measurable it is sufficient to show that for $B \in \mathcal{M}$ the set $\{n \in N \mid \Phi(n)(B) \ltimes q\}$ is a member of $\mathcal{N}$ by Lemma 1.3.2 when $\ltimes$ is taken from the relational operators $\leq, <, >, \geq$; the set B may even be taken from a generator for $\mathcal{M}$. This observation makes handling measurable maps into $\mathfrak{S}(M, \mathcal{M})$ more versatile than meets the eye at first. Second, if $\mathcal{M}_0$ is a generator from $\mathcal{M}$, then we may deduce from Lemma 1.3.2

$$\mathcal{M}^\bullet = \sigma(\{\{\mu \in \mathfrak{S}(M, \mathcal{M}) \mid \mu(B) \ltimes q\} \mid q \in \mathbb{Q}, B \in \mathcal{M}_0\}),$$

so that $\mathcal{M}^\bullet$ is countably generated whenever $\mathcal{M}_0$ is.

Measurability carries over to the integral as a function of the measure.

Proposition 1.6.15. *Let $f \in \mathcal{F}(M, \mathcal{M})$ for the measurable space $(M, \mathcal{M})$ be a nonnegative and bounded measurable function. Then*

$$\mu \mapsto \int_M f \, d\mu$$

constitutes a $\mathcal{M}^\bullet$-$\mathcal{B}(\mathbb{R})$ measurable map.

Proof 1. If $f = \sum_{i=1}^k \alpha_i \cdot \chi_{A_i}$ is a step function with $A_i \in \mathcal{M}$, then

$$\mu \mapsto \int_M f \, d\mu = \sum_{i=1}^k \alpha_i \cdot \mu(A_i),$$

so the assertion is true for step functions f.

2. Because a general measurable function f can be written as $f = f^+ - f^-$ with $f^+ \geq 0$ and $f^- \geq 0$, and because the integral is additive, we may and do assume that $f \geq 0$ holds. From the Monotone Convergence Theorem (Proposition 1.6.1) we obtain a sequence $(f_n)_{n \in \mathbb{N}}$ of measurable step functions $f_n \geq 0$ with $f = \sup_{n \in \mathbb{N}} f_n$, so that

$$\mu \mapsto \int_M f \, d\mu = \lim_{n \to \infty} \int_M f_n \, d\mu.$$

Since the limit of a sequence of measurable functions is measurable again, the assertion follows. □

We obtain from this observation that $\mathcal{M}^\bullet$ is the smallest σ-algebra on $\mathfrak{S}(M, \mathcal{M})$ rendering the integral of a bounded measurable map as a function of the measure a measurable function.

We can say more for separable metric spaces. Then the weak*-σ-algebra constitutes the collection of Borel sets for the topology of weak convergence.

Proposition 1.6.16. $\mathcal{B}(X)^\bullet = \mathcal{B}(\mathfrak{S}(X))$. *for the separable metric space* X.

Proof 1. By the Portmanteau Theorem 1.6.4 the set $\{\mu \in \mathfrak{S}(M, \mathcal{M}) \mid \mu(B) \geq q\}$ is closed whenever F is closed: if $(\mu_n)_{n\in\mathbb{N}}$ is a set with $\mu_n(B) \geq q$ for all n, then

$$q \leq \liminf_{n\to\infty} \mu_n(B) \leq \mu(B).$$

This implies that $\mathcal{B}(X)^\bullet \subseteq \mathcal{B}(\mathfrak{S}(X))$.

2. On the other hand, $\mu \mapsto \int_X f\, d\mu$ is $\mathcal{B}(X)^\bullet$-measurable, whenever $f : X \to \mathbb{R}$ is Borel measurable. This is clear if $f = \chi_A$ by the definition of the weak*-σ-algebra, because $\mu(A) = \int_X \chi_A\, d\mu$. Thus the claim is true if f is a measurable step function. Hence we may deduce it for all f by approximating it through step functions Proposition 1.6.1 (decompose $f = f^+ - f^-$ with $f^+, f^- \geq 0$, and approximate each map separately).

Thus $\mu \mapsto \int_X f\, d\mu$ is also measurable for continuous f. Consequently each element of a base $U(\mu_0, \varepsilon, f_1, \ldots, f_n)$ is an element of $\mathcal{B}(X)^\bullet$; hence each open set, being a countable union of base elements, is in $\mathcal{B}(X)^\bullet$. This implies $\mathcal{B}(\mathfrak{S}(X)) \subseteq \mathcal{B}(X)^\bullet$. □

Stochastic relations

The weak*-σ-algebra is important in characterizing stochastic relations.

Definition 1.6.17. *A* stochastic relation $K : (M, \mathcal{M}) \rightsquigarrow (N, \mathcal{N})$ *between the measurable spaces* $(M, \mathcal{M})$ *and* $(N, \mathcal{N})$ *is a* $\mathcal{M}$-$\mathcal{N}^\bullet$*-measurable map* $K : (M, \mathcal{M}) \to \mathfrak{S}(N, \mathcal{N})$.

Intuitively, $K(m)(T)$ can be interpreted as the probability to hit $T \in \mathcal{N}$ after an input of $m \in M$. The characterization is obvious from the definition of the weak*-σ-algebra. It will be used silently throughout; algebraic properties will be focussed on later, see Chapter 3. Since we will usually omit the σ-algebras from measurable spaces, a stochastic relation K with $K : (M, \mathcal{M}) \rightsquigarrow (N, \mathcal{N})$ will be denoted simply through $\mathsf{K} = (M, N, K)$, unless specific σ-algebras will be addressed.

Proposition 1.6.18. *Given measurable spaces* $(M, \mathcal{M})$ *and* $(N, \mathcal{N})$*, the following is equivalent:*

a. $K : (M, \mathcal{M}) \rightsquigarrow (N, \mathcal{N})$ *is a stochastic relation.*

b. $K(m)$ *is a subprobability measure on* $(N, \mathcal{N})$ *for each* $m \in M$ *such that the map* $m \mapsto K(m)(Q)$ *is* $\mathcal{M}$*-measurable for each measurable set* $Q \in \mathcal{N}$. □

Because the weak*-σ-algebra constitutes the collection of Borel sets for the topology of weak convergence, stochastic relations between separable metric spaces are Borel measurable maps. This follows from Proposition 1.6.16.

Mapping Measures

We have seen that $\mathfrak{S}$ acts on measurable spaces and returns the measurable space of all subprobabilities. But it may be put to work on measurable functions as well, returning a measurable map on subprobabilities. Define for an $\mathcal{A}$-$\mathcal{B}$-measurable map $f : X \to Y$ between two measurable spaces $(X, \mathcal{A})$ and $(Y, \mathcal{B})$ the map $\mathfrak{S}(f) : \mathfrak{S}(X, \mathcal{A}) \to \mathfrak{S}(Y, \mathcal{B})$ upon setting

$$\mathfrak{S}(f)(\mu)(B) := \mu(f^{-1}[B]).$$

Because we have for $B \in \mathcal{B}$

$$\mathfrak{S}(f)^{-1}[\{\nu \in \mathfrak{S}(Y, \mathcal{B}) \mid \nu(B) \geq q\}] = \{\mu \in \mathfrak{S}(X, \mathcal{A}) \mid \mu(f^{-1}[B]) \geq q\},$$

$\mathfrak{S}(f)$ is $\mathcal{A}^\bullet$-$\mathcal{B}^\bullet$-measurable by Lemma 1.3.2. Consequently, $\mathfrak{S}$ is an endofunctor on the category of all measurable spaces with measurable maps; similarly, assigning a measurable space all its probabilities yields the probability functor $\mathfrak{P}$, another endofunctor on that category.

$\mathfrak{S}$ transforms surjective maps between analytic spaces into surjective maps between the corresponding spaces of subprobabilities; to be specific [20, Proposition 1.101]:

Proposition 1.6.19. *Let* X *be an analytic space, and* Y *a separable metric space. If* $f : X \to Y$ *is measurable and onto, so is* $\mathfrak{S}(f) : \mathfrak{S}(X) \to \mathfrak{S}(Y)$. □

The Change of Variable formula is an important tool when elaborating this construction.

Lemma 1.6.20. *Let* $(X, \mathcal{A})$ *and* $(Y, \mathcal{B})$ *be analytic spaces, and assume that* $f : X \to Y$ *is* $\mathcal{A}$*-*$\mathcal{B}$*-measurable. Then*

$$\int_Y g(y)\ \mathfrak{S}(f)(\mu)(dy) = \int_X (g \circ f)(x)\ \mu(dx)$$

holds whenever $g : Y \to \mathbb{R}$ *is* $\mathcal{B}$*-measurable and bounded. In particular*

$$\int_B g(y)\ \mathfrak{S}(f)(\mu)(dy) = \int_{f^{-1}[B]} (g \circ f)(x)\ \mu(dx)$$

whenever $B \subseteq Y$ *is a measurable set.* □

The Change of Variables formula is familiar from calculus with very specific requirements on the transform: if T is a differentiable mapping from an open set $V \subseteq \mathbb{R}^k$ onto a bounded open set $W \subseteq \mathbb{R}^k$ such that T is one-to-one and the inverse of T is continuous, then we have for the integrable map $f : W \to \mathbb{R}$

$$\int_{T[V]} f(w)\, dw = \int_V (f \circ T)(v)|J_T(v)|\, dv$$

with $J_T(x)$ as the Jacobian of T at x, see [71, 8.27]. This is in fact a special case of Lemma 1.6.20, see [45, Chapter 20.2] for a comprehensive discussion.

Some examples are in order.

Example 1.6.21. Let $X = \{1, \ldots, n\}$ be a discrete space, $(Y, \mathcal{B})$ a measurable space, and $f : X \to Y$ be a measurable map. The uniform distribution υ on X assigns to each element probability $1/n$, so that for $B \in \mathcal{B}$

$$\mathfrak{S}\,(f)\,(\upsilon)(B) = \upsilon\big(f^{-1}\,[A]\big) = \frac{\#\{i \in X \mid f(i) \in A\}}{n},$$

$\#$ denoting cardinality. In general, if $p \in \mathfrak{S}\,(X)$ is arbitrary, and if Y is discrete as well, then we obtain for each $y \in Y$

$$\mathfrak{S}\,(f)\,(p)(\{y\}) = \sum_{f(x)=y} p(\{x\}).$$

Thus the probability to hit y equals the cumulative probability for hitting an x with $f(x) = y$. ⊣

Example 1.6.22. Let $X := Y := \mathbb{R}$ with $f(x) := x^2$. The density for a rectangular distribution around $a \in \mathbb{R}$ is given by

$$\mathbf{U}(x \mid a, h) := \frac{1}{2h}\chi_{[a-h,a+h]}(x)$$

$(h > 0)$, see [34, 5.6]. The associated probability μ will be computed for intervals $]-\infty, q]$, because these intervals generate the Borel sets; it is obtained as

$$\mu\big([-\infty, q]\big) = \int_{-\infty}^{q} U(x \mid a, h)\, dx,$$

which evaluates to

$$\begin{cases} 0, & \text{if } q < a - h \\ (q - a + h)/2h, & \text{if } a - h \leq q \leq a + h, \\ 1, & \text{if } q > a + h. \end{cases}$$

Put for simplicity $a = 0, h = 1/2$; then we have for $0 \leq q \leq 1$

$$\begin{aligned}\mathfrak{S}(f)(\mu)([0,q]) &= \mu(f^{-1}[[0,q]])\\ &= 2\mu([0,\sqrt{q}])\\ &= 2\int_0^{\sqrt{q}} \mathbf{U}(x \mid 0, 1/2)\, dx\\ &\overset{(*)}{=} \int_0^q \frac{\mathbf{U}(\sqrt{x} \mid a, 1/2)}{\sqrt{x}}\, dx\\ &= \sqrt{q}.\end{aligned}$$

Equation $(*)$ derives from the Change of Variables formula for $T(v) := \sqrt{v}$ with $T'(v) = 1/(2\sqrt{v})$. Consequently, the image measure has in this case

$$\frac{1}{\sqrt{x}}\mathbf{U}(\sqrt{x} \mid a, 1/2),\ x > 0$$

as its density. ⊣

Example 1.6.23. Let $X := Y := \mathbb{R}$ with $f(x) := 2x + 1$. The density for the normal distribution with mean m and variance σ is given through

$$\mathbf{N}(x \mid m, \sigma) := \frac{1}{\sigma\sqrt{2\pi}} \cdot \exp(-\frac{(x-m)^2}{2\sigma^2})$$

(see [1, 26.2.8], [34, 5.7]) with associated probability measure

$$\mu(A) := \int_A \mathbf{N}(x \mid m, \sigma)\, dx.$$

For calculating $\mathfrak{S}(f)(\mu)$, it again suffices to compute its value for intervals $]-\infty, q]$. We obtain

$$\begin{aligned}\mathfrak{S}(f)(\mu)(]-\infty,q]) &= \mu(f^{-1}[]-\infty,q]])\\ &= \mu(]-\infty, \frac{q-1}{2}])\\ &= \int_{-\infty}^{(q-1)/2} \mathbf{N}(x \mid m, \sigma)\, dx\\ &\overset{(\star)}{=} \frac{1}{2}\int_{-\infty}^{q} \mathbf{N}(\frac{x-1}{2} \mid m, \sigma)\, dx\\ &\overset{(\dagger)}{=} \int_{-\infty}^{q} \mathbf{N}(x \mid 2m+1, 2\sigma)\, dx,\end{aligned}$$

where Equation $(\star)$ derives from the transformation $v \mapsto (v-1)/2$, and Equation $(\dagger)$ is just a simple manipulation of the density.

As a consequence, we see that f transforms a normal distribution into a normal distribution again (which is not particularly surprising). ⊣

Morphisms for Stochastic Relations

While the Change of Variables is a most helpful technical tool, a far more fundamental application of transforming maps through $\mathfrak{S}$ occurs when defining morphisms for stochastic relations.

Definition 1.6.24. *Let* $\mathsf{K} = (X, Y, K)$ *and* $\mathsf{L} = (A, B, L)$ *be stochastic relations over general measurable spaces. Then* $\mathsf{f} : \mathsf{K} \to \mathsf{L}$ *is a* morphism from K to L *iff* $\mathsf{f} = (f, g)$ *with* $f : X \to A$ *and* $g : Y \to B$ *as surjective measurable maps such that this diagram is commutative:*

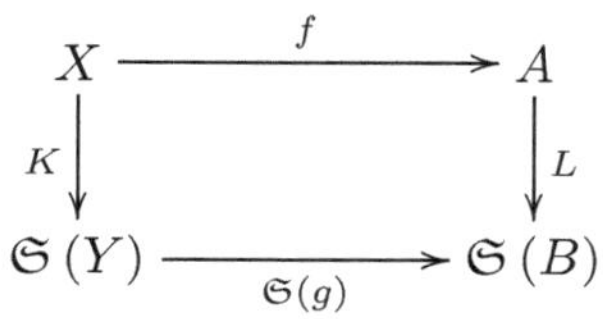

Spelling out the condition implied by the diagram means that we have for each $x \in X$ and each measurable set $D \subseteq B$

$$L\big(f(x)\big)(D) = \big(L \circ f\big)(x)(D) = \big(\mathfrak{S}\,(g) \circ K\big)(x)(D) = K(x)\big(g^{-1}\,[D]\big),$$

so that the probability to hit set D from input $f(x)$ is the same as hitting the set $g^{-1}\,[D]$ from input x.

In the investigation of various forms of bisimilarity we will encounter quite frequently the situation where we have a cospan of morphisms, say

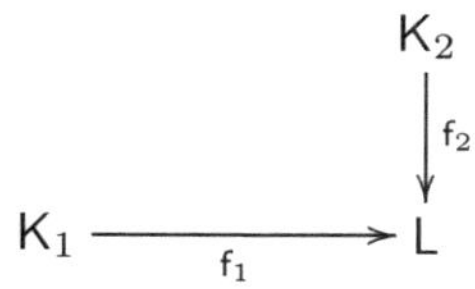

and where we want to construct a semi-pullback, i.e., we want to complete this to a closed square. Hence we want to find a span, i.e., a stochastic relation M and morphisms g_1 and g_2, rendering this diagram commutative.

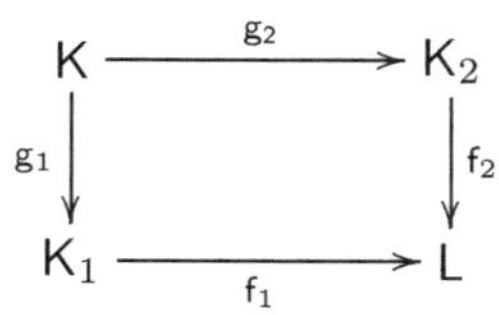

The construction of a span from a cospan is recorded in

Proposition 1.6.25. *Let* K_1 *and* K_2 *be stochastic relations defined over analytic spaces, and assume that* L *is a stochastic relation defined over separable measurable spaces. Given morphisms* $\mathsf{f}_1 : \mathsf{K}_1 \to \mathsf{L}$ *and* $\mathsf{f}_2 : \mathsf{K}_2 \to \mathsf{L}$ *there* *exists a stochastic relation* $\mathsf{K} = (X, Y, K)$ *over Polish spaces* X *and* Y *and morphisms* $\mathsf{g}_1 : \mathsf{K} \to \mathsf{K}_1$ *and* $\mathsf{g}_2 : \mathsf{K} \to \mathsf{K}_2$ *with* $\mathsf{f}_1 \circ \mathsf{g}_1 = \mathsf{f}_2 \circ \mathsf{g}_2$. □

If $\mathsf{K}_i = (X_i, Y_i, K_i)$ is defined over a Polish space, and $\mathsf{f}_i = (\phi_i, \psi_i)$ for $i = 1, 2$, the constituent spaces for K can be described through

$$X = \{\langle x_1, x_2\rangle \mid \phi_1(x_1) = \phi_2(x_2)\},$$
$$Y = \{\langle y_1, y_2\rangle \mid \psi_1(y_1) = \psi_2(y_2)\},$$

and the morphisms can be constructed as the corresponding projections. Neither X or Y nor the resulting stochastic relation need be uniquely determined up to isomorphism, since we do not claim to construct a pullback.

Both the existence and the explicit shape of the constituents of the semi-pullback will be of use later on.

It is sometimes easier to argue with Polish rather than analytic spaces as the base spaces for stochastic relations. This statement helps in such a situation:

Proposition 1.6.26. *Assume that* A *and* B *are separable metric spaces with a stochastic relation* $\mathsf{M} = (A, B, M)$, *and suppose that* $f : X \to A$ *and* $g : Y \to B$ *are Borel measurable and onto, where* X *and* Y *are Polish. Then there exists s stochastic relation* $\mathsf{K} = (X, Y, K)$ *such that* $(f, g) : \mathsf{K} \to \mathsf{M}$ *is a morphism.* □

Closing this section, we illustrate the concept of a morphism through some examples.

Example 1.6.27. Given $\lambda \in \mathbb{R}$, the Poisson distribution with parameter λ is given through

$$\mathbf{P}(n \mid \lambda) = \frac{\lambda^n}{n!}\exp(-\lambda)$$

($n \in \mathbb{N}$). The probability for a random variable with values in $\mathbb{N}$ with this distribution to hit value r is just $\mathbf{P}(r \mid \lambda)$. Let $X := A := \mathbb{R}$, $Y := \mathbb{N}$, and $B := \{n^2 \mid n \in \mathbb{N}\}$; $f : X \to Y$ is defined as $f(x) := x + t$ for some real parameter t, and $g : Y \to B$ is $g(y) := y^2$. Put

$$K(x)(r) := K(x)(\{r\}) := \frac{|x|^r}{r!}\exp(-|x|),$$

so that $K : X \rightsquigarrow Y$ (for discrete measures we will identify elements with their singletons). Assume that

$$L(a)(\ell) = \frac{|x-t|^{\sqrt{\ell}}}{r!}\exp(-|x-t|)$$

for $a \in A, \ell \in B$. Then $L : A \rightsquigarrow B$, and (f, g) is a morphism for K and L. ⊣

Example 1.6.28. The binomial distribution on $n+1$ items with parameter $p, 0 < p < 1$, is given by

$$\mathbf{B}(i \mid n, p) := \binom{n}{i} p^i (1-p)^{n-i}, 0 \leq i \leq n.$$

Let $X := A :=]0,1[, Y := \mathbb{Z}$, and $B := \mathbb{N} \cup \{0\}$, and define $f : X \to A$ as $f(x) := x^2$, $g : Y \to B$ as $g(y) := |y|$. Let

$$K(x)(i) := \frac{1}{2}\mathbf{B}(|i| \mid n, \sqrt{x}),$$

and put

$$L(x)(k) := \frac{1}{2}\big(\mathbf{B}(k \mid n, \sqrt{x}) + \mathbf{B}(-k \mid n, \sqrt{x})\big);$$

then (f, g) is a morphism for $K : X \rightsquigarrow Y$ and $L : A \rightsquigarrow B$. Conversely, given $L : A \rightsquigarrow B$, the relation $K : X \rightsquigarrow Y$ is bound by the relation

$$K(\sqrt{x})(i) + K(\sqrt{x})(-i) = \begin{cases} \mathbf{B}(i \mid n, x) & \text{if } 0 \leq i \leq n, \\ 0 & \text{otherwise.} \end{cases}$$

Consequently, there is a whole family of stochastic relations $\big(]0,1[, \mathbb{Z}, K\big)$ that can be mapped to $\big(]0,1[, \mathbb{N} \cup \{0\}, L\big)$ through a morphism (f, g) with these specific components. ⊣

Example 1.6.29. Let $X := Y := B := \mathbb{R}$ and $A := \mathbb{R}_+$, and define $f : X \to A$ as $x \mapsto x^2$, and $g : Y \to B$ as $g(y) := 2y + 1$. Define $K : X \rightsquigarrow Y$ so that $K(x)$ is distributed according to the normal distribution $\mathbf{N}(\cdot \mid x, \sigma)$ (see Example 1.6.23), and let $L : A \rightsquigarrow B$ be defined so that $L(a)$ has density

$$\mathbf{N}(\cdot \mid 2\sqrt{a} + 1, 2\sigma).$$

Then (f, g) is a morphism from K to L.

Conversely, if $L : A \rightsquigarrow B$ is given through normal distribution $\mathbf{N}(\cdot \mid a, \sigma)$ for $L(a)$, then $K : X \rightsquigarrow Y$ must be given by normal distribution

$$\mathbf{N}(\cdot \mid x^2 - 1/2, \sigma/2)$$

for $K(x)$ in order for (f, g) to be a morphism. The computations follow very closely that in Example 1.6.23. ⊣

1.6.4 Applications of the π-λ-Theorem

The π-λ-Theorem is a fairly important technical tool; it is used typically in the following scenario: we have a property P for which we know the following

i. $P(A)$ holds for all elements A of a generator $\mathcal{A}$ of a σ-algebra $\mathcal{B}$ over the basic set X.
ii. $P(A)$ implies $P(X \setminus A)$.
iii. if $P(A_n)$ holds for all $n \in \mathbb{N}$, and $(A_n)_{n\in\mathbb{N}}$ is mutually disjoint, then $P(\bigcup_{n\in\mathbb{N}} A_n)$ holds.

We then have a look at

$$\mathcal{G} := \{A \in \mathcal{B} \mid P(A)\}$$

($\mathcal{G}$ stands of course for the *good guys*); then $P(A)$ holding for all $A \in \mathcal{A}$ translates into $\mathcal{A} \subseteq \mathcal{G}$, and the other two properties make sure that G is a π-λ-system. We conclude then from the π-λ-Theorem 1.3.1 that

$$\mathcal{G} = \sigma(\mathcal{A}) = \mathcal{B},$$

provided $\mathcal{A}$ is closed under finite intersections. This has been the argumentation for example in the proof of Proposition 1.4.18. It shows that this theorem is rather versatile.

A typical application permits us to infer the equality of two subprobabilities on a σ-algebra from their equality on a generator which is closed under finite intersections. This applies in particular to measures on the reals: here it suffices to investigate the value on intervals of the form $]-\infty, r]$ (we silently did this already when discussing examples for image measures and morphisms in Section 1.6.3).

Lemma 1.6.30. *Let $(X, \mathcal{A})$ be a measurable space, $\mathcal{A} = \sigma(\mathcal{A}_0)$ with $\mathcal{A}_0$ being closed under finite intersections. Assume that $\mu(A) = \mu'(A)$ holds for all $A \in \mathcal{A}_0$, where $\mu, \mu' \in \mathfrak{S}(X, \mathcal{A})$. Then $\mu = \mu'$.*

Proof Consider

$$\mathcal{D} := \{A \in \mathcal{A} \mid \mu(A) = \mu'(A)\},$$

then $\mathcal{D}$ is closed under countable disjoint unions, because measures are σ-additive. Since $X = \bigcap \emptyset \in \mathcal{A}_0$, $\mathcal{D}$ is closed under complementation as well. Because $\mathcal{A}_0 \subseteq \mathcal{D}$, the π-λ-Theorem 1.3.1 says $\mathcal{A} = \sigma(\mathcal{A}_0) \subseteq \mathcal{D} \subseteq \mathcal{A}$. Thus $\mu(A) = \mu'(A)$ for all $A \in \mathcal{A}$. □

We want to demonstrate the application of this theorem for settling questions of measurability, which is also somewhat typical for getting it to work later on.

Define for a set $D \subseteq X \times Y$ the *cut* D_x *of* D *at* $x \in X$ as

$$D_x := \{y \in Y \mid \langle x, y\rangle \in D\},$$

and, symmetrically, the cut D^y of D at $y \in Y$ as $D^y := \{x \in X \mid \langle x, y\rangle \in D\}$.

Lemma 1.6.31. *Let $(X, \mathcal{A})$ and $(Y, \mathcal{B})$ be measurable spaces, and fix $D \in \mathcal{A} \otimes \mathcal{B}$. The map $\langle \nu, x\rangle \mapsto \nu(D_x)$ is a $\mathcal{B}^\bullet \otimes \mathcal{A}$-measurable map on $\mathfrak{S}(Y, \mathcal{B}) \times X$.*

Proof Consider

$$\mathcal{D} := \{D \in \mathcal{A} \otimes \mathcal{B} \mid \langle \nu, x \rangle \mapsto \nu(D_x) \text{ is } \mathcal{B}^\bullet \otimes \mathcal{A} - \text{measurable}\}.$$

Since $((X \times Y) \setminus D)_x = Y \setminus (D_x)$ and

$$(\bigcup_{n \in \mathbb{N}} D_n)_x = \bigcup_{n \in \mathbb{N}} (D_n)_x,$$

it is clear that $\mathcal{D}$ is closed under taking complements and countable disjoint unions. Now let $D = A \times B$ with $A \in \mathcal{A}, B \in \mathcal{B}$. Then $\nu(D_x) = \chi_A(x) \cdot \nu(B)$, thus $\langle \nu, x \rangle \mapsto \nu(D_x)$ is evidently $\mathcal{B}^\bullet \otimes \mathcal{A}$-measurable. But this implies that all measurable rectangles are members of $\mathcal{D}$, and since the set of all these rectangles is closed under finite intersections, $\mathcal{D}$ equals the σ-algebra generated from them, which coincides with $\mathcal{A} \otimes \mathcal{B}$. The assertion is hence true for all measurable subsets of the product. □

Lemma 1.6.31 entails that both $x \mapsto \nu(D_x)$ and $\nu \mapsto \nu(D_x)$ are measurable (but the lemma says considerably more: it establishes joint measurability).

For the interpretation of the continuous time stochastic logics μ**CSL** in Section 2.4 we will require the construction of a projective limit for a projective family of stochastic relations. This construction will be introduced now; it provides among others a fairly typical application of the π-λ-Theorem.

Denote by $X^\infty := \prod_{k \in \mathbb{N}} X$ the infinite product of X with itself. The Borel sets $\mathcal{B}(X^\infty)$ are the smallest σ-algebra that contain all the *cylinder sets*, i.e., sets of the form $\prod_{n \in \mathbb{N}} A_n$ with all but a finite number of factors being equal to X. It is noted that the cylinder sets are closed under finite intersections; this observation will be crucial for some of the proofs that we will sketch below.

Definition 1.6.32. *Let X be a Polish space.*

1. *A sequence $(\mu_n)_{n \in \mathbb{N}}$ of probability measures $\mu_n \in \mathfrak{P}(X^n)$ is called a* projective system *iff $\mu_n(A) = \mu_{n+1}(A \times X)$ for all $n \in \mathbb{N}$ and all Borel sets $A \in \mathcal{B}(X^n)$.*
2. *A probability measure $\mu_\infty \in \mathfrak{P}(X^\infty)$ is called the* projective limit *of the projective system $(\mu_n)_{n \in \mathbb{N}}$ iff*

$$\mu_n(A) = \mu_\infty(A \times \prod_{j>n} X)$$

for all $n \in \mathbb{N}$ and $A \in \mathcal{B}(X^n)$.

Thus a sequence of measures is a projective system iff each measure is the projection of the next one; its projective limit is characterized through the property that its values on cylinder sets coincide with the value of a member of the sequence after taking projections.

It is not immediately obvious that a projective limit exists. The basic idea is to define the limit on the cylinder sets and then to extend this premeasure

— but it has to be established that it is a premeasure indeed. The crucial property is that $\mu_{n_k}(A_k) \to 0$ whenever $(A_n)_{n\in\mathbb{N}}$ is a sequence of cylinder sets A_k (with at most n_k components that do not equal X) that decreases to $\emptyset$. This property is difficult to establish without topological assumptions, but it holds in Polish spaces, and constitutes one of the reasons for our postulating the base space X to be Polish.

The crucial statement is as follows.

Proposition 1.6.33. *Let X be a compact metric space. Then the projective system $(\mu_n)_{n\in\mathbb{N}}$ has a unique projective limit μ_∞.*

Proof (Sketch) 1. Let $A = A'_k \times \prod_{j>k} X$ be a cylinder set with $A'_k \in \mathcal{B}(X^k)$; then define $\mu^*(A) := \mu_k(A'_k)$. Then μ^* is well defined, since the sequence forms a projective system. In order to show that μ^* is a premeasure on the cylinder sets, we have to take a decreasing sequence $(A_n)_{n\in\mathbb{N}}$ of cylinder sets with $\bigcap_{n\in\mathbb{N}} A_n = \emptyset$ and show that $\inf_{n\in\mathbb{N}} \mu^*(A_n) = 0$. In fact, suppose that $(A_n)_{n\in\mathbb{N}}$ is decreasing with $\mu^*(A_n) \geq \delta$ for all $n \in \mathbb{N}$, then we show that $\bigcap_{n\in\mathbb{N}} A_n \neq \emptyset$. We can write $A_n = A'_n \times \prod_{j>k_n} X$ for some $A'_n \in \mathcal{B}(X^{k_n})$. Since X is compact, we get for each n a compact set $K'_n \subseteq A'_n$ such that $\mu_{k_n}(A'_n \setminus K'_n) < \delta/2^n$. Because X^∞ is compact by Tichonov's Theorem, $K''_n := K'_n \times \prod_{j>k_n} X$ is a compact set, and $K_n := \bigcap_{j=1}^n K''_j \subseteq A_n$ is compact as well, with $\mu^*(A_n \setminus K_n) \leq \delta$. Thus $(K_n)_{n\in\mathbb{N}}$ is a decreasing sequence of nonempty compact sets; consequently,

$$\emptyset \neq \bigcap_{n\in\mathbb{N}} K_n \subseteq \bigcap_{n\in\mathbb{N}} A_n.$$

2. Since the cylinder sets generate the Borel sets of X^∞, and since μ^* is a premeasure, we know that there exists a unique extension $\mu_\infty \in \mathfrak{P}(X^\infty)$ to it. Clearly, if $A \subseteq X^n$ is a Borel set, then

$$\mu_\infty(A \times \prod_{j>n} X) = \mu^*(A \times \prod_{j>n} X) = \mu_n(A),$$

so we have constructed a projective limit.

3. Suppose that μ' is another probability measure in $\mathfrak{P}(X^\infty)$ that has the desired property. Consider

$$\mathcal{D} := \{D \in \mathcal{B}(X^\infty) \mid \mu_\infty(D) = \mu'(D)\}.$$

It is clear the $\mathcal{D}$ contains all cylinder sets, that it is closed under complements, and under countable disjoint unions. By the π-λ-Theorem 1.3.1 $\mathcal{D}$ contains the σ-algebra generated by the cylinder sets, hence all Borel subsets of X^∞. This establishes uniqueness of the extension. □

It is important for the proof that compact sets have the finite intersection property: if each finite intersection of a family of compact sets is nonempty, the intersection of the entire family cannot be empty. The construction from above can be made use of when we work in Proposition 1.6.33 in a compact scenario. We liberate ourselves from that restrictive assumption using the Alexandrov embedding of Polish spaces into compact metric spaces.

Proposition 1.6.34. *Let X be a Polish space, and $(\mu_n)_{n\in\mathbb{N}}$ be a projective system on X. Then there exists a unique projective limit $\mu_\infty \in \mathfrak{P}(X^\infty)$ for $(\mu_n)_{n\in\mathbb{N}}$.* □

Our interest in this construction comes from stochastic relations that may form a projective system. We will show now that there exists such a relation which may be thought as the (pointwise) projective limit.

Proposition 1.6.35. *Let X and Y be Standard Borel spaces, and assume that $J^{(n)}$ is a stochastic relation on X and Y^n for each $n \in \mathbb{N}$ such that the sequence $\left(J^{(n)}(x)\right)_{n\in\mathbb{N}}$ forms a projective system on Y for each $x \in X$. Then there exists a unique stochastic relation J_∞ on X and Y^∞ such that $J_\infty(x)$ is the projective limit of $\left(J^{(n)}(x)\right)_{n\in\mathbb{N}}$ for each $x \in X$.*

Proof 0. Assume that the measurable structure on X and on Y is induced by Polish topologies. Let for x fixed $J_\infty(x)$ be the projective limit of the projective system $\left(J^{(n)}(x)\right)_{n\in\mathbb{N}}$. By the definition of a stochastic relation we need to show that the map $x \mapsto J_\infty(x)(B)$ is measurable for every $B \in \mathcal{B}(Y^\infty)$.

1. In fact, consider

$$\mathcal{D} := \{B \in \mathcal{B}(Y^\infty) \mid x \mapsto J_\infty(x)(B) \text{ is measurable}\};$$

then the general properties of measurable functions imply that $\mathcal{D}$ is a σ-algebra on Y^∞. Take a cylinder set $B = B_0 \times \prod_{j>k} Y$ with $B_0 \in \mathcal{B}(Y^k)$ for some $k \in \mathbb{N}$; then, by the properties of the projective limit, we have $J_\infty(x)(B) = J^{(k)}(x)(B_0)$. But $x \mapsto J^{(k)}(x)(B_0)$ constitutes a measurable function on X. Consequently, $B \in \mathcal{D}$, and so $\mathcal{D}$ contains the cylinder sets which generate $\mathcal{B}(Y^\infty)$. Thus measurability is established for each Borel set $B \subseteq Y^\infty$, applying the π-λ-Theorem as in the last part of the proof for Proposition 1.6.33. □

This construction will be put to use when interpreting a path logic over infinite paths in Chapter 2.4.

1.7 Smooth Equivalence Relations

Smooth equivalence relations occur naturally when investigating modal logics with a countable number of formulas. Let $\mathcal{M}$ be a model for such a logic, and denote by $[\![\phi]\!]_\mathcal{M}$ the extension of formula ϕ, i.e, the set of worlds in which ϕ is valid. Assume that these extensions are Borel subsets of some analytic space. Given two worlds w and w', define them as equivalent iff they cannot be separated through a formula, hence iff

$$\mathcal{M}, w \models \phi \Leftrightarrow \mathcal{M}, w' \models \phi$$

holds for all formulas ϕ. Consequently, the two worlds are equivalent iff

$$w \in [\![\phi]\!]_{\mathcal{M}} \Leftrightarrow w' \in [\![\phi]\!]_{\mathcal{M}}$$

holds for each formula ϕ. This equivalence relation has the property that it is determined through the countable family

$$\{[\![\phi]\!]_{\mathcal{M}} \mid \phi \text{ is a formula}\}$$

of Borel sets. Smooth relations are defined in this way through a determining sequence of sets; they occur quite naturally in a variety of contexts, and we will study these relations now in some detail. We first give a definition in terms of a determining sequence of Borel sets and relate other characterizations of smoothness in Lemma 1.7.3. Then we will study invariant sets as an important tool, and finally we will study the relationship of smooth relations on an analytic space X and on the space $\mathfrak{S}(X)$ of its subprobability measures.

Definition 1.7.1. *Let X be an analytic space and ρ an equivalence relation on X. Then ρ is called* smooth *iff there exists a sequence $(A_n)_{n\in\mathbb{N}}$ of Borel sets such that*

$$x \;\rho\; x' \Leftrightarrow \forall n \in \mathbb{N} : [x \in A_n \Leftrightarrow x' \in A_n].$$

The sequence $(A_n)_{n\in\mathbb{N}}$ is said to determine *the relation ρ.*

We obtain immediately from the definition that a smooth equivalence relation — seen as a subset of the Cartesian product — is a Borel set:

Corollary 1.7.2. *Let ρ be a smooth equivalence relation on the analytic space X. Then ρ is a Borel subset of $X \times X$.* □

The following characterization of smooth equivalence relations is sometimes helpful and shows that it is not necessary to look only at sequences of sets. It indicates that the kernels of Borel measurable maps and smooth relations are intimately related.

Lemma 1.7.3. *Let ρ be an equivalence relation on an analytic set X. Then these conditions are equivalent:*

a. ρ is smooth.
b. There exists a sequence $(f_n)_{n\in\mathbb{N}}$ of Borel maps $f_n : X \to Z$ into an analytic space Z such that $\rho = \bigcap_{n\in\mathbb{N}} \mathsf{ker}(f_n)$.
c. There exists a Borel map $f : X \to Y$ into an analytic space Y with $\rho = \mathsf{ker}(f)$.
d. There exists a Borel map $f : X \to [0,1]$ with $\rho = \mathsf{ker}(f)$. □

Thus each smooth equivalence relation may be represented as the kernel of a Borel map, and vice versa.

Example 1.7.4. Consider the space $X := \{0,1\}^\infty$, which is a compact metric space under the product topology when $\{0,1\}$ carries the discrete topology, so it is in particular a Polish space. Define on X the equivalence relation

$$x \; E_0 \; x' \text{ iff } x_n = x'_n \text{ for almost all } n.$$

Thus $x \; E_0 \; x'$ iff there exists $k \in \mathbb{N}$ such that $x_n = x'_n$ for all $n \geq k$. Then E_0 is an equivalence relation. E_0 is not smooth, however. To see this, define for $k \in \mathbb{N}$ the σ-algebra

$$\mathcal{A}_k := \sigma\Big(\bigcup_{n \geq k} \pi_n^{-1}\left[\mathcal{B}(\{0,1\})\right]\Big),$$

and let

$$\mathcal{A}_\infty := \bigcap_{k \in \mathbb{N}} \mathcal{A}_k.$$

This is the σ-algebra of *tail events*. Take moreover as a probability measure $\mu := \bigotimes_{k\in\mathbb{N}} \mu_0$ with $\mu_0 := \frac{1}{2}(\delta_0 + \delta_1)$ as the uniform distribution on $\{0,1\}$. Now suppose that $E_0 = \mathsf{ker}\,(f)$ for some Borel map $f : X \to [0,1]$. Because $f(x)$ does not depend on the first components of x, it is easy to see that f is $\mathcal{A}_\infty$-measurable. By Kolmogorov's famous Zero-One-Law [77, Theorem IV.1.1] we can find $c \in [0,1]$ such that $f(x) = c$ for almost all $x \in X$, which implies that E_0 has only one equivalence class with μ-probability 1, which is impossible. ⊣

Factoring Through a Smooth Relation

The interest in analytic spaces comes from the fact that factoring an analytic space through a smooth equivalence relation will result in an analytic space again. This requires first and foremost the definition of a measurable structure induced by the relation. The natural choice is the structure imposed by the factor map. The final σ-algebra on X/ρ with respect to the Borel sets on X and the natural projection η_ρ will be chosen; it is denoted by $\mathcal{B}(X)/\rho$. Recall that $\mathcal{B}(X)/\rho$ is the largest σ-algebra $\mathcal{C}$ on X/ρ rendering η_ρ a $\mathcal{B}(X)$-$\mathcal{C}$-measurable map. Then it turns out that $\mathcal{B}(X/\rho)$ coincides with $\mathcal{B}(X)/\rho$:

Proposition 1.7.5. *Let X be an analytic space, and assume that α is a smooth equivalence relation on X. Then X/α is an analytic space.*

Proof In accordance with the characterization of smooth relations in Lemma 1.7.3 we assume that α is given through a sequence $(f_n)_{n\in\mathbb{N}}$ of measurable maps $f_n : X \to \mathbb{R}$. The factor map is measurable and onto. Put

$$E_{n,r} := \{[x]_\alpha \mid x \in X, f_n(x) < r\};$$

then

$$\mathcal{E} := \{E_{n,r} \mid n \in \mathbb{N}, r \in \mathbb{Q}\}$$

is a countable set of element of the factor σ-algebra that separates points. The assertion now follows without difficulties from Lemma 1.4.42. □

Usually, factor spaces are analytic but not Standard Borel. Consider this illustrative example.

Example 1.7.6. Let X be an arbitrary uncountable Polish space, and take a subset Q of X which is analytic but not a Borel set (such a set exists by Proposition 1.4.25). From Proposition 1.4.23 we obtain a Borel map $f : \mathbb{N}^\mathbb{N} \to X$ with $f\left[\mathbb{N}^\mathbb{N}\right] = Q$. Thus $\mathsf{ker}(f)$ is a smooth equivalence relation by Lemma 1.7.3, and the factor space $\mathbb{N}^\mathbb{N}/\mathsf{ker}(f)$ is isomorphic to Q by Proposition 1.4.33. ⊣

A smooth relation's invariant Borel sets will be a powerful tool for investigating smooth relations; this σ-algebra will be defined below. In order to appreciate this σ-algebra fully we show that it is often enough to establish a property related to the equivalence relation for the generators, then it will hold on the generated σ-algebra. This will be interesting when we investigate various logics: here the generators are determined by the formulas of the logic, yielding a rather captivating interplay of formulas in the logic and a σ-algebra determined by their semantics.

Lemma 1.7.7. *Assume that an equivalence relation $\sim$ is defined for a set M through*

$$m \sim m' \textit{ iff } \forall G \in \mathcal{G} : [m \in G \Leftrightarrow m' \in G],$$

where the elements of $\mathcal{G}$ are subsets of M. Then

$$m \sim m' \textit{ iff } \forall G \in \sigma(\mathcal{G}) : [m \in G \Leftrightarrow m' \in G].$$

Proof It is not difficult to see that the set

$$\mathcal{H} := \{G \in \sigma(\mathcal{G}) \mid m \in G \Leftrightarrow m' \in G\}$$

is a σ-algebra, where $m, m' \in M$ are fixed. For example, if $G_1, G_2 \in \mathcal{H}$ and $m \in G_1 \cup G_2$, then $m \in G_1$ or $m \in G_2$. Depending on which case applies, $m' \in G_1$ or $m' \in G_2$, thus $m' \in G_1 \cup G_2$, and vice versa. But by assumption $\mathcal{G} \subseteq \mathcal{H}$; thus $\sigma(\mathcal{G}) \subseteq \mathcal{H}$, and the conclusion follows. □

This Lemma tells us that invariance with respect to an equivalence relation, which will be defined below, is carried over from a generator to its σ-algebra. This implies that we have some degrees of freedom when selecting a generator.

Definition 1.7.8. *Let ρ be a smooth equivalence relation on an analytic space X.*

a. A subset $A \subseteq X$ is called ρ-invariant *iff $x \in A$ and $x\ \rho\ x'$ together imply $x' \in A$.*
b. Denote by $\mathbf{\Sigma}\big(\mathcal{B}(X), \rho\big)$ the σ-algebra of ρ-invariant Borel subsets of X.

Thus a ρ-invariant set A can be written as the union of the equivalence classes of its sets, $A = \bigcup\{[x]_\rho \mid x \in A\}$. Invariance can be defined of course also for equivalence relations that are defined on arbitrary sets. Consider a group G and a normal subgroup H. Put $g_1 \sim g_2$ iff $g_1 g_2^{-1} \in H$. Then it is well-known that $\sim$ is an equivalence relation, and that the invariant subsets are exactly the unions of the cosets.

The invariant Borel subsets constitute the σ-algebra of $(A_n)_{n \in \mathbb{N}}$ which determines ρ. This will be investigated further in a moment.

These are quite simple examples:

Lemma 1.7.9. *The identity relation Δ_X and the universal relation U_X are for each analytic space X smooth equivalence relations with*

$$\mathbf{\Sigma}\big(\mathcal{B}(X), \Delta_X\big) = \mathcal{B}(X),$$
$$\mathbf{\Sigma}\big(\mathcal{B}(X), U_X\big) = \{\emptyset, X\}.$$

Proof The assertion is trivial for the universal relation. One argues for the identity relation as follows: the Borel sets of X are countably generated, and one can find such a countable generator $\mathcal{G}$ that separates points. This implies that Δ_X has $\mathcal{G}$ as the determining family, and since $\sigma(\mathcal{G}) = \mathcal{B}(X)$, the assertion follows. □

We have seen in Proposition 1.7.5 that factoring an analytic space with a smooth equivalence yields an analytic space again. This closure property above is fairly fundamental for the development of our algebraic theory, being one of the reasons for our sometimes preferring analytic spaces over Polish ones, since the latter are not closed under factoring through a smooth relation.

1.7.1 Invariant Borel Sets

The invariant Borel sets may be characterized through the factor map by the inverse image of the Borel sets of a factor space. This will give a fairly practical handle on the invariant sets. Lemma 1.7.10 is a bit more general by considering general surjective Borel maps, and we will see that this is helpful indeed.

Lemma 1.7.10. *Let X, Y be analytic spaces, and assume that $f : X \to Y$ is a surjective Borel map. Then $f^{-1}\left[\mathcal{B}(Y)\right] = \mathbf{\Sigma}\big(\mathcal{B}(X), \mathsf{ker}\,(f)\big)$.*

Proof 1. Given $A \in \mathbf{\Sigma}\big(\mathcal{B}(X), \mathsf{ker}\,(f)\big)$, we infer from Lemma 1.4.30 that $f^{-1}\left[f\left[A\right]\right] = A$ holds. Moreover, $f\left[A\right] \subseteq Y$ is analytic. We claim that

$$f[X \setminus A] = Y \setminus f[A]$$

holds. For, if $y \in f[S \setminus A]$, we can find $x \notin B$ with $f(x) = y$. Assuming that $y = f(x')$ for some $x' \in A$, we would infer that $x \in A$ due to the f-invariance of A, and since $f(x) = f(x')$. This is a contradiction. This settles the nontrivial inclusion. From the representation just established we see that $Y \setminus f[A]$ is analytic, and from Souslin's Theorem (Theorem 1.4.29) we infer now that $f[A]$ is Borel in Y.

2. It is clear that for each $B \in \mathcal{B}(Y)$ its inverse image $f^{-1}[B]$ under f is a Borel set which is f-invariant. On the other hand, if $A \in \mathcal{B}(X)$ is f-invariant, we write $A = f^{-1}[f[A]]$ by part 1, and $f[A] \in \mathcal{B}(Y)$ by part 2. This implies the desired equality. □

A by-product of Lemma 1.7.10 yields a characterization of ρ-invariant Borel sets in analytic spaces through the generating sequence $(A_n)_{n\in\mathbb{N}}$. As a consequence, we can describe the ρ-invariant Borel set through the canonical projection η_ρ. We show for later use that the atoms of the σ-algebra generated through the characterization sets for a smooth equivalence relation are exactly the equivalence classes.

Lemma 1.7.11. *Let X be an analytic space with a smooth equivalence relation ρ which is determined through the Borel sets $(A_n)_{n\in\mathbb{N}}$. Then the atoms of $\sigma(\{A_n \mid n \in \mathbb{N}\})$ are exactly the equivalence classes of ρ.*

Proof Given $x \in X$, it is easy to see that

$$[x]_\rho = \bigcap\{A_n \mid x \in A_n\} \cap \bigcap\{X \setminus A_n \mid x \notin A_n\}.$$

Thus each equivalence class is a member of $\sigma(\{A_n \mid n \in \mathbb{N}\})$. It is also an atom: suppose $\emptyset \neq A \in \sigma(\{A_n \mid n \in \mathbb{N}\})$ is a proper subset of $[x]_\rho$, then A is ρ-invariant by Lemma 1.7.7. Let $a \in A, b \in [x]_\rho \setminus A$, then $a \ \rho \ b$, so that $b \in A$ by ρ-invariance, contradicting the choice of b. Consequently $[x]_\rho$ is an atom. If A is an atom, then $A = [a]_\rho$, provided $a \in A$, so that the classes are exactly the atoms. □

Thus we may visualize ρ-invariant sets as a raster image, the single indivisible building blocks being the equivalence classes. This fairly important characterization of the σ-algebra of ρ-invariant sets is obtained as an immediate consequence.

Proposition 1.7.12. *Let X be an analytic space with a smooth equivalence relation ρ. Then the ρ-invariant Borel sets of X are exactly the inverse images of the canonic projection η_ρ, viz.,*

$$\boldsymbol{\Sigma}(\mathcal{B}(X), \rho) = \eta_\rho^{-1}[\mathcal{B}(X/\rho)]$$

holds. Moreover, if ρ is determined by the sequence $(A_n)_{n\in\mathbb{N}}$ of Borel sets $A_n \subseteq X$, then

$$\boldsymbol{\Sigma}(\mathcal{B}(X), \rho) = \sigma(\{A_n \mid n \in \mathbb{N}\}).$$

Proof 1. X/ρ is an analytic space, and $\eta_\rho : X \to X/\rho$ is surjective and onto. Thus the first assertion follows from Lemma 1.7.10 upon observing that $\rho = \mathsf{ker}(\eta_\rho)$ holds.

2. Since each A_n is a Borel set, $\sigma(\{A_n \mid n \in \mathbb{N}\})$ is contained in $\mathbf{\Sigma}(\mathcal{B}(X), \rho)$. Take conversely $B \in \mathbf{\Sigma}(\mathcal{B}(X), \rho)$, then B is a Borel set which is a union of atoms by Lemma 1.7.11, so that $B \in \sigma(\{A_n \mid n \in \mathbb{N}\})$ follows from the Blackwell-Mackey Theorem 1.4.43. □

This yields a fairly convenient way of determining the Borel sets $\mathcal{B}(X/\rho)$ of the factor space, which will be used over and over again.

Corollary 1.7.13. *Let X be an analytic space, and assume that the equivalence relation ρ is determined by the sequence $(A_n)_{n\in\mathbb{N}}$ of Borel sets. Then $\mathcal{B}(X/\rho) = \sigma(\{\eta_\rho[A_n] \mid n \in \mathbb{N}\})$.*

Proof Put

$$\mathcal{E} := \sigma(\{\eta_\rho[A_n] \mid n \in \mathbb{N}\});$$

then, given a measurable space $(P, \mathcal{P})$, a map $g : X/\rho \to P$ is $\mathcal{E}$-$\mathcal{P}$-measurable iff $g^{-1}[C] \in \mathcal{E}$ for each $C \in P$ iff $(\eta_\rho^{-1} \circ g^{-1})[C] \in \mathbf{\Sigma}(\mathcal{B}(X), \rho) \subseteq \mathcal{B}(X)$. From the representation of $\mathbf{\Sigma}(\mathcal{B}(X), \rho)$ in Proposition 1.7.12 and from Lemma 1.3.3 we infer that $\mathcal{E}$ is the final σ-algebra with respect to the factor map η_ρ, hence $\mathcal{E} = \mathcal{B}(X)$. □

A second corollary states that a smooth equivalence relation is determined uniquely by its invariant sets:

Corollary 1.7.14. *If $\mathcal{C} \subseteq \mathcal{B}(X)$ is a countably generated sub-σ-algebra of the Borel sets of X, then there exists a unique smooth equivalence relation $\rho_\mathcal{C}$ on X with $\mathcal{C} = \mathbf{\Sigma}(\mathcal{B}(X), \rho_\mathcal{C})$.* □

1.7.2 Operations on Smooth Relations

We will study briefly operations with smooth equivalence relations which will represent useful tools for constructions on coalgebraic logics.

Proposition 1.7.15. *Let $(\alpha_n)_{n\in\mathbb{N}}$ be a sequence of smooth equivalence relations on the analytic space X.*

a. The infimum $\bigwedge_{n\in\mathbb{N}} \alpha_n$ with respect to inclusion is a smooth equivalence relation; it is given by $\bigcap_{n\in\mathbb{N}} \alpha_n$.
b. $\mathbf{\Sigma}(\mathcal{B}(X), \bigwedge_{n\in\mathbb{N}} \alpha_n) = \sigma(\{A_{n,m} \mid n, m \in \mathbb{N}\})$, provided α_n is generated by the sequence $(A_{n,m})_{m\in\mathbb{N}} \subseteq \mathcal{B}(X)$.

Consequently, smooth equivalence relations on an analytic space are closed under countable infima.

Proof It is trivial that $\bigwedge_{n\in\mathbb{N}} \alpha_n = \bigcap_{n\in\mathbb{N}} \alpha_n$ holds for each sequence $(\alpha_n)_{n\in\mathbb{N}}$ of equivalence relations. The assertions then follow easily from Corollary 1.7.14 □

One may wonder whether the set of smooth equivalence relations on analytic spaces is closed under finite suprema. This is equivalent to the question of whether the intersection of two countably generated σ-algebras is always countably generated, or whether the category of analytic spaces with measurable maps is closed under pushouts. See the Bibliographic Notes for a brief discussion.

The interplay between smooth relations and measurable maps is illustrated by the technique of transporting a smooth relation backwards along a measurable map.

Lemma 1.7.16. *Let α be a smooth equivalence relation on the analytic space A so that $\alpha = \mathsf{ker}\,(h)$ for some measurable map $h : A \to W$, W being an analytic space. Define for the Polish space X and the Borel map $f : X \to A$ on X the smooth relation $\alpha_f := \mathsf{ker}\,(h \circ f)$. If $E \subseteq X$ is an α_f-invariant Borel set, then*

a. $f\,[E]$ is an α-invariant Borel set in A,
b. $E = f^{-1}\,[f\,[E]]$.

The invariant Borel sets of α_f are just the inverse images of the invariant Borel set of α under f, viz.,

$$\mathbf{\Sigma}\big(\mathcal{B}(X), \alpha_f\big) = f^{-1}\left[\mathbf{\Sigma}\big(\mathcal{B}(A), \alpha\big)\right].$$

Proof 1. Let $E_0 := f^{-1}\,[F]$ be the inverse image of an α-invariant set $F \subseteq A$, and assume that $x \in E_0$ with $x\ \alpha_f\ x'$. Since $f(x) \in F$, and since $h(f(x)) = h(f(x'))$, we have $f(x') \in F$; thus $x' \in E_0$. Consequently, E_0 is α_f-invariant, and we have shown that

$$\mathbf{\Sigma}\big(\mathcal{B}(X), \alpha_f\big) \supseteq f^{-1}\left[\mathbf{\Sigma}\big(\mathcal{B}(A), \alpha\big)\right]$$

holds.

2. Let $E \in \mathbf{\Sigma}\big(\mathcal{B}(X), \alpha_f\big)$; then we assert that

$$E' := f\,[E] \in \mathbf{\Sigma}\big(\mathcal{B}(A), \alpha\big).$$

Since E is α_f-invariant, $f\,[E]$ is α-invariant by construction. The hard part is showing that E' is a Borel set. First it is clear that E' is an analytic set, because it is the image of a Borel set under a Borel map. We claim that $f\,[X \setminus E] = A \setminus f\,[E]$. From this we may conclude that E' is also co-analytic, and thus is a Borel set by Souslin's Theorem 1.4.29.

The argumentation in the proof of Lemma 1.7.10 shows that $f\,[X \setminus E] \subseteq A \setminus f\,[E]$ holds. This establishes the desired equality and shows that E' is in fact a Borel set. But we can say more: $E = f^{-1}\,[f\,[E]]$ will be shown to hold.

Let $x \in f^{-1}[f[E]]$, thus $f(x) \in f[E]$, hence $f(x) = f(x')$ for some $x' \in E$. But this implies $x \in E$ since the latter set is α_f-invariant, and $x \; \alpha_f \; x'$. The other inclusion is trivial again.

3. The argument shows that each element of $\mathbf{\Sigma}(\mathcal{B}(X), \alpha_f)$ can be represented as the inverse image of an element from $\mathbf{\Sigma}(\mathcal{B}(A), \alpha)$ under f; thus

$$\mathbf{\Sigma}(\mathcal{B}(X), \alpha_f) \subseteq f^{-1}\left[\mathbf{\Sigma}(\mathcal{B}(A), \alpha)\right]$$

is established. □

Recall that the image of a Borel set under a Borel map is usually not a Borel set. This is an immediate consequence of the observation in Proposition 1.4.25 that there are strictly more analytic sets than Borel sets. Hence the first property in Lemma 1.7.16 indicates that invariance paired with smoothness is fairly strong a property.

Sums And Products

We turn now to sums and products. Knowing that smooth equivalence relations are closed under the sum operation is helpful when discussing congruences on the sum of stochastic relations. It is clear that the sum and the product of a countable number of analytic spaces is analytic again; this can easily be established through Proposition 1.4.23.

Lemma 1.7.17. *Let X and Y be analytic spaces with smooth equivalence relations α resp. β. Then $\alpha + \beta$ is a smooth equivalence relation on $X + Y$.*

Proof If $(A_n)_{n\in\mathbb{N}}$ and $(B_n)_{n\in\mathbb{N}}$ determine α resp. β, then the countable set of Borel sets $\{A_n + B_m \mid n, m \in \mathbb{N}\}$ determines $\alpha + \beta$. □

Smooth equivalence relations are closed under intersections and under countably infinite products.

Lemma 1.7.18. *If ρ, ρ' are smooth equivalence relations on the analytic space X, then $\rho \cap \rho'$ is smooth, and*

$$\mathbf{\Sigma}(\mathcal{B}(X), \rho \cap \rho') = \sigma\big(\mathbf{\Sigma}(\mathcal{B}(X), \rho) \cup \mathbf{\Sigma}(\mathcal{B}(X), \rho')\big).$$

Proof Assume that $(A_n)_{n\in\mathbb{N}}$ and $(A'_n)_{n\in\mathbb{N}}$ determine ρ resp. ρ'. Then the countable set $\{A_n \cap A'_m \mid n, m \in \mathbb{N}\}$ determines $\rho \cap \rho'$. The representation for the invariant Borel sets for $\rho \cap \rho'$ follows then easily from Proposition 1.7.12. □

The closure under countably infinite products will of course have finite products as a special case.

Lemma 1.7.19. *Assume that $(X_n)_{n\in\mathbb{N}}$ is a sequence of analytic spaces, and let ρ_n be a smooth equivalence relation on X_n for each $n \in \mathbb{N}$. Define*

$$(a_n)_{n\in\mathbb{N}}\,(\times_{n\in\mathbb{N}}\,\rho_n)\,(a'_n)_{n\in\mathbb{N}} \text{ iff } \forall n\in\mathbb{N}: a_n\ \rho_n\ a'_n.$$

Then

a. $\times_{n\in\mathbb{N}}\rho_n$ *is a smooth equivalence relation on* $\prod_{n\in\mathbb{N}} X_n$.
b. $\mathbf{\Sigma}\big(\mathcal{B}(\prod_{n\in\mathbb{N}} X_n), \times_{n\in\mathbb{N}}\rho_n\big) = \bigotimes_{n\in\mathbb{N}}\mathbf{\Sigma}\big(\mathcal{B}(X_n),\rho_n\big)$.

Proof 1. Abbreviate the equivalence relation $\times_{n\in\mathbb{N}}\rho_n$ by ρ^∞. Assume that ρ_n is determined by the sequence $(Z_{n,m})_{m\in\mathbb{N}}$ of Borel sets $Z_{n,m}\subseteq X_n$. Define for $n_1\dots n_k\in\mathbb{N}^*$ the cylinder set

$$W_{n_1\dots n_k} := Z_{1,n_1}\times\dots\times Z_{k,n_k}\times\prod_{j>k} X_j.$$

Then it is easy to see that

$$(a_n)_{n\in\mathbb{N}}\,\rho^\infty\,(a'_n)_{n\in\mathbb{N}} \Leftrightarrow \forall v\in\mathbb{N}^*: \big[\langle a_1,\dots,a_k\rangle\in W_v \iff \langle a'_1,\dots,a'_k\rangle\in W_v\big].$$

This implies that ρ^∞ is generated through a countable family of Borel sets.

2. Since for each index $v\in\mathbb{N}^*$ the set W_v is a ρ^∞-invariant Borel set which is constructed from ρ_n-invariant factors, we have

$$\mathbf{\Sigma}\big(\mathcal{B}(\prod_{n\in\mathbb{N}} X_n),\rho^\infty\big)\subseteq\bigotimes_{n\in\mathbb{N}}\mathbf{\Sigma}\big(\mathcal{B}(X_n),\rho_n\big);$$

on the other hand,

$$\bigotimes_{n\in\mathbb{N}}\mathbf{\Sigma}\big(\mathcal{B}(X_n),\rho_n\big)$$

is generated by cylinder sets of the form

$$B_1\times\ldots\times B_n\times\prod_{j>n} X_j,$$

which are ρ^∞-invariant. This implies the other inclusion. □

A finite version is available here as well: the product of two smooth equivalence relations is smooth again, and the invariant Borel sets for the product are just the product of the Borel sets for the factors.

Amalgamation

A special case of equivalence relation on the sum of two sets will be provided by the amalgamated sum of two equivalence relations. We do not require for the time being a topological or a measurable structure on the sets.

Suppose that α and β are equivalence relations on the sets M resp. N, and assume that we can find a bijection $g: M/\alpha\to N/\beta$ between the factor spaces. Consequently, given a class $[m]_\alpha\in M/\alpha$ there exists a unique class

$[n]_\beta \in N/\beta$ to which it relates via g, and vice versa. Thus the factor spaces are like two copies of one another.

The *amalgamation* $\alpha \diamond \beta$ *of* α *and* β is defined on the sum $M + N$ of M and N (with injections inj_M resp. inj_N) in the following way: injections of elements of M and N are related to each other through α resp. β, and if $w = inj_M(m)$, and $w' = inj_N(n)$, then $w\ \alpha \diamond \beta\ w'$ iff $n \in g\big([m]_\alpha\big)$. Formally, the amalgamation $\alpha \diamond \beta$ of α and β is defined through the partition

$$\{inj_M\,[[m]_\alpha] \mid m \in M\} \cup \{inj_M\,\big[g\big([m]_\alpha\big)\big] \mid m \in M\}$$

of $M + N$ (usually, we will omit the injections). Consequently, we we have for $w, w' \in M + N$

$$w\ \alpha \diamond \beta\ w' \iff \begin{cases} m\ \alpha\ m', & w = inj_M(m), w' = inj_M(m'),\\ n\ \beta\ n', & w = inj_N(n), w' = inj_N(n'),\\ [n]_\beta = g\big([m]_\alpha\big), & w = inj_M(m), w' = inj_N(n),\\ g\big([m]_\alpha\big) = [n]_\beta\,, & w = inj_N(n), w' = inj_M(m) \end{cases}$$

The amalgamation will permit us to relate Kripke models; it will be constructed in different ways.

1.7.3 Congruences for Stochastic Relations

Given a stochastic relation $\mathsf{K} = (X, Y, K)$ with analytic spaces X and Y, a pair $\mathsf{c} = (\alpha, \beta)$ of smooth equivalence relations α and β on X resp. Y is called a *congruence for* K iff $K(x)(B) = K(x')(B)$, whenever $x\ \alpha\ x'$, and whenever $B \in \mathbf{\Sigma}\big(\mathcal{B}(Y), \beta\big)$ is an invariant Borel subset of Y. This means that if α cannot distinguish x and x', and if β cannot distinguish the elements of B, then the probability $K(x)(B)$ equals the probability $K(x')(B)$.

Now let $\mathsf{c} = (\alpha, \beta)$ be a congruence. Because $\{x \in X \mid K(x)(B) \geq q\}$ is an α-invariant Borel set in X, provided $B \in \mathbf{\Sigma}\big(\mathcal{B}(Y), \beta\big)$, we obtain the following.

Proposition 1.7.20. *Let* $\mathsf{K} = (X, Y, K)$ *be a stochastic relation with analytic spaces* X *and* Y, *and assume* $\mathsf{c} = (\alpha, \beta)$ *is a pair of smooth equivalence relations on* X *resp.* Y. *Denote by* $K^{\upharpoonright}(x)$ *the restriction of the measure* $K(x)$ *to the* σ*-algebra* $\mathbf{\Sigma}\big(\mathcal{B}(Y), \beta\big)$. *These statements are equivalent*

a. c *is a congruence for* K.
b. $K^{\upharpoonright} : (X, \mathbf{\Sigma}\big(\mathcal{B}(X), \alpha\big)) \rightsquigarrow (Y, \mathbf{\Sigma}\big(\mathcal{B}(Y), \beta\big))$ *is a stochastic relation.* □

The *factor relation* K/c is defined fairly canonically through

$$\mathsf{K}/\mathsf{c} = (X/\alpha, Y/\beta, K_{\mathsf{c}})$$

with

$$K_{\mathsf{c}}([x]_\alpha)(B) := K(x)(\eta_\beta^{-1}[B])\big(= (\mathfrak{S}(\eta_\beta) \circ K)(x)(B)\big)$$

whenever $x \in X$ and $B \in \mathcal{B}(Y/\beta)$ is a Borel set. Consequently, the diagram

$$\begin{array}{ccc} X & \xrightarrow{\ \eta_\alpha\ } & X/\alpha \\ {\scriptstyle K}\big\downarrow & & \big\downarrow{\scriptstyle K_{\mathsf{c}}} \\ Y & \xrightarrow[\ \mathfrak{S}(\eta_\beta)\ ]{} & Y/\beta \end{array} \tag{1.1}$$

commutes, so that $\eta_{\mathsf{c}} := (\eta_\alpha, \eta_\beta)$ is a morphism $\mathsf{K} \to \mathsf{K}/\mathsf{c}$ with kernel c.

This establishes partly the following characterization of congruences.

Proposition 1.7.21. *Let X and Y be analytic spaces with smooth equivalence relations α resp. β. Then $\mathsf{c} = (\alpha, \beta)$ is a congruence for $K : X \rightsquigarrow Y$ iff there exists a stochastic relation $K_{\mathsf{c}} : X/\alpha \rightsquigarrow Y/\beta$ which renders the diagram (1.1) commutative.*

Proof If c is a congruence, we just have established that we can construct a stochastic relation $K_{\mathsf{c}} : X/\alpha \rightsquigarrow Y/\beta$. Assume conversely that the diagram (1.1) commutes. If $D \in \mathbf{\Sigma}\big(\mathcal{B}(Y), \beta\big)$ is a β-invariant Borel set, then we know from Lemma 1.7.16 that $D = \eta_\beta^{-1}[\eta_\beta[D]]$, and that $\eta_\beta[D] \in \mathcal{B}(Y/\beta)$. Now assume $x\ \alpha\ x'$; then we obtain

$$\begin{aligned} K(x)(D) &= K(x)\big(\eta_\beta^{-1}[\eta_\beta[D]]\big) \\ &= \big(\mathfrak{S}(\eta_\beta) \circ K\big)(x)\big(\eta_\beta[D]\big) \\ &= K_{\mathsf{c}}([x]_\alpha)\big(\eta_\beta[D]\big) \\ &= K_{\mathsf{c}}([x']_\alpha)\big(\eta_\beta[D]\big) \\ &= K(x')(D), \end{aligned}$$

the latter equality just unwinding for $K(x')$ what we wound up for $K(x)$. Consequently, $x\ \alpha\ x'$ and $D \in \mathbf{\Sigma}\big(\mathcal{B}(Y), \beta\big)$ together imply that $K(x)(D) = K(x')(D)$; hence c is a congruence for K. $\square$

We will occasionally restrict our attention to nontrivial congruences: call a congruence (α, β) for the stochastic relation K *nontrivial* iff β is not the universal relation U_Y on Y (consequently, there exists $y, y' \in Y$ such that $\langle y, y' \rangle \notin \beta$). Thus the invariant Borel sets $\mathbf{\Sigma}\big(\mathcal{B}(Y), \beta\big)$ will contain more elements than the trivial σ-algebra $\{\emptyset, Y\}$. This property will be of interest for investigating bisimilarity when it comes to excluding trivial cases, as the discussion in Section 2.2 will show.

If two congruences on K are related through inclusion, the factor spaces are related through a morphism.

Corollary 1.7.22. *Let K and c be as above, and assume that $\mathsf{d} = (\gamma, \delta)$ is a congruence for K with $\alpha \subseteq \gamma$ and $\beta \subseteq \delta$. Then there exists a unique morphism*

$\mathsf{f} : \mathsf{K}/\mathsf{c} \to \mathsf{K}/\mathsf{d}$ *rendering the diagram*

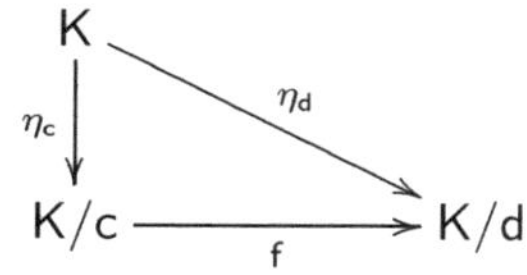

commutative.

Proof (Sketch) Let $\mathsf{K} = (X, Y, K)$; then $[x]_\alpha \mapsto [x]_\gamma$ and $[y]_\beta \mapsto [y]_\delta$ are shown to induce the desired morphism f. □

This is a version of the Correspondence Theorem from Universal Algebra [13, Theorem 6.20] which relates congruences on a quotient to a sublattice of relations on the base algebra.

1.8 Grounded Relations on Subprobabilities

We will study the interplay of equivalence relations on $\mathfrak{S}(X)$ and X, respectively by investigating the trace that an equivalence relation ξ on $\mathfrak{S}(X)$ leaves on the base space X, and by lifting (or randomizing) an equivalence relation on X to $\mathfrak{S}(X)$. The trace is constructed through the Dirac measures on the points of the base space, and thus, technically speaking, from the unit for the Giry monad (see Chapter 3.2), and lifting is done through comparing measures on invariant sets. Both approaches appear natural when it comes to transforming an equivalence relation from one space to another. The question arises about whether constructing the trace of a lifted equivalence relation yields the original relation, and, conversely, whether lifting the trace changes anything. These questions will present themselves when randomized morphisms are considered, so we deal with them here in suitable generality.

The central rôle is played in the present discussions by the invariant sets for a smooth equivalence relation. On the one hand, they are defined on the base space; on the other hand they are part of the domain of a subprobability measure; hence they permit comparing these measures. Thus they serve as a kind of hub in the discussion that will follow. We will often take an equivalence relation on $\mathfrak{S}(X)$ and see what we can say about the invariant sets with respect to its trace. Thus a relation of the subprobabilities is interlocked with those on the base space.

Randomization

We begin with the randomization of an equivalence relation on a measurable space, and with the converse process, a kind of de-randomization, which has a look at the trace on the base space. We will compare the relations that arise, putting particular emphasis on those relations on $\mathfrak{S}(X)$ that are finer than the randomization of their trace, and on those which equal this randomization; we call the latter ones grounded. A complete characterization of grounded relations will be given in terms of positive convex structures.

Definition 1.8.1. *Given a measurable space $(A, \mathcal{M})$ and an equivalence relation ξ on A, define on $\mathfrak{S}(A, \mathcal{M})$ the* lifting $\overline{\xi}$ of ξ *through $\mu \ \overline{\xi} \ \mu'$ iff $\mu(B) = \mu'(B)$ for all $B \in \mathbf{\Sigma}(\mathcal{M}, \xi)$.*

Thus the subprobabilities μ and μ' are $\overline{\xi}$-equivalent iff they behave in the same way on the ξ-invariant sets in $\mathcal{M}$. The randomization of an equivalence relation on A is clearly an equivalence relation on $\mathfrak{S}(A, \mathcal{M})$.

It will be helpful to know that equality on a sub-σ-algebra carries over to the integrals of functions which are measurable with respect to this σ-algebra.

Lemma 1.8.2. *Let $\mathcal{M}$ be a σ-algebra on a set A, and assume that $\mu(B) = \mu'(B)$ holds for all $B \in \mathcal{M}$. Then*

$$\int_X f \, d\mu = \int_X f \, d\mu'$$

holds for all $f : X \to \mathbb{R}$ that are $\mathcal{M}$-$\mathcal{B}(\mathbb{R})$-measurable.

Proof It is no loss of generality to assume that f is nonnegative, for f can be written as $f = f^+ - f^-$ with positive part $f^+ := \max\{f, 0\}$ and negative part $f^- := \max\{-f, 0\}$, and the integral is additive.

Since f is $\mathcal{M}$-$\mathcal{B}(\mathbb{R})$-measurable, the set

$$\{f > t\} := \{x \in X \mid f(x) > t\}$$

is a member of $\mathcal{M}$ for every $t \in \mathbb{R}$; thus, using the Choquet representation from Proposition 1.6.2,

$$\int_X f \, d\mu = \int_0^\infty \mu(f > t)dt = \int_0^\infty \mu'(f > t)dt = \int_X f \, d\mu'.$$

□

Let X be an analytic space with Borel sets $\mathcal{B}(X)$, and let α be smooth; we will fix X and α throughout. An elementary consequence of Corollary 1.7.14 is that we can recover α from $\overline{\alpha}$.

Corollary 1.8.3. *$x \ \alpha \ x'$ iff $\delta_x \ \overline{\alpha} \ \delta_{x'}$.* □

Here are two examples.

Example 1.8.4. We extend Lemma 1.7.9. Let X be an analytic space. Since the identity Δ_X is a smooth equivalence relation, and since $\mathbf{\Sigma}(\mathcal{B}(X), \Delta_X) = \mathcal{B}(X)$, we have

$$\overline{\Delta_X} = \Delta_{\mathfrak{S}(X)}.$$

Furthermore, the universal relation $U_X = X \times X$ is smooth, and $\mathbf{\Sigma}\big(\mathcal{B}(X), U_X\big)$ equals $\{\emptyset, X\}$. Thus

$$\overline{U_X} = \{\langle \mu, \mu' \rangle \mid \mu(X) = \mu'(X)\}.$$

$\dashv$

Lemma 1.8.5. *If $\alpha = \mathsf{ker}\,(f)$ for the Borel map $f : X \to T$ and the analytic space T, then $\overline{\alpha} = \mathsf{ker}\,(\mathfrak{S}\,(f))$. Hence $\overline{\alpha}$ is a smooth equivalence relation on $\mathfrak{S}\,(X)$.*

Proof 0. It is no loss of generality to assume that f is onto, since the image of an analytic set under a Borel map is again an analytic set by Proposition 1.4.26.

1. Note that $f^{-1}\,[B]$ is α-invariant for $B \in \mathcal{B}(T)$. Thus $\langle \mu, \mu' \rangle \in \mathsf{ker}\,(\mathfrak{S}\,(f))$ implies

$$\mu(f^{-1}\,[B]) = \mathfrak{S}\,(f)\,(\mu)(B) = \mathfrak{S}\,(f)\,(\mu')(B) = \mu'(f^{-1}\,[B])$$

for all Borel sets $B \in \mathcal{B}(T)$. Because each invariant Borel set $A \in \mathbf{\Sigma}\big(\mathcal{B}(X), \alpha\big)$ may be represented as the inverse image under f of a Borel set in T by Lemma 1.7.10, we see that $\langle \mu, \mu' \rangle \in \overline{\alpha}$.

2. Conversely, let $\mu(A) = \mu'(A)$ for each $A \in \mathbf{\Sigma}\big(\mathcal{B}(X), \alpha\big) = f^{-1}\,[\mathcal{B}(T)]$. Thus

$$\mathfrak{S}\,(f)\,(\mu)(B) = \mu(f^{-1}\,[B]) = \mu'(f^{-1}\,[B]) = \mathfrak{S}\,(f)\,(\mu')(B)$$

is established for each Borel set $B \in \mathcal{B}(T)$. This in turn implies that the pair $\langle \mu, \mu' \rangle$ is in the kernel of $\mathfrak{S}\,(f)$. □

For investigating the converse, assume that γ is a smooth equivalence relation on $\mathfrak{S}\,(X)$. We ask under which conditions there exists a smooth equivalence relation α on X such that $\gamma = \overline{\alpha}$. Define the equivalence relation $\lfloor \gamma \rfloor$ on X upon setting

$$x \ \lfloor \gamma \rfloor\ x' \Leftrightarrow \delta_x \ \gamma \ \delta_{x'}.$$

Then $\lfloor \gamma \rfloor$ is smooth: if $\gamma = \mathsf{ker}\,(H)$, then $\lfloor \gamma \rfloor = \mathsf{ker}\,(H \circ \mathfrak{e}_X)$, where $\mathfrak{e}_X : X \ni x \mapsto \delta_x \in \mathfrak{S}\,(X)$ maps each point into its Dirac measure.

We are interested in those smooth relations on $\mathfrak{S}\,(X)$ that are related to their traces on X, so they deserve a special name.

Definition 1.8.6. *A smooth equivalence relation γ on $\mathfrak{S}\,(X)$ is called*

1. near-grounded *iff* $\gamma \subseteq \overline{\lfloor \gamma \rfloor}$,
2. grounded *iff* $\gamma = \overline{\lfloor \gamma \rfloor}$.

Suppose that γ is near-grounded. Then $\mu \ \gamma \ \mu'$ implies $\mu(B) = \mu'(B)$ for all Borel sets that are $\lfloor \gamma \rfloor$-invariant; in this way γ, which is defined on $\mathfrak{S}\,(X)$, is related to the $\lfloor\gamma\rfloor$-invariant Borel sets of X. Hence we have at least some hints at how μ and μ' relate to each other on the base set.

Constructing $\overline{\lfloor\cdot\rfloor}$ is a bit akin to forming the closure of the interior of a set in a topological space. If γ is grounded, then it is uniquely determined by its restriction on $\{\delta_x \mid x \in X\}$, and thus essentially by its behavior on the base set. This is intuitively clear and will be formally substantiated in Proposition 1.8.15.

We give an example for a grounded equivalence relation, and provide an example indicating that there are near-grounded relations which are not grounded.

Example 1.8.7. Let $X := [-1, 1]$ and put

$$\mu \; \gamma \; \mu' \Leftrightarrow \forall A \in \mathcal{B}(X) : \int_A t^2 \; \mu(dt) = \int_A t^2 \; \mu'(dt);$$

then γ is a smooth equivalence relation. In fact, let $\mathcal{A}$ be a countable generator for $\mathcal{B}(X)$, and define

$$H(\mu) := \left(\int_A t^2 \; \mu(dt)\right)_{A \in \mathcal{A}};$$

then $H : \mathfrak{S}(X) \to [0,1]^{\mathbb{N}}$ is continuous with $\gamma = \mathsf{ker}(H)$.

It is clear that $x_1 \lfloor\gamma\rfloor \; x_2$ iff $x_1^2 = x_2^2$, so that a Borel set $S \in \mathcal{B}(X)$ is $\lfloor\gamma\rfloor$-invariant iff S is symmetric, thus iff $r \in S$ implies $-r \in S$. Consequently,

$$\mu \; \overline{\lfloor\gamma\rfloor} \; \mu' \Leftrightarrow \int_S t^2 \; \mu(dt) = \int_S t^2 \; \mu'(dt) \text{ for all symmetric Borel sets } S.$$

This implies that $\gamma = \overline{\lfloor\gamma\rfloor}$, so γ is grounded. ⊣

Example 1.8.8. Let $K : X \rightsquigarrow Y$ be a stochastic relation with X, Y Polish, and define for $\mu \in \mathcal{B}(X)$ and $E \in \mathcal{B}(X \times Y)$

$$(\mu \otimes K)(E) := \int_X K(x)(E_x) \; \mu(dx),$$

where $E_x := \{y \in Y \mid \langle x, y\rangle \in E\}$ is the vertical cut of E at x. Put

$$\mu \; \gamma_K \; \mu' \Leftrightarrow \mu \otimes K = \mu' \otimes K.$$

Then γ_K is smooth on account of $\mu \mapsto \mu \otimes K$ defining a Borel map $\mathfrak{S}(X) \to \mathfrak{S}(X \times Y)$.

Then $\lfloor\gamma_K\rfloor = \mathsf{ker}(K)$. Now let $K = \nu$ be constant; hence $\mu \otimes K$ equals the product measure $\mu \otimes \nu$. Furthermore $\mathsf{ker}(K) = X \times X$, so that $\boldsymbol{\Sigma}(\mathcal{B}(X), \lfloor\gamma_\nu\rfloor) = \{\emptyset, X\}$; see Example 1.8.4. Consequently, $\mu \; \overline{\lfloor\gamma_\nu\rfloor} \; \mu'$ iff $\mu(X) = \mu'(X)$. Now select μ, μ', ν so that $\mu(X) = \mu'(X)$ but $\mu \otimes \nu \neq \mu' \otimes \nu$, indicating $\overline{\lfloor\gamma_\nu\rfloor} \not\subseteq \gamma_\nu$. Thus γ_ν is near-grounded but not grounded. ⊣

1.8.1 A Borel Isomorphism

The dependence of a grounded equivalence relation γ on its trace $\lfloor\gamma\rfloor$, which it leaves as a footprint on the base space is strong, because in this case γ is completely determined by $\lfloor\gamma\rfloor$. We will show that this is reflected on the Borel structure of its factor space $\mathfrak{S}(X)/\gamma$, which equals $\mathfrak{S}(X/\lfloor\gamma\rfloor)$ up to a Borel isomorphism. Thus this factor is essentially a space of probability measures.

It turns out that the construction under consideration can be carried out already for a near-grounded relation, yielding a surjective Borel map $\partial_\gamma : \mathfrak{S}(X)/\gamma \to \mathfrak{S}(X/\lfloor\gamma\rfloor)$. Grounded relations are different from their near-grounded cousins: this is witnessed by the fact that γ is grounded iff $\mathfrak{S}(X)/\gamma$ and $\mathfrak{S}(X/\lfloor\gamma\rfloor)$ are isomorphic. We study in this section the map ∂_γ and its properties, providing some insight into the inner working of the transformations between equivalence relations on different, albeit closely related spaces.

Structural Properties

Define for a smooth and near-grounded equivalence relation γ on $\mathfrak{S}(X)$ with X analytic the map

$$\partial_\gamma : \begin{cases} \mathfrak{S}(X)/\gamma \to \mathfrak{S}(X/\lfloor\gamma\rfloor) \\ [\mu]_\gamma \qquad \mapsto \mathfrak{S}\left(\eta_{\lfloor\gamma\rfloor}\right)(\mu). \end{cases}$$

Thus we have $\partial_\gamma([\mu]_\gamma)(C) = \mu(\eta_{\lfloor\gamma\rfloor}^{-1}[C])$ for the Borel set $C \in \mathcal{B}(X/\lfloor\gamma\rfloor)$ and $\mu \in \mathfrak{S}(X)$.

Lemma 1.8.9. *Whenever X is analytic, and γ is a smooth and near-grounded equivalence relation on $\mathfrak{S}(X)$, ∂_γ is $\mathcal{B}(\mathfrak{S}(X)/\gamma)$-$\mathcal{B}(\mathfrak{S}(X/\lfloor\gamma\rfloor))$-measurable and onto.*

Proof 0. Suppose $\mu\ \gamma\ \mu'$; then $\mu(B) = \mu'(B)$ whenever $B \in \Sigma(\mathcal{B}(X), \lfloor\gamma\rfloor)$, since γ is near-grounded. Thus $\mathfrak{S}\left(\eta_{\lfloor\gamma\rfloor}\right)(\mu) = \mathfrak{S}\left(\eta_{\lfloor\gamma\rfloor}\right)(\mu')$; consequently, ∂_γ is well defined. Given $\mu \in \mathfrak{S}(X)$, it is clear that $\partial_\gamma([\mu]_\gamma) \in \mathfrak{S}(X/\lfloor\gamma\rfloor)$.

1. Let $\nu \in \mathfrak{S}(X/\lfloor\gamma\rfloor)$; then there exists $\mu \in \mathfrak{S}(X)$ such that $\nu = \mathfrak{S}\left(\eta_{\lfloor\gamma\rfloor}\right)(\mu)$ by Proposition 1.6.19. Thus $\partial_\gamma([\mu]_\gamma) = \nu$.

2. In order to establish measurability of ∂_γ, we need to show that $\partial_\gamma^{-1}[G]$ is a Borel set in $\mathfrak{S}(X)/\gamma$ whenever $G \subseteq \mathfrak{S}(X/\lfloor\gamma\rfloor)$ is a Borel set. Since the Borel sets of $\mathfrak{S}(X/\lfloor\gamma\rfloor)$ are exactly the elements of the weak*-σ-algebra with respect to the Borel sets on $X/\lfloor\gamma\rfloor$ by Proposition 1.6.16, we may assume that there exist $H \in \mathcal{B}(X/\lfloor\gamma\rfloor)$ and a Borel set $F \subseteq \mathbb{R}$ such that

$$G = \{\tau \in \mathfrak{S}(X/\lfloor\gamma\rfloor) \mid \tau(H) \in F\}.$$

Because $H \in \mathcal{B}(X/\lfloor\gamma\rfloor)$, we know that $\eta_{\lfloor\gamma\rfloor}^{-1}[H] \in \mathbf{\Sigma}(\mathcal{B}(X), \lfloor\gamma\rfloor)$. Thus we want to show that

$$\partial_\gamma^{-1}[G] = \{[\mu]_\gamma \mid \mu(\eta_{\lfloor\gamma\rfloor}^{-1}[H]) \in F\} \in \mathcal{B}(\mathfrak{S}(X)/\gamma),$$

or equivalently, that $\eta_\gamma^{-1}\left[\partial_\gamma^{-1}[G]\right]$ is a γ-invariant member of $\mathcal{B}(\mathfrak{S}(X))$. Now

$$\eta_\gamma^{-1}\left[\partial_\gamma^{-1}[G]\right] = (\partial_\gamma \circ \eta_\gamma)^{-1}[G] = \{\mu \in \mathfrak{S}(X) \mid \mu(\eta_{\lfloor\gamma\rfloor}^{-1}[H]) \in F\},$$

and this is a γ-invariant Borel set in $\mathfrak{S}(X)$. □

If γ is grounded, we can even say a bit more:

Lemma 1.8.10. *If γ is a near-grounded equivalence relation on $\mathfrak{S}(X)$ with X analytic, then ∂_γ is a bijection iff γ is grounded.*

Proof Take $\mu_1, \mu_2 \in \mathfrak{S}(X)$ which lie in different γ-classes; then there exists for grounded γ a $\lfloor\gamma\rfloor$-invariant Borel set $B \in \mathbf{\Sigma}(\mathcal{B}(X), \lfloor\gamma\rfloor)$ such that $\mu_1(B) \neq \mu_2(B)$; consequently, because

$$C = \eta_{\lfloor\gamma\rfloor}^{-1}\left[\eta_{\lfloor\gamma\rfloor}[C]\right]$$

holds for each $\lfloor\gamma\rfloor$-invariant Borel set C, we obtain

$$\partial_\gamma([\mu_1]_\gamma)(\eta_{\lfloor\gamma\rfloor}[B]) = \mu_1(B) \neq \mu_2(B) = \partial_\gamma([\mu_2]_\gamma)(\eta_{\lfloor\gamma\rfloor}[B]).$$

Conversely, it is easy to see that γ is grounded if ∂_γ is one-to-one. □

It will be shown now, moreover, that ∂_γ is a Borel isomorphism. This requires a bit of preparatory work. We claim that the image of certain Borel sets under $\mathfrak{S}(\eta_{\lfloor\gamma\rfloor})$ are Borel sets again. Recall that Borel sets are usually not preserved under Borel maps.

Lemma 1.8.11. *Under the assumptions of Lemma 1.8.10, if $G_0 \subseteq \mathfrak{S}(X)$ is an $\lfloor\gamma\rfloor$-invariant Borel set, then $\mathfrak{S}(\eta_{\lfloor\gamma\rfloor})[G_0]$ is a Borel set in $\mathfrak{S}(X/\lfloor\gamma\rfloor)$.*

Proof 0. Since G_0 is a Borel set, we know that $\mathfrak{S}(\eta_{\lfloor\gamma\rfloor})[G_0]$ is an analytic set in $\mathfrak{S}(X/\lfloor\gamma\rfloor)$. We show that the complement of this set is analytic as well. If we would know that $\mathfrak{S}(\eta_{\lfloor\gamma\rfloor})[\mathfrak{S}(X) \setminus G_0]$ is disjoint from $\mathfrak{S}(\eta_{\lfloor\gamma\rfloor})[G_0]$, we would be done by Souslin's Theorem 1.4.29, so the crucial and non-obvious property is disjointness. We actually establish a bit more by showing

$$\mathfrak{S}(\eta_{\lfloor\gamma\rfloor})[\mathfrak{S}(X) \setminus G_0] = \mathfrak{S}(X/\lfloor\gamma\rfloor) \setminus \mathfrak{S}(\eta_{\lfloor\gamma\rfloor})[G_0].$$

1. Because $\mathfrak{S}(\eta_{\lfloor\gamma\rfloor}) : \mathfrak{S}(X) \to \mathfrak{S}(X/\lfloor\gamma\rfloor)$ is onto, we obtain

$$\begin{aligned}\mathfrak{S}(X/\lfloor\gamma\rfloor) \setminus \mathfrak{S}(\eta_{\lfloor\gamma\rfloor})[G_0] &= \mathfrak{S}(\eta_{\lfloor\gamma\rfloor})[\mathfrak{S}(X)] \setminus \mathfrak{S}(\eta_{\lfloor\gamma\rfloor})[G_0] \\ &\subseteq \mathfrak{S}(\eta_{\lfloor\gamma\rfloor})[\mathfrak{S}(X) \setminus G_0].\end{aligned}$$

2. Now assume that the converse inclusion is false, hence we can find $\tau \in \mathfrak{S}\left(\eta_{\lfloor\gamma\rfloor}\right)[\mathfrak{S}(X) \setminus G_0]$ such that $\tau \in \mathfrak{S}\left(\eta_{\lfloor\gamma\rfloor}\right)[G_0]$ also holds. By the first membership there exists $\nu \notin G_0$ with $\tau = \mathfrak{S}\left(\eta_{\lfloor\gamma\rfloor}\right)(\nu)$, by the second property there exists $\kappa \in G_0$ with $\tau = \mathfrak{S}\left(\eta_{\lfloor\gamma\rfloor}\right)(\kappa)$. Thus we find

$$\mu(C) = \mathfrak{S}\left(\eta_{\lfloor\gamma\rfloor}\right)(\nu)(C) = \nu(\eta_{\lfloor\gamma\rfloor}^{-1}[C]),$$

and similarly

$$\mu(C) = \mathfrak{S}\left(\eta_{\lfloor\gamma\rfloor}\right)(\kappa)(C) = \kappa(\eta_{\lfloor\gamma\rfloor}^{-1}[C])$$

for all $C \in \mathcal{B}(X/\lfloor\gamma\rfloor)$. We know that

$$\mathbf{\Sigma}\big(\mathcal{B}(X), \lfloor\gamma\rfloor\big) = \{\eta_{\lfloor\gamma\rfloor}^{-1}[C] \mid C \in \mathcal{B}(X/\lfloor\gamma\rfloor)\};$$

hence we have found that $\kappa(D) = \nu(D)$ for all $D \in \mathbf{\Sigma}\big(\mathcal{B}(X), \lfloor\gamma\rfloor\big)$. This implies $\nu \ \overline{\lfloor\gamma\rfloor}\ \kappa$. Since $\kappa \in G_0$, and G_0 is γ-invariant, we conclude $\nu \in G_0$, arriving at a contradiction.

3. Hence we have established that

$$\mathfrak{S}(X/\lfloor\gamma\rfloor) \setminus \mathfrak{S}\left(\eta_{\lfloor\gamma\rfloor}\right)[G_0] = \mathfrak{S}\left(\eta_{\lfloor\gamma\rfloor}\right)[\mathfrak{S}(X) \setminus G_0],$$

so that $\mathfrak{S}\left(\eta_{\lfloor\gamma\rfloor}\right)[G_0]$ is the complement of an analytic set. An application of the Souslin Theorem 1.4.29 then yields that the set under consideration is a Borel set. □

Proposition 1.8.12. *Let X be an analytic space, and γ a grounded equivalence relation on $\mathfrak{S}(X)$. Then $\partial_\gamma : \mathfrak{S}(X)/\gamma \to \mathfrak{S}(X/\lfloor\gamma\rfloor)$ is a Borel isomorphism.*

Proof We know from Lemma 1.8.9 that ∂_γ is bijective and measurable; from Lemma 1.8.11 it is inferred that the image of a Borel set under ∂_γ is Borel as well, so that ∂_γ^{-1} is also a Borel map. □

We will use this characterization of the factor space $\mathfrak{S}(X)/\gamma$ through essentially a space of probability measures when dealing with the factor map, which in turn is associated with a randomized morphism through a congruence.

1.8.2 Characterizing Groundedness

For the investigations of the conditions under which a smooth relation on $\mathfrak{S}(X)$ is grounded, we put

$$\Omega := \big\{\langle\alpha_1, \ldots, \alpha_k\rangle \mid k \in \mathbb{N}, \alpha_i \geq 0, \sum_{i=1}^{k} \alpha_i \leq 1\big\}$$

for the rest of the chapter, the elements of Ω being called *positive convex tuples* or simply *positive convex*.

Definition 1.8.13. *An equivalence relation γ on $\mathfrak{S}(X)$ is said to be* positive convex *iff $\mu_i \ \gamma \ \mu_i'$ for $1 \leq i \leq n$ and $\langle \alpha_1, \ldots, \alpha_n \rangle \in \Omega$ together imply*

$$(\alpha_1 \cdot \mu_1 + \ldots + \alpha_n \cdot \mu_n) \ \gamma \ (\alpha_1 \cdot \mu_1' + \ldots + \alpha_n \cdot \mu_n')$$

for each $n \in \mathbb{N}$. A partition of $\mathfrak{S}(X)$ is called positive convex *iff its associated equivalence relation is.*

It is clear that $\gamma = \overline{\alpha}$ entails γ being positive convex. But it works the other way as well: A positive convex and smooth equivalence relation $\gamma = \mathsf{ker}(H)$ with a surjective Borel map $H : \mathfrak{S}(X) \to T$ and a Polish space T induces a positive convex structure on T.

But first, positive convexity will be described abstractly, following Pumplün's approach [67].

Definition 1.8.14. *A* positive convex structure $\mathcal{P}$ *on a set T assigns to each $\alpha = \langle \alpha_1, \ldots, \alpha_n \rangle \in \Omega$ a map $\alpha_{\mathcal{P}} : T^n \to T$, which we write as*

$$\alpha_{\mathcal{P}}(t_1, \ldots, t_n) = \sum_{1 \leq i \leq n}^{\mathcal{P}} \alpha_i \cdot t_i,$$

such that

i. $\sum_{1 \leq i \leq n}^{\mathcal{P}} \delta_{i,k} \cdot t_i = t_k$, *where $\delta_{i,j}$ is Kronecker's δ (thus $\delta_{i,j} = 1$ if $i = j$, and $\delta_{i,j} = 0$, otherwise),*

ii. the identity

$$\sum_{1 \leq i \leq n}^{\mathcal{P}} \alpha_i \cdot \Big(\sum_{1 \leq k \leq m}^{\mathcal{P}} \beta_{i,k} \cdot t_k \Big) = \sum_{1 \leq k \leq m}^{\mathcal{P}} \Big(\sum_{i=1}^{n} \alpha_i \beta_{i,k} \Big) \cdot t_k$$

holds whenever $\langle \alpha_1, \ldots, \alpha_n \rangle, \langle \beta_{i,1}, \ldots, \beta_{i,m} \rangle \in \Omega, 1 \leq i \leq n$.

Property i looks quite trivial when written down this way. Rephrasing, it states that the map

$$\langle \delta_{1,k}, \ldots, \delta_{n,k} \rangle_{\mathcal{P}} : T^n \to T,$$

which is assigned to the n-tuple $\langle \delta_{1,k}, \ldots, \delta_{n,k} \rangle$ through $\mathcal{P}$, acts as the projection to the k-th component for $1 \leq k \leq n$. Similarly, property ii may be re-coded in a formal but less concise way. One usually uses the notation from vector spaces rather freely, omitting the explicit reference to the structure whenever possible. Hence simple addition $\alpha_1 \cdot x_1 + \alpha_2 \cdot x_2$ is written rather than $\sum_{1 \leq i \leq 2}^{\mathcal{P}} \alpha_i \cdot x_i$, with the understanding that it refers to a fixed positive convex structure $\mathcal{P}$ on T.

It can be shown that for a positive convex structure the usual rules for manipulating sums in vector spaces apply, e.g., [67],

$$1 \cdot t = t,$$
$$\sum_{i=1}^{n} \alpha_i \cdot t_i = \sum_{i=1, \alpha_i \neq 0}^{n} \alpha_i \cdot t_i,$$
$$(\alpha_1 \cdot t_1 + \alpha_2 \cdot t_2) + \alpha_3 \cdot t_3 = \alpha_1 \cdot t_1 + (\alpha_2 \cdot t_2 + \alpha_3 \cdot t_3).$$

Nevertheless, care should be taken because not all rules apply: we cannot in general conclude $t = t'$ from $\alpha \cdot t = \alpha \cdot t'$, even if $\alpha \neq 0$.

Proposition 1.8.15. *Let γ be a positive convex and smooth equivalence relation on $\mathfrak{S}(M)$ such that $\gamma = \mathsf{ker}(H)$ for some surjective Borel map $H : \mathfrak{S}(M) \to T$ with M, T measurable. Then*

$$\sum_{1 \leq i \leq n}^{H} \alpha_i \cdot H(\mu_i) := H\Big(\sum_{i=1}^{n} \alpha_i \cdot \mu_i\Big)$$

with $\langle \alpha_1, \ldots, \alpha_n \rangle \in \Omega$ and $\mu_1, \ldots, \mu_n \in \mathfrak{S}(M)$ defines a positive convex structure on T.

Proof 1. Because γ is positive convex, and $\gamma = \mathsf{ker}(H)$, we may infer from $H(\mu_i) = H(\mu_i')$ for $1 \leq i \leq n$ and $\alpha_1, \ldots, \alpha_n \in \Omega$ that

$$H\Big(\sum_{i=1}^{n} \alpha_i \cdot \mu_i\Big) = H\Big(\sum_{i=1}^{n} \alpha_i \cdot \mu_i'\Big);$$

thus $\sum^H$ is well defined. It is also easy to see that property i in Definition 1.8.14 is satisfied.

2. Now let $\langle \alpha_1, \ldots, \alpha_n \rangle, \langle \beta_{i,1}, \ldots, \beta_{i,m} \rangle \in \Omega, 1 \leq i \leq n$ and assume that $t_i = H(\mu_i)$ holds. Then

$$\begin{aligned}
\sum_{1 \leq i \leq n}^{H} \alpha_i \cdot \Big(\sum_{1 \leq k \leq m}^{H} \beta_{i,k} \cdot t_k\Big) &= \sum_{1 \leq i \leq n}^{H} \alpha_i \cdot H\Big(\sum_{k=1}^{m} \beta_{i,k} \cdot \mu_k\Big) \\
&= H\Big(\sum_{i=1}^{n} \alpha_i \cdot \Big(\sum_{k=1}^{m} \beta_{i,k} \cdot \mu_k\Big)\Big) \\
&= H\Big(\sum_{k=1}^{m} \Big(\sum_{i=1}^{n} \alpha_i \cdot \beta_{i,k}\Big) \cdot \mu_k\Big) \\
&= \sum_{1 \leq k \leq m}^{H} \Big(\sum_{i=1}^{n} \alpha_i \cdot \beta_{i,k}\Big) \cdot t_k.
\end{aligned}$$

□

Assume that $T = \mathfrak{S}(Y)$ for some Polish space Y. This space carries a natural positive convex structure nat which assigns to $\langle \alpha_1, \ldots, \alpha_n \rangle \in \Omega$ the map $\langle \alpha_1, \ldots, \alpha_n \rangle_{\mathsf{nat}}$ with

$$\langle \alpha_1, \ldots, \alpha_n \rangle_{\mathsf{nat}}(\mu_1, \ldots, \mu_n) := \sum_{i=1}^{n} \alpha_i \cdot \mu_i,$$

so that $\sum^{\mathsf{nat}}_{1 \leq i \leq n} \alpha_i \cdot \mu_i = \sum^{n}_{i=1} \alpha_i \cdot \mu_i$. It obviously has all the properties of a positive convex structure.

The observation yields a characterization of those smooth equivalence relations on $\mathfrak{S}(X)$ that are generated from their cousins on the base space in terms of point-affine maps.

Definition 1.8.16. *Call a surjective map $H : \mathfrak{S}(M) \to \mathfrak{S}(T)$ for the measurable spaces M and T* point-affine *iff*

i. $\sum^{H} = \sum^{\mathsf{nat}}$,
ii. for each $m \in M$ there exists $t \in T$ such that $H(\delta_m) = \delta_t$.

The condition i in Definition 1.8.16 is tantamount to saying that H is affine, i.e., distributes over positive convex combinations. A positive convex and smooth equivalence relation $\gamma = \mathsf{ker}(H)$ with $H : \mathfrak{S}(X) \to T$ induces on T a positive convex structure by Proposition 1.8.15.

Proposition 1.8.17. *Let $\gamma = \mathsf{ker}(H)$ be a positive convex and smooth equivalence relation on $\mathfrak{S}(X)$ with $H : \mathfrak{S}(X) \to \mathfrak{S}(Y)$ surjective and continuous for the Polish space X and the separable metric space Y. Then these conditions are equivalent*

a. γ is grounded.
b. H is point-affine.

Proof 1. The proof for $a \Rightarrow b$ is straightforward; thus $b \Rightarrow a$ needs to be taken care of.

2. Define $f : X \to Y$ through $\delta_{f(x)} = H(\delta_x)$; then f is well defined and continuous. This is so since $x_n \to x$ in X implies $\delta_{x_n} \rightharpoonup_w \delta_x$ in the weak topology on $\mathfrak{S}(X)$, so that

$$\delta_{f(x_n)} = H(\delta_{x_n}) \to H(\delta_x) = \delta_{f(x)},$$

from which continuity of f follows, because $\{\delta_y \mid y \in Y\}$ is homeomorphic to Y. This implies that $\mathfrak{S}(f) : \mathfrak{S}(X) \to \mathfrak{S}(Y)$ is continuous as well.

We show that $H = \mathfrak{S}(f)$, from which the assertion will follow through Lemma 1.8.5. In fact, from the definition of f we see that

$$\mathfrak{S}(f)(\delta_x) = \delta_{f(x)} = H(\delta_x)$$

holds, and we infer from property ii that for $(\alpha_1, \ldots, \alpha_n) \in \Omega$

$$\mathfrak{S}(f)\big(\sum_{i=1}^{n}\alpha_i\cdot\delta_{x_i}\big)=\sum_{i=1}^{n}\alpha_i\cdot\delta_{f(x_i)}=$$
$$\sum_{1\le i\le n}^{\mathsf{nat}}\alpha_i\cdot H(\delta_{x_i})=\sum_{1\le i\le n}^{H}\alpha_i\cdot H(\delta_{x_i})=H\big(\sum_{i=1}^{n}\alpha_i\cdot\delta_{x_i}\big).$$

This implies that $\mathfrak{S}(f)(\mu)=H(\mu)$ holds for discrete measures μ. Since each measure can be approximated by a discrete one in the weak topology by Proposition 1.6.7, and since both H and $\mathfrak{S}(f)$ are continuous, the assertion follows. □

The proof of Proposition 1.8.17 makes essential use of the weak topology on $\mathfrak{S}(X)$, the space on which the relation is defined. This seems to indicate that installing a Polish topology on the base space X is essential, but this is not the case: we are going to generalize this result to analytic spaces now, using the technique to move a smooth relation from a Polish space to an analytic one, developed when investigating simple relations in Section 1.7.2.

Lemma 1.8.18. *Let A and X be analytic spaces, and $f : X \to A$ be Borel and onto. Let γ be an equivalence relation on $\mathfrak{S}(A)$, and define the equivalence relation $\gamma_{\mathfrak{S}(f)}$ on $\mathfrak{S}(X)$ through*

$$\mu\ \gamma_{\mathfrak{S}(f)}\ \mu'\Leftrightarrow\mathfrak{S}(f)(\mu)\ \gamma\ \mathfrak{S}(f)(\mu')$$

whenever $\mu,\mu'\in\mathfrak{S}(X)$. Then the following holds

a. If γ is smooth, so is $\gamma_{\mathfrak{S}(f)}$.
b. $\lfloor\gamma_{\mathfrak{S}(f)}\rfloor=\lfloor\gamma\rfloor_f$.
c. γ is near-grounded iff $\gamma_{\mathfrak{S}(f)}$ is near-grounded.
d. γ is grounded iff $\gamma_{\mathfrak{S}(f)}$ is grounded.

Proof 1. The first part is trivial, because γ is assumed to be smooth. The second part follows immediately from the observation that $\mathfrak{S}(f)(\delta_x)=\delta_{f(x)}$ and from the definition of $\lfloor\gamma\rfloor_f$; see Lemma 1.7.16.

2. Assume that γ is near-grounded, and take $\mu,\mu'\in\mathfrak{S}(X)$. Then

$$\begin{aligned}\mu\ \gamma_{\mathfrak{S}(f)}\ \mu'&\Leftrightarrow\mathfrak{S}(f)(\mu)\ \gamma\ \mathfrak{S}(f)(\mu')\\&\overset{(*)}{\Rightarrow}\forall D\in\mathbf{\Sigma}(\mathcal{B}(A),\lfloor\gamma\rfloor):\mu(f^{-1}[D])=\mu'(f^{-1}[D])\\&\overset{(\natural)}{\Leftrightarrow}\forall E\in f^{-1}\left[\mathbf{\Sigma}(\mathcal{B}(A),\lfloor\gamma\rfloor)\right]:\mu(E)=\mu'(E)\\&\overset{(\sharp)}{\Leftrightarrow}\forall E\in\mathbf{\Sigma}(\mathcal{B}(A),\lfloor\gamma\rfloor_f):\mu(E)=\mu'(E)\\&\Leftrightarrow\forall E\in\mathbf{\Sigma}(\mathcal{B}(A),\lfloor\gamma_{\mathfrak{S}(f)}\rfloor):\mu(E)=\mu'(E)\\&\Leftrightarrow\mu\ \overline{\lfloor\gamma_{\mathfrak{S}(f)}\rfloor}\ \mu'.\end{aligned}$$

Implication $(*)$ makes use of the assumption that γ is near-grounded, i.e., that $\gamma\subseteq\overline{\lfloor\gamma\rfloor}$ holds, equivalence $(\natural)$ shifts the inverse operator f^{-1} from sets to the σ-algebra, and the next equivalence ($\sharp$) applies the characterization of invariant sets from Lemma 1.7.16. Thus we see that the near-groundedness of γ implies the near-groundedness of $\gamma_{\mathfrak{S}(f)}$.

In a similar way the converse statement is proved. Let $\mu, \mu' \in \mathfrak{S}(A)$; then we can find $\mu_0, \mu_0' \in \mathfrak{S}(X)$ such that $\mu = \mathfrak{S}(f)(\mu_0), \mu' = \mathfrak{S}(f)(\mu_0')$. Now assume that $\gamma_{\mathfrak{S}(f)}$ is near-grounded, then

$$\begin{aligned}\mu\ \gamma\ \mu' &\Leftrightarrow \mu_0\ \gamma_{\mathfrak{S}(f)}\ \mu_0' \\ &\overset{(\circledast)}{\Rightarrow} \mu_0\ \overline{\lfloor\gamma\rfloor_f}\ \mu_0' \\ &\overset{(\spadesuit)}{\Leftrightarrow} \forall E \in f^{-1}\left[\mathbf{\Sigma}(\mathcal{B}(A), \lfloor\gamma\rfloor)\right] : \mu_0(E) = \mu_0'(E) \\ &\Leftrightarrow \forall G \in \mathbf{\Sigma}(\mathcal{B}(A), \lfloor\gamma\rfloor) : \mu(G) = \mu'(G) \\ &\Leftrightarrow \forall G \in \mathbf{\Sigma}(\mathcal{B}(A), \lfloor\gamma\rfloor) : \mathfrak{S}(f)(\mu_0)(G) = \mathfrak{S}(f)(\mu_0')(G).\end{aligned}$$

Implication ($\circledast$) uses the inclusion $\gamma_{\mathfrak{S}(f)} \subseteq \overline{\lfloor\gamma_{\mathfrak{S}(f)}\rfloor} = \overline{\lfloor\gamma\rfloor_f}$, and equivalence ($\spadesuit$) applies the characterization of invariant sets from Lemma 1.7.16. The other arguments are straightforward. Thus the near-groundedness of $\gamma_{\mathfrak{S}(f)}$ implies the near-groundedness of γ. This establishes part c.

3. For the proof of part d, it is noted that for grounded relations the implications ($*$) resp. ($\circledast$) can be reversed. □

Using this characterization of a relation induced by the inverse image of a Borel map we can generalize Proposition 1.8.17 to analytic base spaces.

Proposition 1.8.19. *Let $\gamma = \mathsf{ker}(H)$ be a positive convex and smooth equivalence relation on $\mathfrak{S}(A)$ with $H : \mathfrak{S}(A) \to \mathfrak{S}(Y)$ surjective and Borel for the analytic space A and the analytic space Y. Then γ is grounded iff H is point-affine.*

Proof 1. Let first Y be a separable metric space. Assume that $\gamma = \mathsf{ker}(H)$ is grounded. Since A is an analytic space, we can find a surjective Borel map $f : X \to A$ that defines the measurable structure on A. Because of Proposition 1.4.19 we may assume that $x \mapsto \big(H \circ \mathfrak{S}(f)\big)(\delta_x)$ is continuous. Thus the assertion follows from the observation that $\sum^{H\circ\mathfrak{S}(f)} = \sum^H$ together with Lemma 1.8.18, part d, and the characterization of grounded relations in Proposition 1.8.17.

2. If Y is an analytic space, then we can find a separable metric space Z such that $(Y, \mathcal{B}(Y))$ is Borel isomorphic to $(Z, \mathcal{B}(Z))$ by Corollary 1.4.35. Thus $\mathfrak{S}(Y)$ and $\mathfrak{S}(Z)$ are Borel isomorphic, and the assertion follows from part 1. □

1.9 Bibliographic Notes

Polish Spaces

The basic measure-theoretic constructions as well as the discussion of Polish spaces are folklore. Apart from the *Gentle Introduction* in [20, Chapter 1] where most of these results are collected, and where everything is proved, we point to some references that have been used rather frequently. Halmos [43] is the first, albeit somewhat outdated reference to Measure Theory; the book by Hewitt and Stromberg [45] is a textbook which focusses on connections to Real Analysis; both Parthasarathy [65] and Billingsley [6] deal with measure-theoretic aspects (while [7] is more down-to-earth and application-oriented). Fremlin's comprehensive survey [36] should be mentioned as well. [33] and [69] are rich sources of information related to topological spaces. The reader finds an extensive discussion of the topology of weak convergence in the monographs by Parthasarathy [65] and by Billingsley [6] with emphasis on the Prohorov metric; the Hutchinson metric, and its variants (as well as the naming problems associated with it) find a careful review in Edgar's treatise [32].

Borel Sets

The study of Borel sets cannot be undertaken nowadays without the monographs by Kechris [49] and by Srivastava [79]; the towering early reference in this area is Kuratowski [55], which contains many examples; see also [59]. Special aspects of analytic spaces are discussed in Arveson [2]. The relationship between the weak topology and the weak*-σ-algebra has been rediscovered many times in the area of Markov transition systems; to the best of the author's knowledge Rieder [70, Lemma 6.1] first stated and proved that the Borel sets for the weak topology are just the members of the weak*-σ-algebra; the compact case has been treated without a proof in [30, 3.1]. Measurable selections are an important topic in areas like stochastic dynamic programming; overviews are given in [82, 46] and in [14]. The existence of semi-pullbacks was established first for a special case in [31] and in the general form recorded here in [19].

Smooth Equivalence Relations

The application of smooth equivalence relations starts — like many interesting developments in this area — with the paper [17] by Desharnais, Edalat, and Panangaden; the reference was, however, implicit. Later developments in the investigation of stochastic Kripke models for various types of logics and in the investigation of the coalgebraic structure of stochastic relations indicated that these relations are fairly helpful. They permit the definition of congruences; the kernel of a morphism is a congruence, and vice versa, each congruence is the kernel of a morphism [20, Chapter 5], just as in other structures investigated in Universal Algebra [42, 13]. Factoring stochastic relations was first investigated in [24]. When investigating the question about whether this classical result can be established also for weak morphisms, near-grounded relations were introduced in [25]; the proofs in Section 1.8 are from that paper.

It is shown in Proposition 1.7.15 that the set of all smooth equivalence relations is closed under countable infima. It can be shown that the same does not hold for suprema: the supremum of two smooth equivalence relations needs not be smooth; an example has been proposed by S. Thomas. It is discussed in [28], and it has as a consequence the facts that the category of analytic spaces with Borel maps is not closed under pushouts, and that the intersection of two countably generated σ-algebras is not necessarily countably generated. Example 1.7.4 exhibits a non-smooth equivalence relation with interesting structural properties; it is taken from [73].

Measurable Selectors

Measurable selection theory constitutes an important technique in Operations Research; see, e.g., [82] for a survey. It is also a helpful tool in the theory of Borel sets; see [79, Chapter 5]. It has been applied to stochastic Kripke models specifically to construct semi-pullbacks in categories of stochastic relations [19] and for establishing the existence of certain bisimulations for simple stochastic relations [20, Section 5.7]. An overview as well as some samples for this powerful technique is provided in [20]. The approach can be used when a direct construction is not available, and when the set of candidates for a solution is sufficiently well behaved.

Chapter 2
Modal Logics

2.1 Introduction

This chapter introduces modal and continuous time stochastic logics. We discuss these logics here in some detail since we want to define the classical notions of bisimilarity, logical and behavioral equivalence for logics the interpretation of which is well understood. This will enable us later on to draw from this source knowledge as well as experience when defining these terms for coalgebraic logics, and when investigating them. We indicate the results and give sketches for the more interesting proofs, which we will need in later chapters as well. These considerations make substantial use of the techniques that have been developed in the context of stochastic relations, about which we reported in Chapter 1, in particular in Sections 1.6 and 1.7. We will return to the topics dealt with in the present chapter mainly in Chapter 4, when coalgebraic logics are investigated. It will turn out that nearly all results are special cases of a coalgebraic scenario. This might advise against dealing with, e.g., modal logics and continuous time stochastic logics on this level of detail. On the other hand it turns out that the approaches for the general, coalgebraic case are essentially motivated by these specific logics, and that these logics serve as excellent examples for the somewhat abstract treatment in Chapter 4. For instance, we will study residence times in some detail in the present chapter, and this will help in our appreciating the properties of some important functors when dealing with stochastic right coalgebras in Section 4.4.

We first deal with modal logics which are introduced fairly generally through modal similarity types. Kripke models are introduced in both their non-deterministic and their stochastic versions, and we give examples from well-known logics illustrating these models and highlighting their different approaches. Before jumping into this discussion in Section 2.3, however, we define in Section 2.2 bisimilar stochastic relations in terms of spans of morphisms and give a criterion for them to be bisimilar; this is based on simu-

E.-E. Doberkat, *Stochastic Coalgebraic Logic*, Monographs in Theoretical Computer Science. An EATCS Series, DOI 10.1007/978-3-642-02995-0_2,

lation equivalent congruences. Because each modal logic defines congruences on the state space of a Kripke model, we put this criterion to work when looking into the relationship of bisimilarity, behavioral equivalence — which is defined in terms of a cospan of morphisms — and logical equivalence in Section 2.3.2. Investigating congruences like that is also at the heart of looking into the same relationship when discussing continuous time stochastic logic in Section 2.4. This is a logic akin to the well-known logic **CTL** used for model checking which takes time explicitly into consideration. We augment the logic by adding a fixed-point operator, partly for indicating that the techniques we present here are so flexible that adding a substantially different operator does not affect them, but also to prepare for things to come when discussing coalgebraic stochastic logics which contain infinitesimal operators as abstractions for fixed point operators. The discussion of the continuous time logics is technically a bit more involved than the one for the conceptually simpler modal logic because we need to argue on two levels, viz., on the level of states — here we will define state formulas — and on the level of infinite paths, which will be captured through path formulas; see Section 2.4.5. This foreshadows the development for some coalgebraic generalization as well; see, e.g., Section 4.3.

The models are defined —nearly by default — over analytic spaces, and sometimes even over Polish ones. We show in an appendix that weaker results can be obtained already in Kripke models that are defined over general measurable spaces. It will be shown that two such Kripke models are logically equivalent iff they are behaviorally equivalent; this will be established for a very simple negation-free Hennessy-Milner logic.

2.2 Bisimulations

Bisimulations are introduced as spans of morphisms such that common events exist. They relate two systems in terms of their elements, and hence in terms of nondeterministic relations of their state spaces.

In fact, assume that $(S, (\rightarrow_a)_{a\in A})$ and $(T, (\rightarrow'_a)_{a\in A})$ are two labeled transition systems; then a relation $R \subseteq S \times T$ is called a bisimulation iff

- Whenever $\langle s, t\rangle \in R$ and $s \rightarrow_a s_1$, then there exists t_1 with $t \rightarrow'_a t_1$ and $\langle s_1, t_1\rangle \in R$.
- Whenever $\langle s, t\rangle \in R$ and $t \rightarrow'_a t_1$, then there exists s_1 with $s \rightarrow_a s_1$ and $\langle s_1, t_1\rangle \in R$.

We interpret a labeled transition system as a coalgebra (S, α) for the functor $\mathfrak{F} := \mathcal{P}(A \times -)$ which sends the set S to $\mathcal{P}(A \times S)$ and the map $f : S \rightarrow T$ to $\mathcal{P}(A \times S) \ni Q \mapsto \{\langle a, f(s)\rangle \mid \langle a, s\rangle \in Q\} \in \mathcal{P}(A \times T)$. It is an easy exercise to show that R is a bisimulation iff there exists a coalgebraic structure xi on R such that this diagram commutes:

$$\begin{array}{ccccc} S & \xleftarrow{\pi_S} & R & \xrightarrow{\pi_T} & T \\ {\scriptstyle\alpha}\downarrow & & \downarrow{\scriptstyle xi} & & \downarrow{\scriptstyle\beta} \\ \mathfrak{F}(S) & \xleftarrow[\mathfrak{F}(\pi_S)]{} & \mathfrak{F}(R) & \xrightarrow[\mathfrak{F}(\pi_T)]{} & \mathfrak{F}(T) \end{array}$$

Definition 2.2.1. *The stochastic relations* $\mathsf{K} = (X, Y, K)$ *and* $\mathsf{L} = (V, W, L)$ *are called* bisimilar *iff there exist a stochastic relation* $\mathsf{M} = (A, B, M)$ *and morphisms* $\mathsf{f} = (\phi, \psi) : \mathsf{M} \to \mathsf{L}, \mathsf{g} = (\gamma, \delta) : \mathsf{M} \to \mathsf{L}$ *such that*

a. the diagram

$$\begin{array}{ccccc} X & \xleftarrow{\phi} & A & \xrightarrow{\gamma} & V \\ {\scriptstyle K}\downarrow & & \downarrow{\scriptstyle M} & & \downarrow{\scriptstyle L} \\ \mathfrak{S}(Y) & \xleftarrow[\mathfrak{S}(\psi)]{} & \mathfrak{S}(B) & \xrightarrow[\mathfrak{S}(\delta)]{} & \mathfrak{S}(W) \end{array}$$

is commutative,

b. the σ*-algebra* $\psi^{-1}[\mathcal{B}(Y)] \cap \delta^{-1}[\mathcal{B}(W)]$ *is nontrivial, i.e., contains not only* $\emptyset$ *and* B*.*

The relation M *is called* mediating.

The first condition on bisimilarity is in accordance with the general definition of bisimilarity of coalgebras in Section 1.2.1; it requests that f and g form a span of morphisms

$$\mathsf{K} \xleftarrow{\mathsf{f}} \mathsf{M} \xrightarrow{\mathsf{g}} \mathsf{L},$$

so that we have for each $a \in A, D \in \mathcal{B}(Y), E \in \mathcal{B}(W)$ the equalities

$$K\big(\phi(a)\big)(D) = \big(\mathfrak{S}(\psi) \circ M\big)(a)(D) = M(a)\big(\psi^{-1}[D]\big)$$

and

$$L\big(\gamma(a)\big)(E) = \big(\mathfrak{S}(\delta) \circ M\big)(a)(E) = M(a)\big(\delta^{-1}[E]\big).$$

The second condition, however, states that we can find an event $C^* \in \mathcal{B}(B)$ which is common to both K and L in the sense that

$$\psi^{-1}[D] = C^* = \delta^{-1}[E]$$

for some $D \in \mathcal{B}(Y)$ and $E \in \mathcal{B}(W)$ such that both $C^* \neq \emptyset$ and $C^* \neq B$ hold (note that for $C^* = \emptyset$ or $C^* = W$ we can always take the empty and the full set, respectively). Given such a C^* with D and E from above we get for each $a \in A$

$$\begin{aligned} K(\phi(a))(D) &= M(a)(\psi^{-1}[D]) \\ &= M(a)(C^*) \\ &= M(a)(\delta^{-1}[E]) \\ &= L(\gamma(a))(E); \end{aligned}$$

thus the event C^* ties K and L together. Loosely speaking, $\psi^{-1}[\mathcal{B}(Y)] \cap \delta^{-1}[\mathcal{B}(W)]$ can be described as the σ-algebra of common events, which is required to be nontrivial. Note that without the second condition two relations K and L which are strictly probabilistic (i.e., for which the entire space is always assigned probability 1) would always be bisimilar: Put $A := X \times V, B := Y \times W$ and set for $\langle x, v\rangle \in A$ as the mediating relation $M(x,v) := K(x) \otimes L(v)$; then the projections will make the diagram commutative. It is also clear that this argument does not work for the subprobabilistic case. The second condition in Definition 2.2.1 serves to prevent this somewhat anomalous behavior; it is technically not too restrictive, as we will see below.

A criterion for stochastic relations to be bisimilar is derived from simulation-equivalent congruences. They will be introduced now, and the relation to bisimilarity is indicated as well.

Simulation Equivalence

Simulation-equivalent relations behave on their classes in exactly the same fashion. This requires the equivalence classes, in particular the Borel structure on the respective factor spaces, to be related in a suitable way: knowing one factor space and its Borel structure entails detailed knowledge about the Borel structure of the other one, in particular about its generators. This is captured through the idea that one equivalence relation *spawns* the other one — if it is known how to generate the Borel structure on one factor space, then this knowledge is carried over to the other one.

Definition 2.2.2. *Let α and β be smooth equivalence relations on the analytic spaces X resp. Y, and assume that $\Upsilon : X/\alpha \to Y/\beta$ is a map between the equivalence classes. We say that α* spawns *β* via *$(\Upsilon, \mathcal{A}_0)$ iff $\mathcal{A}_0$ is a countable generator of $\mathbf{\Sigma}(\mathcal{B}(X), \alpha)$ such that*

a. $\mathcal{A}_0$ is closed under finite intersections,
b. $\{\Upsilon_A \mid A \in \mathcal{A}_0\}$ is a generator of $\mathbf{\Sigma}(\mathcal{B}(Y), \beta)$, where $\Upsilon_A := \bigcup\{\Upsilon([x]_\alpha) \mid x \in A\}$.

Thus if α spawns β, then the measurable structure induced by α on X is all we need for constructing the measurable structure induced by β on Y: the map Υ can be made to carry over the generator $\mathcal{A}_0$ from $\mathbf{\Sigma}(\mathcal{B}(X), \alpha)$ to $\mathbf{\Sigma}(\mathcal{B}(Y), \beta)$ and to transport the atoms from one σ-algebra to the other. This is of particular interest since the atoms are just the equivalence classes by Lemma 1.7.11. Hence α together with Υ and the generator $\mathcal{A}_0$ is all we may care to know or to learn about β.

Spawning is used to model congruences that behave in the same fashion; this requires that at least the Borel structures of the underlying factor spaces be comparable. The definition of simulation-equivalent congruences then reads as follows.

Definition 2.2.3. *Let* $\mathsf{K} = (X, Y, K)$ *and* $\mathsf{K}' = (X', Y', K')$ *be stochastic relations over Standard Borel spaces with congruences* $\mathsf{c} = (\alpha, \beta)$ *and* $\mathsf{c}' = (\alpha', \beta')$*, respectively.*

a. Congruence c simulates c' *(symbolically* $\mathsf{c} \propto \mathsf{c}'$*) iff* α *spawns* α' *via* $(\Upsilon, \mathcal{A}_0)$ *and* β *spawns* β' *via* $(\Theta, \mathcal{B}_0)$ *such that*

$$\forall x \in X \forall x' \in \Upsilon([x]_\alpha) \forall B \in \mathcal{B}_0 : K(x)(B) = K'(x')(\Theta_B).$$

b. Call these congruences simulation-equivalent *iff both* $\mathsf{c} \propto \mathsf{c}'$ *and* $\mathsf{c}' \propto \mathsf{c}$ *hold.*

Simulation-equivalent congruences behave in exactly the same way. The same behavior is exhibited on each equivalence class, as far as the input is concerned, and on the respective invariant output sets. It becomes plain at this point that a characterization of equivalent behavior through congruences exhibits the double face of congruences: it is certainly necessary to use the equivalence relation on the input spaces; but since the behavior on the output spaces is modelled through probabilities, we need also the invariant Borel sets for a characterization.

Simulation-equivalent congruences on stochastic relations give rise to a factor object built on their sum, as we have seen in Section 1.7.3. This construction will be used for investigating the bisimilarity of stochastic relations, and later on for a closer discussion of the bisimilarity of Kripke models for a multitude of logics, ranging from general modal logics (Section 2.3.2) to continuous time logics (Section 2.4.5) and to coalgebraic logics (Section 4.3.3).

Assume that c and c' are simulation-equivalent congruences on the Polish objects $\mathsf{K} = (X, Y, K)$ and $\mathsf{K}' = (X', Y', K')$, respectively. Assume that α spawns α' via $(\Upsilon, \{(C_n)_{n\in\mathbb{N}}\})$, and that β spawns β' via $(\Theta, \{(D_n)_{n\in\mathbb{N}}\})$. Construct for K and K' the direct sum

$$\mathsf{K} \oplus \mathsf{K}' := \big(X + X', Y + Y', K \oplus K'\big);$$

where the only non-obvious construction is $K \oplus K'$: put for the Borel set $E \subseteq Y + Y'$

$$(K \oplus K')(z)(E) := \begin{cases} K(z)(E \cap Y), & \text{if } z \in X \\ K'(z)(E \cap Y'), & \text{if } z \in X', \end{cases}$$

then clearly $K \oplus K' : X + X' \rightsquigarrow Y + Y'$ (we omit the injections). Define respectively on $X + X'$ and $Y + Y'$ the σ-algebras

$$\mathcal{G} := \{C + \Upsilon_{C'} \mid C \in \mathbf{\Sigma}(\mathcal{B}(X), \alpha), C' \in \mathbf{\Sigma}(\mathcal{B}(X'), \alpha')\}$$
$$\mathcal{H} := \{D + \Theta_{D'} \mid D \in \mathbf{\Sigma}(\mathcal{B}(Y), \beta), D' \in \mathbf{\Sigma}(\mathcal{B}(Y'), \beta')\}.$$

Then $\mathcal{G}$ and $\mathcal{H}$ are countably generated sub-σ-algebras of the respective Borel sets. Because the σ-algebras in question are countably generated, so is their sum. Both $\mathcal{G}$ and $\mathcal{H}$ respectively define smooth equivalence relations on $X+X'$ and $Y + Y'$ by Corollary 1.7.14. By simulation-equivalence it follows that these equivalences are just the amalgamation of the participating relations (the amalgamation is defined on page 52).

Because the congruences are simulation-equivalent, we see that $z\ (\alpha \diamond \alpha')\ z'$ implies $(K \oplus K')(z)(F) = (K \oplus K')(z')(F)$ for all $F \in \mathcal{H}$. Hence,

$$\mathcal{G} = \mathbf{\Sigma}(\mathcal{B}(X + X'), \alpha \diamond \alpha')$$
$$\mathcal{H} = \mathbf{\Sigma}(\mathcal{B}(Y + Y'), \beta \diamond \beta'),$$

and $\mathsf{c} \diamond \mathsf{c}' := (\alpha \diamond \alpha', \beta \diamond \beta')$ is a congruence on $\mathsf{K} \oplus \mathsf{K}'$.

Bisimilarity

The factor object $(\mathsf{K} \oplus \mathsf{K}')/(\mathsf{c} \diamond \mathsf{c}')$ constructed in this way will be of interest when helping us establish the bisimilarity of K and K', provided they have simulation-equivalent nontrivial congruences.

Proposition 2.2.4. *If there exists nontrivial congruences c_i on the Polish objects K_i for $i = 1, 2$ that are simulation-equivalent, then*

a. there exist morphisms $\mathsf{f}_1 : \mathsf{K}_1 \to (\mathsf{K}_1 \oplus \mathsf{K}_2)/(\mathsf{c}_1 \diamond \mathsf{c}_2)$ and $\mathsf{f}_2 : \mathsf{K}_2 \to (\mathsf{K}_1 \oplus \mathsf{K}_2)/(\mathsf{c}_1 \diamond \mathsf{c}_2)$.
b. K_1 and K_2 are bisimilar.

Proof (Sketch) 1. Assume $\mathsf{K}_i = (X_i, Y_i, K_i)$ and $\mathsf{c}_i = (\alpha_i, \beta_i)$ for $i = 1, 2$. Construct the sum $\mathsf{K}_1 \oplus \mathsf{K}_2$ as above, and let (κ_i, λ_i) be the corresponding injections, which are, however, no morphisms. Let

$$(\eta_{\alpha_1 \diamond \alpha_2}, \eta_{\beta_1 \diamond \beta_2}) : \mathsf{K}_1 \oplus \mathsf{K}_2 \to (\mathsf{K}_1 \oplus \mathsf{K}_2)/(\mathsf{c}_1 \diamond \mathsf{c}_2)$$

be the factor map. Then $(\eta_{\alpha_1 \diamond \alpha_2} \circ \kappa_i, \eta_{\beta_1 \diamond \beta_2} \circ \lambda_i)$ constitutes a morphism $\mathsf{K}_i \to (\mathsf{K}_1 \oplus \mathsf{K}_2)/(\mathsf{c}_1 \diamond \mathsf{c}_2)$, as will be shown now. Surjectivity has to be established, and we have to show that the σ-algebra of common events is nontrivial.

2. Each equivalence class $a \in (X_1 + X_2)/(\alpha_1 \diamond \alpha_2)$ can be represented as $a = [x_1]_{\alpha_1} + [x_2]_{\alpha_2}$ for some suitably chosen $x_1 \in X_1, x_2 \in X_2$. Similarly, each equivalence class $b \in (Y_1 + Y_2)/(\beta_1 \diamond \beta_2)$ can be written as $b = [y_1]_{\beta_1} + [y_2]_{\beta_2}$ for some $y_1 \in Y_1, y_2 \in Y_2$. Conversely, the sum of classes is a class again.

3. Now we have this diagram:

$$\begin{array}{ccc} & & \mathsf{K}_1 \\ & & \Big\downarrow{\scriptstyle (\eta_{\alpha_1\diamond\alpha_2}\circ\kappa_1, \eta_{\beta_1\diamond\beta_2}\circ\lambda_1)} \\ \mathsf{K}_2 & \xrightarrow[(\eta_{\alpha_1\diamond\alpha_2}\circ\kappa_2, \eta_{\beta_1\diamond\beta_2}\circ\lambda_2)]{} & (\mathsf{K}_1 \oplus \mathsf{K}_2)/(\mathsf{c}_1 \diamond \mathsf{c}_2) \end{array}$$

This yields part *a*.

4. The semi-pullback of the pair of morphisms with a joint target constructed in the first step exists by Proposition 1.6.25. It is a Polish object (A, B, M), where

$$A := \{\langle x_1, x_2\rangle \in X_1 \times X_2 \mid [x_1]_{\alpha_1\diamond\alpha_2} = [x_2]_{\alpha_1\diamond\alpha_2}\},$$
$$B := \{\langle y_1, y_2\rangle \in Y_1 \times Y_2 \mid [y_1]_{\beta_1\diamond\beta_2} = [y_2]_{\beta_1\diamond\beta_2}\}.$$

We finally are required to establish that there are indeed nontrivial common events. Since c is nontrivial, we can find an invariant Borel set $D \in \mathbf{\Sigma}(\mathcal{B}(Y_1), \beta_1)$ with $\emptyset \neq D \neq Y_1$. Assume that β_1 spawns β_2 via $(\Theta, \{D_n \mid n \in \mathbb{N}\})$, then $\emptyset \neq \Theta_D \neq Y_2$ also holds. Because D is β_1-invariant,

$$\pi_{1,Y_1}^{-1}[D] = \{\langle y_1, y_2\rangle \mid y_1 \in D\} = \{\langle y_1, y_2\rangle \mid y_2 \in \Theta_D\} = \pi_{2,Y_2}^{-1}[\Theta_D];$$

thus

$$\pi_{1,Y_1}^{-1}[D] \in \pi_{1,Y_1}^{-1}[\mathcal{B}(Y_1)] \cap \pi_{2,Y_2}^{-1}[\mathcal{B}(Y_2)],$$

and we are done once it is shown that $\pi_{1,Y_1}^{-1}[D] \neq B$. Since $D \neq Y_1$ is invariant, there exists y_1 with

$$[y_1]_{\beta_1\diamond\beta_2} \cap D = [y_1]_{\beta_1} \cap D = \emptyset.$$

Let $[y_2]_{\beta_2} := \Theta([\psi_1]_{\beta_1})$, then $[y_2]_{\beta_1\diamond\beta_2} \cap \Theta_D = [y_2]_{\beta_2} \cap \Theta_D = \emptyset$. Consequently, $\langle y_1, y_2\rangle \in B \setminus \pi_{1,Y_1}^{-1}[D]$. This shows that $\pi_{1,Y_1}^{-1}[\mathcal{B}(Y_1)] \cap \pi_{2,Y_2}^{-1}[\mathcal{B}(Y_2)]$ is nontrivial. □

We note the following for later use.

Corollary 2.2.5. *Under the conditions of Proposition 2.2.4, the stochastic relation* $(\mathsf{K}_1 \oplus \mathsf{K}_2)/(\mathsf{c}_1 \diamond \mathsf{c}_2)$ *is isomorphic both to* $\mathsf{K}_1/\mathsf{c}_1$ *and to* $\mathsf{K}_2/\mathsf{c}_2$. □

The strategy of the proof to Proposition 2.2.4 has been to make sure that the classes associated with the congruences are distributed evenly among the summands in the sense that each class in the sum is the sum of appropriate classes. This then implies that we can construct surjective maps, and from them morphisms through some general mechanisms. The idea works in particular with isomorphic factor spaces.

Proposition 2.2.6. *Let* K *and* K' *be analytic objects such that* K/c *is isomorphic to* K'/c' *for some nontrivial congruences* c *and* c'*. Then*

a. c *and* c' *are simulation-equivalent,*
b. K *and* K' *are bisimilar.*

Proof (Sketch) 0. Let $\mathsf{K} = (X, Y, K)$ with $\mathsf{c} = (\alpha, \beta)$; similarly for K' and c'. Assume that $\mathsf{f} = (\Phi, \Psi)$ is the isomorphism $\mathsf{K}/\mathsf{c} \to \mathsf{K}'/\mathsf{c}'$ which is composed of the Borel isomorphisms $\Phi : X/\alpha \to X'/\alpha'$ and $\Psi : Y/\beta \to Y'/\beta'$. Let moreover $\mathcal{A}$ and $\mathcal{B}$ be countable generators of $\mathbf{\Sigma}(\mathcal{B}(X), \alpha)$ and $\mathbf{\Sigma}(\mathcal{B}(Y), \beta)$ which are closed under finite intersections. Then α spawns α' via $(\Phi, \mathcal{A})$, and β spawns β' via $(\Psi, \mathcal{B})$. Hence we have to establish for each $x \in X, x' \in \Phi([x]_\alpha)$ and for each β-invariant Borel subset $B \subseteq Y$ that $K(x)(B) = K'(x')(\Psi_B)$ holds. This will imply that c simulates c'; interchanging the rôles of c and c' then will yield simulation-equivalence.

1. Given $B \in \mathbf{\Sigma}(\mathcal{B}(Y), \beta)$ we know from Lemma 1.7.10 that we can find a Borel set $B_1 \in \mathcal{B}(Y/\beta)$ such that $B = \eta_\beta^{-1}[B_1]$. Since Ψ is a Borel isomorphism, we find $B_2 \in \mathcal{B}(Y'/\beta')$ with $B_1 = \Psi^{-1}[B_2]$. A routine calculation shows that $\Psi_B = \eta_{\beta'}^{-1}[B_2]$. Now assume that $x \in X, x' \in \Phi([x]_\alpha)$; then the following chain of equations is obtained from the argumentation above, and from the assumption that f is an isomorphism

$$\begin{aligned}
K(x)(B) &= K(x)(\eta_\beta^{-1}\left[\Psi^{-1}[B_2]\right]) \\
&= K_{\alpha,\beta}([x]_\alpha)(\Psi^{-1}[B_2]) \\
&= K'_{\alpha',\beta'}(\Phi([x]_\alpha)(B_2) \\
&= K'(x')(\eta_{\beta'}^{-1}[B_2]) \\
&= K'(x')(\Psi_B).
\end{aligned}$$

This establishes the desired relation $\mathsf{c} \propto \mathsf{c}'$ and completes the proof for the first part.

2. Bisimilarity now follows through Proposition 2.2.4. □

This will be a helpful tool and construction for the investigations to follow.

2.3 Modal Logics: Syntax and Semantics

We did establish a criterion for bisimilarity through simulation-equivalent congruences and discussed bisimilarity in terms of isomorphic factor spaces. We will now apply this to modal logics.

This section defines modal logic, and Kripke models are defined in their usual nondeterministic and their stochastic versions, together with their satisfaction relation. Some examples are given in order to exhibit probabilistic models for specific logics. Then we discuss bisimilarity and the related notions of logical and behavioral equivalence for these Kripke models.

Let P be a countable set of propositional letters which is fixed throughout; $O \neq \emptyset$ is a set of modal operators. $\tau = (O, \mathfrak{ar})$ is called a *modal similarity type* iff $O \neq \emptyset$, and if $\mathfrak{ar} : O \to \mathbb{N}$ is a map, assigning each modal operator $\triangle$ its arity $\mathfrak{ar}(\triangle) \geq 1$. We will not deal with modal operators of arity 0, since they do not have to be dealt with as modal constants in an interpretation. The similarity type τ will be fixed.

We define three modal languages based on τ and P. The formulas of the *basic modal language* $\mathfrak{M}_{\mathfrak{b}}(\tau, P)$ are given by the syntax

$$\phi ::= p \mid \top \mid \phi_1 \wedge \phi_2 \mid \neg\phi \mid \triangle(\phi_1, \ldots \phi_{\mathfrak{ar}(\triangle)}),$$

where $p \in P$ is a propositional letter. If we have $O = \{\diamond\}$ with $\mathfrak{ar}(\diamond) = 1$, we obtain the formulas of the well-known *basic modal language* with negation. Omitting negation in $\mathfrak{M}_{\mathfrak{b}}(\tau, P)$ defines the formulas in the *negation-free basic modal language* $\mathfrak{M}_{\mathfrak{1}}(\tau, P)$. Finally the *extended modal language* $\mathfrak{M}_{\mathfrak{s}}(\tau, P)$ is defined through the syntax

$$\phi ::= p \mid \top \mid \phi_1 \wedge \phi_2 \mid \neg\phi \mid \triangle_q(\phi_1, \ldots \phi_{\mathfrak{ar}(\triangle)}),$$

where $q \in \mathbb{Q} \cap [0, 1]$ is a rational number, and $p \in P$ is a propositional letter. Again, if we deal with $O = \{\diamond\}$ as the similarity type, then we get an entire line of new formulas through $(\diamond_q)_{q \in \mathbb{Q} \cap [0,1]}$.

Nondeterministic Kripke Models

A *nondeterministic τ-Kripke model* $\mathcal{R} = (S, R_\tau, V)$ consists of a state space S, a family $R_\tau = (R_\triangle)_{\triangle \in O}$ of set-valued maps $R_\triangle : S \to \mathcal{P}\left(S^{\mathfrak{ar}(\triangle)}\right)$, and a set-valued map $V : P \to \mathcal{P}(S)$.

The satisfaction relation $\models$ for a nondeterministic τ-Kripke model $\mathcal{R}$ is defined as usual for $\mathfrak{M}_{\mathfrak{b}}(\tau, P)$:

$$\begin{aligned}
\mathcal{R}, s \models p &\Leftrightarrow s \in V(p) \\
\mathcal{R}, s \models \neg\phi &\Leftrightarrow \mathcal{R}, s \not\models \phi \\
\mathcal{R}, s \models \phi_1 \wedge \phi_2 &\Leftrightarrow \mathcal{R}, s \models \phi_1 \text{ and } \mathcal{R}, s \models \phi_2 \\
\mathcal{R}, s \models \triangle(\phi_1, \ldots, \phi_{\mathfrak{ar}(\triangle)}) &\Leftrightarrow \exists \langle s_1, \ldots, s_{\mathfrak{ar}(\triangle)} \rangle \in R_\triangle(s) : \mathcal{R}, s_i \models \phi_i \\
&\qquad \text{for } 1 \leq i \leq \mathfrak{ar}(\triangle).
\end{aligned}$$

Denote by

$$[\![\phi]\!]_{\mathcal{R}} := \{s \in S \mid \mathcal{R}, s \models \phi\}$$

the set of states for which formula ϕ is valid (the *extension of formula* ϕ), and by

$$Th_{\mathcal{R}}(s) := \{\phi \in \mathfrak{M}_{\mathfrak{b}}(\tau, P) \mid \mathcal{R}, s \models \phi\}$$

the *theory of state* s in $\mathcal{R}$.

Stochastic Kripke Models

In analogy, a *stochastic τ-Kripke model* $\mathcal{K} = (S, K_\tau, V)$ has a state space S which is endowed with a σ-algebra $\mathcal{A}$, a family $K_\tau = (K_\triangle)_{\triangle \in O}$ of stochastic relations $K_\triangle : S \rightsquigarrow S^{\mathfrak{ar}(\triangle)}$, and a set-valued map $V : P \to \mathcal{A}$. The stochastic relation $K_\triangle : S \rightsquigarrow S^{\mathfrak{ar}(\triangle)}$ is denoted by $F_\triangle(\mathcal{K})$. We will usually assume that S is a Polish space, and that the σ-algebra are the Borel sets.

The interpretation of formulas in $\mathfrak{M}_\mathfrak{s}(\tau, P)$ for a stochastic τ-Kripke model $\mathcal{K}$ is fairly straightforward, the interesting case arising when a modal operator is involved:

$$\mathcal{K}, s \models \triangle_q(\phi_1, \ldots, \phi_{\mathfrak{ar}(\triangle)})$$

holds iff there exists measurable subsets $A_1, \ldots, A_{\mathfrak{ar}(\triangle)}$ of S such that $\mathcal{K}, s_i \models \phi_i$ holds for all $s_i \in A_i$ for $1 \leq i \leq \mathfrak{ar}(\triangle)$, and

$$K_\triangle(s)(A_1 \times \ldots \times A_{\mathfrak{ar}(\triangle)}) \geq q.$$

Arguing with state transition systems in mind, this interpretation of validity reflects that upon the move indicated by $\triangle_q$, a state s satisfies $\triangle_q(\phi_1, \ldots, \phi_{\mathfrak{ar}(\triangle)})$ iff we can find states s_i satisfying ϕ_i with a $K_\triangle$-probability not smaller than q. Note that the usual operators $\triangle$ and ∇ are replaced by a whole spectrum of operators $\triangle_q$ which permit a finer and probabilistically more adequate notion of satisfaction.

Again, let $[\![\phi]\!]_\mathcal{K}$ be the set of all states for which $\phi \in \mathfrak{M}_\mathfrak{s}(\tau, P)$ is satisfied under $\mathcal{K}$, and

$$Th_\mathcal{K}(s) := \{\phi \in \mathfrak{M}_\mathfrak{s}(\tau, P) \mid \mathcal{K}, s \models \phi\}$$

the theory for state $s \in S$.

An easy inductive argument shows that the sets $[\![\phi]\!]_\mathcal{K}$ are measurable, so that they may be used as arguments for the stochastic relations we are working with:

Lemma 2.3.1. *$[\![\phi]\!]_\mathcal{K}$ is a measurable subset of S for each $\phi \in \mathfrak{M}_\mathfrak{s}(\tau, P)$.* □

As in the case of stochastic relations we need to exclude trivial cases.

Definition 2.3.2. *A τ-Kripke model $\mathcal{K}$ with state space S is called* degenerate *iff $[\![\phi]\!]_\mathcal{K} = S$ or $[\![\phi]\!]_\mathcal{K} = \emptyset$ holds for each formula $\phi \in \mathfrak{M}_\mathfrak{s}(\tau, P)$.*

A degenerate model does not usually carry useful information. The restriction is quite similar to not permitting the universal relation as a part of a congruence, and of requesting the existence of nontrivial common events for bisimulations. We will see that these constraints are closely related.

2.3.1 Examples

We show how some well-known logics may be interpreted through Kripke models, indicating that specific logics require specific probabilistic arguments. We introduce first the logic associated with labeled transition systems. This example is of historic significance [58]. It is shown also that the basic temporal language may be interpreted stochastically by reversing a relation. Arrow logic as a popular logic modelling simple programming constructs is interpreted through a simple transformation of a distribution. In presenting these examples we follow essentially the representation of the respective logics in [8].

Example 2.3.3. Suppose that the set Act of labels is a countable set; it is thought of as an alphabet of actions. Each action $a \in \mathsf{Act}$ is associated with a unary modal operator $\langle a \rangle$; so put $\tau := (O, \mathfrak{ar})$ with $O := \{\langle a \rangle \mid a \in \mathsf{Act}\}$ and $\mathfrak{ar}(\langle a \rangle) := 1$.

1. A nondeterministic τ-Kripke model is based on a labeled transition system $(S, (\rightarrow_a)_{a \in \mathsf{Act}})$ which associates a binary relation $\rightarrow_a \subseteq S \times S$ with each action a. Thus
$$s \models \langle a \rangle \phi \Leftrightarrow \exists s' : s \rightarrow_a s' \wedge s' \models \phi.$$
2. A stochastic τ-Kripke model is based on a *labeled Markov transition system*, say $(S, (k_a)_{a \in \mathsf{Act}})$, which associates with each action a a stochastic relation $k_a : S \rightsquigarrow S$. Thus
$$s \models \langle a \rangle_q \phi \Leftrightarrow k_a(s)([\![\phi]\!]) \geq q;$$
hence a transition is replaced by the probability with which it can occur.

Variants of the logic $\mathfrak{M}_\mathfrak{s}(\tau, P)$ with $P = \emptyset$ were investigated in the literature by Larsen and Skou, and by Desharnais, Edalat and Panangaden with a reference to the logic investigated by Hennessy and Milner [44]; we refer to them also as *Hennessy-Milner logic* $\mathfrak{L}(\mathsf{Act}, \mathbb{Q} \cap [0,1])$. Consequently, this logic's formulas are given through
$$\phi ::= \top \mid \phi_1 \wedge \phi_2 \mid \langle a \rangle_r \phi.$$
Here $a \in \mathsf{Act}$ is an action, and the threshold r is a rational number from the unit interval. $\dashv$

Example 2.3.4. The basic temporal language has two unary modal operators **F** (forward) and **B** (backward), so that $O = \{\mathbf{F}, \mathbf{B}\}$.

1. A nondeterministic τ-Kripke model interprets the forward operator **F** through a relation $R \subseteq S \times S$ and the backward operator **B** through the converse $R^\smile$ of relation R; thus $R^\smile := \{\langle s', s \rangle \mid \langle s, s' \rangle \in R\}$. Consequently, we have

$$s \models \mathbf{B}\phi \Leftrightarrow \exists t \in S : \langle t, s\rangle \in R \wedge t \models \phi.$$

2. A probabilistic interpretation interprets **F** through a stochastic relation $K : S \rightsquigarrow S$, so that

$$s \models \mathbf{F}_q\phi \Leftrightarrow K(s)([\![\phi]\!]) \geq q.$$

The backward operator **B** is interpreted through the converse $K_\mu^\smile : S \rightsquigarrow S$, provided the state space S is Standard Borel and an initial probability μ is given. The converse $K_\mu^\smile$ of a stochastic relation K given an initial probability μ is a stochastic relation $L : S \rightsquigarrow S$ such that

$$\int_S K(s)(B_s)\ \mu(ds) = \int_S L(t)(B^t)\ \mu(dt)$$

holds for each Borel set $B \subseteq S \times S$. It is known that the converse relation exists whenever the state space S is a Polish space [23]; this carries over obviously to Standard Borel spaces. Thus

$$s \models \mathbf{B}_q\phi \Leftrightarrow K_\mu^\smile(s)([\![\phi]\!]) \geq q.$$

An easy calculation shows that

$$\begin{aligned} s \models \mathbf{B}_1\mathbf{F}_1\phi &\Leftrightarrow K_\mu^\smile(s)\left(\{s' \mid K(s')([\![\phi]\!]) = 1\}\right) = 1 \\ &\Leftrightarrow \int_S K(s')([\![\phi]\!])\ K_\mu^\smile(s)(ds') = 1. \end{aligned}$$

Note that the definition of the converse requires an initial probability (this is intuitively clear: if the probability for a backward running process is described, one has to say where to start). It is also noteworthy that a topological assumption has been made; if the state space is not a Polish space, then the technical arguments permitting the definition of the converse are not available. $\dashv$

Example 2.3.5. Arrow logic has three modal operators modelling reversal, composition, and skip respectively. Thus $O = \{\mathbf{1}, \otimes, \circ\}$ with respective arities $\mathfrak{ar}(\mathbf{1}) = 0, \mathfrak{ar}(\otimes) = 1,$ and $\mathfrak{ar}(\circ) = 2$.

1. The usual nondeterministic interpretation of arrow logic is done over a world of pairs; so the base state space is $S \times S$ for some S, with associated relations

$$\begin{aligned} R_1 &= \Delta_S = \{\langle s, s\rangle \mid s \in S\}, \\ R_\otimes &= \{\langle\langle s_0, s_1\rangle, \langle s_1, s_0\rangle\rangle \mid s_0, s_1 \in S\}, \\ R_\circ &= \{\langle\langle s_0, s_1\rangle, \langle s_0, s\rangle, \langle s, s_1\rangle\rangle \mid s, s_0, s_1 \in S\}. \end{aligned}$$

Thus, e.g.,

$$\langle s, s'\rangle \models \phi \circ \psi \Leftrightarrow \exists s_0 : \langle s, s_0\rangle \models \phi \wedge \langle s_0, s'\rangle \models \psi$$

and

$$\langle s, s'\rangle \models \otimes\phi \Leftrightarrow \langle s', s\rangle \models \phi.$$

2. Now assume again that S is a Polish space, and let $\mu \in \mathfrak{P}(S)$ be a probability. Put for $A \in \mathcal{B}(S \times S)$

$$\hat{\mu}(A) := \mu(\{s \in S \mid \langle s, s\rangle \in A\});$$

thus $\hat{\mu}$ transports a Borel set in S to a Borel set in the diagonal of $S \times S$. Interpret the composition operator $\circ_q$ through the stochastic relation

$$K_\circ(s, s') := \delta_s \otimes \hat{\mu} \otimes \delta_{s'}.$$

Note that the operator $\otimes$ is somewhat overloaded: it denotes the modal operator for reversal, and the product operator for measures. The context should make it clear which version is meant.
We obtain then

$$\begin{aligned} K_\circ(s, s')([\![\phi]\!] \times [\![\psi]\!]) &= (\delta_s \otimes \hat{\mu} \otimes \delta_{s'})\,([\![\phi]\!] \times [\![\psi]\!]) \\ &= \hat{\mu}(\{\langle s_1, s_2\rangle \mid \langle s, s_1\rangle \in [\![\phi]\!], \langle s_2, s'\rangle \in [\![\psi]\!]\}) \\ &= \mu(\{s_1 \mid \langle s, s_1\rangle \in [\![\phi]\!], \langle s_1, s'\rangle \in [\![\psi]\!]\}). \end{aligned}$$

Consequently,

$$\langle s, s'\rangle \models \phi \circ_1 \psi \Leftrightarrow \langle s, s_1\rangle \models \phi \wedge \langle s_1, s'\rangle \models \psi \text{ for } \mu\text{-almost all } s_1$$

(here *μ-almost all* s_1 means as usual that the set of all s_1 for which the property does not hold has μ-measure 0). More generally, $\langle s, s'\rangle \models \phi \circ_q \psi$ iff $\langle s, s_1\rangle \models \phi \wedge \langle s_1, s'\rangle \models \psi$ for all s_1 from a Borel set S_0 with $\mu(S_0) \geq q$. Finally, put $K_\otimes(s, s') := \delta_{\langle s', s\rangle}$; then $\langle s, s'\rangle \models \otimes_q\phi \Leftrightarrow \langle s', s\rangle \models \phi$, for all rational q with $0 \leq q \leq 1$ (which is evidently independent of q), and let

$$K_1(s, s') := \begin{cases} 0, & s \neq s' \\ \delta_{\langle s, s\rangle}, & s = s' \end{cases}$$

(here 0 is the null measure); then

$$\langle s, s'\rangle \models \mathbf{1} \Leftrightarrow s = s'.$$

Note that in general we did exclude modal constants, i.e., modal operators of arity 0, when defining modal similarity types. The example shows that it is possible to include them nevertheless without much ado. ⊣

2.3.2 Bisimulations for Kripke Models

This section investigates morphisms for stochastic τ-Kripke models. Bisimilarity and logical equivalence are related to each other.

Definition 2.3.6. *The stochastic τ-Kripke models $\mathcal{K}$ and $\mathcal{K}'$ are said to be* logical equivalent *iff* $\{Th_{\mathcal{K}}(s) \mid s \in S\} = \{Th_{\mathcal{K}'}(s') \mid s' \in S'\}$.

Thus $\mathcal{K}$ and $\mathcal{K}'$ are logically equivalent iff given $s \in S$ there exists $s' \in S'$ such that $Th_{\mathcal{K}}(s) = Th_{\mathcal{K}'}(s')$, and vice versa.

Morphisms for stochastic Kripke models should be based on morphisms for the underlying stochastic relations, and they should take the propositional constants into account.

Definition 2.3.7. *Let $\mathcal{K}$ and $\mathcal{K}'$ be stochastic τ-Kripke models with $\mathcal{K} = \big(S, (K_\triangle)_{\triangle\in O}), V\big)$ and $\mathcal{K}' = \big(S', (K'_\triangle)_{\triangle\in O}), V'\big)$. A* strong morphism *$\Phi : \mathcal{K} \to \mathcal{K}'$ is determined through a measurable and surjective map $\Phi : S \to S'$ so that these conditions are satisfied:*

a. $V(p) = \Phi^{-1}\left[V'(p)\right]$ holds for all $p \in P$,
b. $K'_\triangle \circ \Phi = \mathfrak{S}\left(\Phi^{\mathfrak{ar}(\triangle)}\right) \circ K_\triangle$ holds for each modal operator $\triangle$.

Here $\Phi^{\mathfrak{ar}(\triangle)} : \langle x_1, \ldots, x_{\mathfrak{ar}(\triangle)}\rangle \mapsto \langle \Phi(x_1), \ldots, \Phi(x_{\mathfrak{ar}(\triangle)})\rangle$ distributes the map Φ into the components. Consequently, if $\Phi : \mathcal{K} \to \mathcal{K}'$ is a strong morphism, then

$$(\Phi, \Phi^{\mathfrak{ar}(\triangle)}) : F_\triangle(\mathcal{K}) \to F_\triangle(\mathcal{K}')$$

is a morphism between the corresponding stochastic relations for each modal operator $\triangle \in O$. In addition we know for each propositional letter p that $\mathcal{K}, s \models p$ iff $\mathcal{K}', \Phi(s) \models p$.

Bisimulations are defined again as spans of — strong — morphisms. Similarly, we define behavioral equivalence through a cospan of morphisms, essentially mimicking the corresponding definition for stochastic relations.

Definition 2.3.8. *Let $\mathcal{K}_1$ and $\mathcal{K}_2$ be stochastic τ-Kripke models.*

a. $\mathcal{K}_1$ and $\mathcal{K}_2$ are called strongly bisimilar *iff there exists a stochastic τ-Kripke model $\mathcal{M}$ and strong morphisms*

$$\mathcal{K}_1 \xleftarrow{\Phi_1} \mathcal{M} \xrightarrow{\Phi_2} \mathcal{K}_2,$$

such that the σ-algebra of common events $\Phi_1^{-1}\left[\mathcal{B}(S_1)\right] \cap \Phi_2^{-1}\left[\mathcal{B}(S_2)\right]$ is nontrivial (here S_i is the state space of $\mathcal{K}_i, i = 1, 2$).

b. $\mathcal{K}_1$ and $\mathcal{K}_2$ are called behaviorally equivalent *iff there exists a stochastic τ-Kripke model $\mathcal{L}$ and strong morphisms*

$$\mathcal{K}_1 \xrightarrow{\Psi_1} \mathcal{L} \xleftarrow{\Psi_2} \mathcal{K}_2.$$

We relate logical equivalence, strong bisimilarity and behavioral equivalence of Kripke models $\mathcal{K}$ and $\mathcal{K}'$, provided the models are based on Polish spaces. Fix the stochastic τ-Kripke models $\mathcal{K} := \big(S, (K_\triangle)_{\triangle \in O}), V\big)$ and $\mathcal{K}' := \big(S', (K'_\triangle)_{\triangle \in O}, V'\big)$.

It is well known that morphisms preserve theories for logics of the Hennessy-Milner type. This is also true for stochastic relations:

Lemma 2.3.9. *If $\Phi : \mathcal{K} \to \mathcal{K}'$ is a strong morphism, then $Th_{\mathcal{K}}(s) = Th_{\mathcal{K}'}(\Phi(s))$ holds for all states $s \in S$.* □

Define the equivalence relation α on state space S through

$$s_1 \ \alpha \ s_2 \Leftrightarrow Th_{\mathcal{K}}(s_1) = Th_{\mathcal{K}}(s_2);$$

thus two states are α-equivalent iff they satisfy exactly the same formulas in $\mathfrak{M}_{\mathfrak{s}}(\tau, P)$. Thus they are equivalent iff they cannot be separated through a formula. In a similar way α' is defined on S'. Because we have at most countably many formulas, α and α' are smooth equivalence relations. Define the equivalence relation $\beta_\triangle$ on $S^{\mathfrak{ar}(\triangle)}$ through

$$\langle s_1, \dots, s_{\mathfrak{ar}(\triangle)} \rangle \ \beta_\triangle \ \langle t_1, \dots, t_{\mathfrak{ar}(\triangle)} \rangle \Leftrightarrow s_1 \ \alpha \ t_1 \wedge \dots \wedge s_{\mathfrak{ar}(\triangle)} \ \alpha \ t_{\mathfrak{ar}(\triangle)};$$

then $\beta_\triangle = \times_{i=1}^{\mathfrak{ar}(\triangle)} \beta$ is smooth, and we know that the σ-algebra of β-invariant sets can be written in terms of the α-invariant sets, viz.,

$$\mathbf{\Sigma}\big(\mathcal{B}(S^{\mathfrak{ar}(\triangle)}), \beta_\triangle\big) = \bigotimes_{i=1}^{\mathfrak{ar}(\triangle)} \mathbf{\Sigma}\big(\mathcal{B}(S), \alpha\big)$$

(see Lemma 1.7.19). The relation $\beta'_\triangle$ is defined in the same way for α'.

The equivalence of $\mathcal{K}$ and $\mathcal{K}'$ makes these relations into simulation-equivalent congruences.

Lemma 2.3.10. *If the nondegenerate Kripke models $\mathcal{K}$ and $\mathcal{K}'$ are logically equivalent, then $(\alpha, \beta_\triangle)$ and $(\alpha', \beta'_\triangle)$ are simulation-equivalent and nontrivial congruences for the stochastic relations $F_\triangle(\mathcal{K})$ and $F_\triangle(\mathcal{K}')$.* □

Accordingly, we know from Proposition 2.2.4 that for logical equivalent Kripke models $\mathcal{K}$ and $\mathcal{K}'$ and for each modal operator $\triangle$ the stochastic relations $F_\triangle(\mathcal{K})$ and $F_\triangle(\mathcal{K}')$ are bisimilar. All the mediating relations can be collected to form a mediating Kripke model. This yields the following.

Theorem 2.3.11. *Assume that $\mathcal{K}$ and $\mathcal{K}'$ are nondegenerate stochastic τ-Kripke models over analytic spaces. Then the following statements are equivalent:*

a. $\mathcal{K}$ and $\mathcal{K}'$ are strongly bisimilar,
b. $\mathcal{K}$ and $\mathcal{K}'$ are logically equivalent,

c. $\mathcal{K}$ and $\mathcal{K}'$ are behaviorally equivalent. □

Commenting on the development, it is noted that Theorem 2.3.11 is derived from Proposition 2.2.4, and hence from a condition that arose from the consideration of stochastic relations alone. This is in marked contrast to the proof carried out in [27] which starts from the logic and develop the properties of simulation-equivalent congruences implicitly. Analyzing the proof, it becomes clear that the model constructed there will usually not be defined over a Standard Borel space. This is so since factoring destroys the property of being a Polish space, rendering the factor space analytic instead, see Example 1.7.6.

Logical equivalence appears here as some sort of catalyst which permits proving that bisimilarity and behavioral equivalence describe the same phenomenon, a link that is missing in the general development of stochastic relations; see Section 2.2. There we have simulation-equivalent congruences at our disposal, which are always tied to a relation, while the logic serves here as an arbitrator which is completely independent of the Kripke model interpreting it.

2.4 Temporal Logics: μCSL

We will define as a further illustration *continuous time stochastic logic with fixed points operators*, abbreviated as μ**CSL**. The logic will be introduced formally first; then models and their morphisms are introduced and the interpretation of μ**CSL** is given. Some standard properties like Borel measurability are established, and logical equivalence is defined. We deal with properties on states and on paths; the equivalence relations defined by the logic on these sets are related to each other. The relations obtained from these constructions are modified so that they fit into the mold of the models for the logic. The main result is that logical equivalence and bisimilarity are equivalent, and that this holds also for behavioral equivalence, provided the factor space induced by the theory of states is a Standard Borel space again (this is so since the projective limit construction, on which interpretations are based, does not seem to work for general analytic spaces, but only for their Standard Borel brethren, see Proposition 1.6.35).

2.4.1 The Logic μCSL

State formulas and path formulas for $\mu\mathbf{CSL} = \mu\mathbf{CSL}(\mathsf{AP}, \mathsf{SV}, \mathsf{PV})$ are given through this syntax (with mutually disjoint and countable sets AP, SV, and PV of atomic proposition, state variables, and path variables, respectively): The intuitive idea is that the logic works over infinite paths which have as a component alternating a state and a time; the time is interpreted as residence time for the state.

The formal definition reads as follows.

- *State formulas* are defined through the syntax

$$\phi ::= \top \mid a \mid Z \mid \neg\phi \mid \phi \wedge \phi' \mid \mathcal{S}_{\ltimes p}(\phi) \mid \mathcal{P}_{\ltimes p}(\psi).$$

  Here $a \in \mathsf{AP}$ is an atomic proposition, $Z \in \mathsf{SV}$ is a state variable, ψ is a path formula, $\ltimes$ is one of the relational operators $<$, $\leq$, $>$, $\geq$, and $p \in [0,1]$ is a rational number.
- *Path formulas* are defined through

$$\psi ::= \tilde{\top} \mid P \mid \neg\psi \mid \psi \wedge \psi' \mid \mathcal{X}^I\,\psi \mid \phi\,\mathcal{U}^I\,\phi' \mid \mu P.\psi$$

  with $P \in \mathsf{PV}$ as a path variable, ϕ, ϕ' as state formulas, $I \subseteq \mathbb{R}_+$ a closed interval of the real numbers with rational bounds (including $I = \mathbb{R}_+$); these intervals will be called *rational intervals*. The operator μ describes the smallest fixed point; it binds variables in the usual sense. We assume that the variable bound by it is in the range of an even number of negations.

  The informal interpretation of the operators is as follows.

1. The operator $\mathcal{S}_{\ltimes p}(\phi)$ gives the *steady state probability* for ϕ to hold with the boundary condition $\ltimes p$, where ϕ is a state formula. This is a state formula again.
2. The *path quantifier* formula $\mathcal{P}_{\ltimes p}(\psi)$ holds for a state iff the probability of all paths starting in this state and satisfying path formula ψ is specified by $\ltimes p$. Thus, e.g., ψ holds on almost all paths starting from that state iff it satisfies $\mathcal{P}_{\geq 1}(\psi)$. Of course, $\mathcal{P}_{\ltimes p}(\psi)$ is a state formula.
3. The *next operator* $\mathcal{X}^I\,\phi$ is assumed to hold on an infinite path of states and residence times iff the residence time for the first state is an element of interval I, and if the second state satisfies ϕ.
4. The *until-operator* $\phi_1\,\mathcal{U}^I\,\phi_2$ holds on path σ iff we can find a point in time $t \in I$ such that the state $\sigma@t$ which the infinite path σ denotes at time t satisfies ϕ_2, and for all times t' before that, $\sigma@t'$ satisfies ϕ_1 (the notation $\sigma@t$ will be defined formally on page 89).

2.4.2 Defining Models and Their Morphisms

We are ready for the definition of models for μ**CSL** and their morphisms. We will work with projective limits (see Section 1.6.4) for interpreting path formulas; models will be based on Polish spaces rather than more generally on analytic spaces.

Definition 2.4.1. $\mathcal{M} = (S, M, \mathcal{I}, V)$ *is called a* model for μ**CSL** *iff*

1. *S is a Polish space, the* state space *of $\mathcal{M}$,*
2. *$M : S \rightsquigarrow \mathbb{R}_+ \times S$ is a stochastic relation with $M(s)(\mathbb{R}_+ \times S) = 1$ for all $s \in S$; the stochastic relation M is the* law of change *of $\mathcal{M}$,*
3. *$\mathcal{I} = (\Sigma, \Pi)$ interprets the variables,*
   a. *$\Sigma : \mathsf{SV} \to \mathcal{B}(S)$ assigns each state variable a Borel set in S,*
   b. *$\Pi : \mathsf{PV} \to \mathcal{B}((S \times \mathbb{R}_+)^\infty)$ assigns each path variable a Borel set of paths,*
4. *$V : \mathsf{AP} \to \mathcal{B}(S)$ maps each atomic proposition to a Borel set of states.*

Thus a model says how residence times and state changes are to be handled: if $s \in S$ is the present state, then $M(s)(I \times B)$ gives the probability that after $t \in I$ time units a state change will happen, and that the new state will be a member of Borel set $B \subseteq S$. Each model says how the variables are to be interpreted; this is written down through the maps Σ and Π, and we say what sets the atomic propositions are taken from. Note that we assume in each case that the sets under consideration are Borel. Otherwise we could not assign them any probability directly or indirectly; hence this assumption is made for keeping the model within the realm of our probabilistic reasoning. We postulate that the law of change assigns probability 1 to $\mathbb{R}_+ \times S$ for each state. Without this assumption, mass along infinite paths might vanish too fast; from a technical point of view, this assumption is necessary because we will construct from this law a projective limit which requires its components to assign probability 1 to the base space (see Section 1.6.4). Consequently, we will use the probability functor $\mathfrak{P}$ rather than its cousin $\mathfrak{S}$.

Morphisms

We define a morphism $\Phi : \mathcal{M} \to \mathcal{N}$ for the models $\mathcal{M}$ and $\mathcal{N}$. It is based on a map $\Phi : S \to S'$ between state spaces, which is extended to a map $\Phi_\infty : (S \times \mathbb{R}_+)^\infty \to (S' \times \mathbb{R}_+)^\infty$ upon setting

$$\Phi_\infty(\langle s_0, t_0, s_1, t_1, \ldots \rangle) := \langle \Phi(s_0), t_0, \Phi(s_1), t_1, \ldots \rangle;$$

thus we transform the states according to Φ but leave the residence times alone; define additionally $id_{\mathbb{R}_+} \times \Phi : \langle t, s \rangle \mapsto \langle t, \Phi(s) \rangle$, and similarly, $\Phi \times id_{\mathbb{R}_+}$.

Definition 2.4.2. *Let $\mathcal{M} = (S, M, \mathcal{I}, V)$ and $\mathcal{N} = (S', N, \mathcal{I}', V')$ be models for μ**CSL**. Then $\Phi : \mathcal{M} \to \mathcal{N}$ is called a* morphism *from $\mathcal{M}$ to $\mathcal{N}$ iff*

a. *$\Phi : S \to S'$ is a surjective and Borel measurable map between the state spaces,*
b. *$(\Phi, id_{\mathbb{R}_+} \times \Phi) : M \to N$ is a morphism for the associated stochastic relations M and N,*
c. *$\Phi^{-1}[\Sigma'(Z)] = \Sigma(Z)$ for each state variable Z,*
d. *$\Phi_\infty^{-1}[\Pi'(P)] = \Pi(P)$ for each path variable P,*

e. $\Phi^{-1}[V'(a)] = V(a)$ *for each atomic proposition* a.

We require the map underlying a morphism to be onto since we want to be able to trace each state in S' back to a state in S, inheriting the corresponding property from the basic stochastic relations. Condition b says that this diagram is commutative:

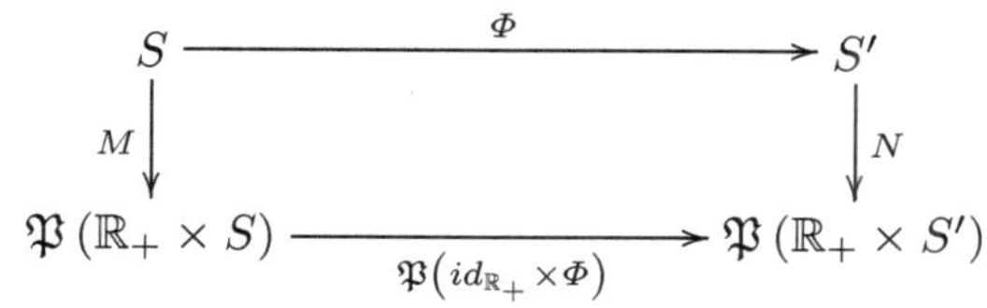

Thus we have in particular

$$N(\Phi(s))(I \times B) = M(s)\left(I \times \Phi^{-1}[B]\right)$$

for every state $s \in S$, every rational interval I, and every Borel set $B \in \mathcal{B}(S)$. Conditions c to e relate the interpretations of variables and atomic propositions. For example, condition c says that for a state s and a state variable Z we have $s \in \Sigma(Z)$ iff $\Phi(s) \in \Sigma'(Z)$.

Projective Limits

We model the one-step behavior of a model through its transition law, but we are not yet able to say how this relates to the behavior along paths. This is done iteratively for paths of finite length, and by passing to a limit (in this case to the projective limit) for infinite paths.

The construction works like this.

Let $M : S \rightsquigarrow \mathbb{R}_+ \times S$ be the stochastic relation underlying a model. Fix a state $s \in S$, and proceed inductively along finite paths: Put

$$M_1(s) := M(s),$$

and set in the inductive step for the Borel set $D \subseteq (\mathbb{R}_+ \times S)^{n+1}$

$$\begin{aligned} M_{n+1}(s)(D) :=& \\ &\int_{(\mathbb{R}_+ \times S)^n} M(s_n)\left(\{\langle t, s\rangle \mid \langle t_0, s_1, \dots, t_{n-1}, s_n, t, s\rangle \in D\}\right) \times \\ &\qquad \times M_n(s)(d\langle t_0, s_1, \dots, t_{n-1}, s_n\rangle) = \\ &\qquad\qquad \int_{(\mathbb{R}_+ \times S)^n} M(\hbar_S(\mathbf{w}))(D_{\mathbf{w}})\ M_n(s)(d\mathbf{w}), \end{aligned}$$

where we have set $\hbar_S(t_0, s_1, \ldots, t_{n-1}, s_n) := s_n$ for simplifying the notation. Thus the argument to $M(s_n) = M(\hbar_S(\mathbf{w}))$ is the set of all times and states $\langle t, s\rangle$ such that $\langle \mathbf{w}, t, s\rangle = \langle t_0, s_1, \ldots, t_{n-1}, s_n, t, s\rangle$ is a member of D. Analyzing the expression further, we see that at step $n+1$ the probability for the pair that consists of timing a transition and changing a state is an element of $\{\langle t, s\rangle \mid \langle t_0, s_1, \ldots, t_{n-1}, s_n, t, s\rangle \in D\}$ equals

$$M(\hbar_S(\mathbf{w}))(D_{\mathbf{w}}) = M(s_n)\left(\{\langle t, s\rangle \mid \langle t_0, s_1, \ldots, t_{n-1}, s_n, t, s\rangle \in D\}\right),$$

provided the corresponding times and states that have been run through during steps $1, \ldots, n$ are given by $\mathbf{w} = \langle t_0, s_1, \ldots, t_{n-1}, s_n\rangle$ which in turn is captured through $M_n(s)(d\mathbf{w})$.

Standard arguments show that $M_n : S \rightsquigarrow (\mathbb{R}_+ \times S)^n$ is a stochastic relation. For each state $s \in S$ the sequence $(M_n(s))_{n\in\mathbb{N}}$ forms a projective system (Definition 1.6.32), provided $M(s)(\mathbb{R}_+ \times S) = 1$ holds for each $s \in S$: for each Borel set $B \subseteq (\mathbb{R}_+ \times S)^n$ the equality

$$M_{n+1}(B \times (\mathbb{R}_+ \times S)) = M_n(s)(B)$$

holds. Consistency of this family has as a consequence the fact that the measures can be extended to Borel sets of infinite sequences. We obtain from Proposition 1.6.35 the existence of the projective limit.

Proposition 2.4.3. *Given a stochastic relation $M : S \rightsquigarrow \mathbb{R}_+ \times S$ such that $M(s)(\mathbb{R}_+ \times S) = 1$ for all $s \in S$, there exists a unique stochastic relation $M_\infty : S \rightsquigarrow (\mathbb{R}_+ \times S)^\infty$ such that*

$$M_\infty(s)\big(B \times \prod_{j>n}(\mathbb{R}_+ \times S)\big) = M_n(s)(B)$$

for each Borel set $B \in \mathcal{B}((\mathbb{R}_+ \times S)^n)$ and each state $s \in S$. M_∞ is the projective limit of $(M_n)_{n\in\mathbb{N}}$. □

Intuitively, this equation means that the behavior of the infinite paths up to horizon n is uniquely determined by the transition law M_n. This construction entails that a morphism Φ between models may be interpreted as a morphism (Φ, Φ_∞) between these projective limits. To be specific:

Proposition 2.4.4. *Let $\mathcal{M}$ and $\mathcal{N}$ be models, $\Phi : \mathcal{M} \to \mathcal{N}$ be a morphism from $\mathcal{M}$ to $\mathcal{N}$. Then $(\Phi, \Phi_\infty) : M_\infty \to N_\infty$ is a morphism between the stochastic relations M_∞ and N_∞.* □

2.4.3 Interpreting μCSL

We are now ready for an interpretation of μ**CSL**. Fix a model $\mathcal{M} = (S, M, \mathcal{I}, V)$ over the Standard Borel space S and let $M_\infty : S \rightsquigarrow \mathbb{R}_+ \times (S \times$

$\mathbb{R}_+)^\infty = (\mathbb{R}_+ \times S)^i nfty$ be the associated stochastic relation that relates (initial) states to paths.

The Semantics

The semantics of μ**CSL** is then described recursively through relation $\models$ between states respectively paths, and formulas as described below. We will need to describe in what state the model is at a given time t. Assume the behavior is given through a path $\sigma = \langle s_0, t_0, s_1, t_1, \ldots \rangle$ which may more graphically be written through $s_0 \xrightarrow{t_0} s_1 \xrightarrow{t_1} s_2 \xrightarrow{t_2} s_3 \ldots$. Put $\sigma[1\ldots] := s_1 \xrightarrow{t_1} s_2 \xrightarrow{t_2} s_3 \ldots$ as the path after step 1, and define $\delta(\sigma, 0) := t_0$ as the first residence time. Given $t \in \mathbb{R}_+$, assume that there is a smallest index k such that $t < \sum_{i=0}^{k} t_i$; then we would intuitively say that σ is at time k in state s_k; define accordingly $\sigma@t := s_k$. It can be shown that the set $\{\langle \sigma, t \rangle \in (S \times \mathbb{R}_+)^\infty \mid \sigma@t \text{ is defined}\}$ is a Borel subset of $(S \times \mathbb{R}_+)^\infty$, and that $\langle \sigma, t \rangle \mapsto \sigma@t$ constitutes a measurable map from the latter set to S.

Let $\mathcal{M} = (S, M, (\Sigma, \Pi), V)$ be a model. Given a state variable Z and a Borel set $Q \in \mathcal{B}(S)$, denote by $\mathcal{M}[Z\backslash Q]$ the model $(S, M, (\Sigma', \Pi), V)$ with $\Sigma'(Z) := Q$; otherwise Σ' coincides with Σ. Similarly, the model $\mathcal{M}[P\backslash U]$ is defined for the path variable P and the Borel set $U \in \mathcal{B}((S \times \mathbb{R}_+)^\infty)$. Substituting values in this way may be iterated.

We again denote for model $\mathcal{M}$ the extensions of formulas through

$$[\![\phi]\!]_{\mathcal{M}} := \{s \in S \mid \mathcal{M}, s \models \phi\}$$

and

$$[\![\psi]\!]_{\mathcal{M}} := \{\sigma \in (S \times \mathbb{R}_+)^\infty \mid \mathcal{M}.\sigma \models \psi\}$$

Hence these sets denote all states or paths for which the respective formula holds. Relation $\models$ is defined inductively; let $s \in S$ be a state, and $\sigma \in (S \times \mathbb{R}_+)^\infty$ be an infinite path alternating between states and residence times.

$\mathcal{M}, s \models \top$ is true for all $s \in S$.
$\mathcal{M}, s \models a \Leftrightarrow s \in V(a)$.
$\mathcal{M}, s \models Z \Leftrightarrow s \in \Sigma(Z)$ for $Z \in \mathsf{SV}$.
$\mathcal{M}, s \models \phi_1 \wedge \phi_2 \Leftrightarrow \mathcal{M}, s \models \phi_1$ and $\mathcal{M}, s \models \phi_2$.
$\mathcal{M}, s \models \neg\phi \Leftrightarrow \mathcal{M}, s \models \phi$ is false.
$\mathcal{M}, s \models \mathcal{S}_{\ltimes p}(\phi) \Leftrightarrow \lambda := \lim_{t\to\infty} M_\infty(s)(\{\tau \mid \langle s, \tau\rangle @t \models \phi\})$ exists, and $\lambda \ltimes p$.
$\mathcal{M}, s \models \mathcal{P}_{\ltimes p}(\psi) \Leftrightarrow M_\infty(s)(\{\tau \mid \langle s, \tau\rangle \models \psi\}) \ltimes p$.

$\mathcal{M}, \sigma \models \tilde{\top}$ is true for all $\sigma \in (S \times \mathbb{R}_+)^\infty$
$\mathcal{M}, \sigma \models P \Leftrightarrow \sigma \in \Pi(P)$ for $P \in \mathsf{PV}$.
$\mathcal{M}, \sigma \models \psi_1 \wedge \psi_2 \Leftrightarrow \mathcal{M}, \sigma \models \psi_1$ and $\mathcal{M}, \sigma \models \psi_2$.
$\mathcal{M}, \sigma \models \neg\psi \Leftrightarrow \mathcal{M}, \sigma \models \psi$ is false.
$\mathcal{M}, \sigma \models \mathcal{X}^I\, \psi \Leftrightarrow \mathcal{M}, \sigma[1\ldots] \models \psi$ and $\delta(\sigma, 0) \in I$.
$\mathcal{M}, \sigma \models \phi_1\, \mathcal{U}^I\, \phi_2 \Leftrightarrow \exists t \in I : \mathcal{M}, \sigma@t \models \phi_2$ and $\forall t' \in [0, t[\colon \mathcal{M}, \sigma@t' \models \phi_1$.
$\mathcal{M}, \sigma \models \mu P.\psi \Leftrightarrow \sigma \in \bigcup_{i\geq 0} R_i$ with $R_0 := [\![\psi]\!]_{\mathcal{M}[P\backslash\emptyset]}, R_{i+1} := [\![\psi]\!]_{\mathcal{M}[P\backslash R_i]}$.

Define the theory $Th_{\mathcal{M}}(s)$ of a state s as above as the formulas which hold in s:

$$Th_{\mathcal{M}}(s) := \{\phi \mid \phi \text{ is a state formula}, \mathcal{M}, s \models \phi\}.$$

Similarly, the theory $Th_{\mathcal{M}}(\sigma)$ of a path σ is defined:

$$Th_{\mathcal{M}}(\sigma) := \{\psi \mid \psi \text{ is a path formula}, \mathcal{M}, \sigma \models \psi\}.$$

The extensions of the formulas are Borel-measurable. This is established through induction on the structure of a formula. Those formulas that contain the μ-operator need special consideration.

Proposition 2.4.5. *$[\![\phi]\!]_{\mathcal{M}} \in \mathcal{B}(S)$ for all state formulas ϕ, and $[\![\psi]\!]_{\mathcal{M}} \in \mathcal{B}((S \times \mathbb{R}_+)^\infty)$ for all path formulas ψ.* □

Of course it is important to know that the sets under consideration are Borel, for otherwise the corresponding sets are not in the range of the corresponding probability, and one cannot compute probabilities like $M_\infty(s)(\{\tau \mid \langle s, \tau\rangle \models \psi\})$.

We note that the μ-operator plays a special rôle: intuitively, it models the smallest fixed point. This is noted just for the sake of completeness.

Proposition 2.4.6. *$[\![\mu P.\psi]\!]_{\mathcal{M}}$ is the smallest fixed point of $R \mapsto [\![\psi]\!]_{\mathcal{M}[P\backslash R]}$.*
□

Theories are invariant under model morphisms; to be specific:

Proposition 2.4.7. *Let $\mathcal{M}$ and $\mathcal{M}'$ be models for μ**CSL**, and assume that $\Phi : \mathcal{M} \to \mathcal{M}'$ is a morphism. Then*

a. $[\![\phi]\!]_{\mathcal{M}} = \Phi^{-1}\left[[\![\phi]\!]_{\mathcal{M}'}\right]$ *for all state formulas* ϕ.
b. $[\![\psi]\!]_{\mathcal{M}} = \Phi_{\infty}^{-1}\left[[\![\psi]\!]_{\mathcal{M}'}\right]$ *for all state formulas* ψ. □

2.4.4 Congruences

We will define two equivalence relations on states respectively on paths. These relations are fundamental for discussing bisimilarity and behavioral as well as logical equivalence later on.

Fix the model $\mathcal{M} = (S, M, \mathcal{I}, V)$ and define

$$s \;\zeta_{\mathcal{M}}\; s' \Leftrightarrow Th_{\mathcal{M}}(s) = Th_{\mathcal{M}}(s'),$$
$$\sigma \;\omega_{\mathcal{M}}\; \sigma' \Leftrightarrow Th_{\mathcal{M}}(\sigma) = Th_{\mathcal{M}}(\sigma').$$

Then both $\zeta_{\mathcal{M}}$ and $\omega_{\mathcal{M}}$ are smooth equivalence relations on S resp. on $(S \times \mathbb{R}_+)^\infty$. This is so since there are only countably many formulas, and because we have

$$s \;\zeta_{\mathcal{M}}\; s' \Leftrightarrow \big[\mathcal{M}, s \models \phi \Leftrightarrow \mathcal{M}, s' \models \phi\big] \text{ for all state formulas } \phi,$$
$$\Leftrightarrow \big[s \in [\![\phi]\!]_{\mathcal{M}} \Leftrightarrow s' \in [\![\phi]\!]_{\mathcal{M}}\big] \text{ for all state formulas } \phi.$$

From this it is clear that the countable set $\big\{[\![\phi]\!]_{\mathcal{M}} \mid \phi \text{ is a state formula}\big\}$ determines the relation $\zeta_{\mathcal{M}}$, and that

$$\mathbf{\Sigma}\big(\mathcal{B}(S), \zeta_{\mathcal{M}}\big) = \sigma\big(\{[\![\phi]\!]_{\mathcal{M}} \mid \phi \text{ is a state formula}\}\big).$$

In a similar way we see that $\omega_{\mathcal{M}}$ is smooth, and that

$$\mathbf{\Sigma}\big(\mathcal{B}((S \times \mathbb{R}_+)^\infty), \omega_{\mathcal{M}}\big) = \sigma\big(\{[\![\psi]\!]_{\mathcal{M}} \mid \psi \text{ is a path formula}\}\big)$$

holds as well. These two relations will be studied now in some detail, and it will turn out that the relationship of $\zeta_{\mathcal{M}}$ and $\omega_{\mathcal{M}}$ is closer than meets the eye.

In a first step it is established that $\zeta_{\mathcal{M}}$ and $\omega_{\mathcal{M}}$ form essentially a congruence for M_∞.

Proposition 2.4.8. *The pair $(\zeta_{\mathcal{M}}, \Delta_{\mathbb{R}_+} \times \omega_{\mathcal{M}})$ of smooth equivalence relations is a congruence for $M_\infty : S \rightsquigarrow \mathbb{R}_+ \times (S \times \mathbb{R}_+)^\infty$.* □

Relating the Relations

Two infinite paths are $\omega_{\mathcal{M}}$-equivalent iff their state components are $\zeta_{\mathcal{M}}$-equivalent (and the timing information is identical). This will support the investigation of logical equivalence later on, mainly since the information available for states is easier to handle than that for infinite paths. We obtain through a detailed analysis.

Proposition 2.4.9. $\omega_{\mathcal{M}} = (\zeta_{\mathcal{M}} \times \Delta_{\mathbb{R}_+})^{\infty}$. □

This has two interesting consequences:

Corollary 2.4.10. *We have*

$$\mathbf{\Sigma}\big(\mathcal{B}((S \times \mathbb{R}_+)^{\infty}), (\zeta_{\mathcal{M}} \times \Delta_{\mathbb{R}_+})^{\infty}\big) = \mathbf{\Sigma}\big(\mathcal{B}((S \times \mathbb{R}_+)^{\infty}), \omega_{\mathcal{M}}\big)$$

and

$$M_{\infty} : (S, \mathbf{\Sigma}(\mathcal{B}(S), \zeta_{\mathcal{M}})) \rightsquigarrow \Big(\mathbb{R}_+ \times (S \times \mathbb{R}_+)^{\infty}, \mathbf{\Sigma}\big(\mathcal{B}(\mathbb{R}_+ \times (S \times \mathbb{R}_+)^{\infty}), \big(\Delta_{\mathbb{R}_+} \times \zeta_{\mathcal{M}}\big)^{\infty}\big)\Big)$$

is a stochastic relation. □

The consequence of the equality in Proposition 2.4.9 is that we may check the equivalence of paths locally, i.e., through the equivalence of states. This represents a considerable reduction in complexity, because the equivalence relation $\omega_{\mathcal{M}}$ that operates on infinite paths is uniquely determined through the relation $\zeta_{\mathcal{M}}$ which in turn operates on states. It will be reflected in the representation of the equivalence classes, as we will see in Corollary 2.4.11. The reduction makes checking some properties of course much easier, and it has also technical advantages when it comes to checking the semi-pullback of two models, as we will in the next section.

We give a first consequence of this equality in terms of a representation of the equivalence classes.

Corollary 2.4.11. *Given* $\sigma \in (S \times \mathbb{R}_+)^{\infty}$, *the* $\omega_{\mathcal{M}}$*-class of* $\sigma = s_0 \xrightarrow{t_0} s_1 \xrightarrow{t_1} \dots$ *can be represented as*

$$[\sigma]_{\omega_{\mathcal{M}}} = \prod_{j \geq 0} \Big([s_j]_{\zeta_{\mathcal{M}}} \times \{t_j\}\Big).$$

Moreover, we have Borel isomorphisms between these analytic spaces:

$$(S \times \mathbb{R}_+)^{\infty}/\omega_{\mathcal{M}} \cong \big((S \times \mathbb{R}_+)/(\zeta_{\mathcal{M}} \times \Delta_{\mathbb{R}_+})\big)^{\infty} \cong \big((S/\zeta_{\mathcal{M}}) \times \mathbb{R}_+\big)^{\infty}. \ \square$$

We are now in a position to define the logical equivalence of models, and to relate it to spans of morphisms.

2.4.5 Logical Equivalence and Bisimilarity

Logical equivalence between two models says roughly that, given a state in one model, there exists a state in the other model so that in both exactly the same formulas are valid; similarly for paths. This equivalence is modelled after the corresponding equivalence that has been investigated in modal logics; see Definition 2.3.6. We have seen there that it is closely tied to the notion of bisimulation through the Hennessy-Milner Theorem. The relationship of this equivalence to bisimulations will be discussed now.

Let $\mathcal{M} = (S, M, \mathcal{I}, V)$ and $\mathcal{N} = (S', N, \mathcal{J}, W)$ be models for μ**CSL**. We assume that $\mathcal{M}$ is *nondegenerate*, i.e., that there exists a state formula ϕ with $\emptyset \neq [\![\phi]\!]_{\mathcal{M}} \neq S$. Being nondegenerate implies that the factor space $S/\zeta_{\mathcal{M}}$ is not trivial. Corollary 2.4.11 entails the existence of a path formula ψ such that $\emptyset \neq [\![\psi]\!]_{\mathcal{M}} \neq (S \times \mathbb{R}_+)^\infty$.

Basic Definitions

Define the models $\mathcal{M}$ and $\mathcal{N}$ as logically equivalent iff they accept exactly the same formulas. This is similar to logical equivalence for Kripke models, as the discussion in Section 2.3.2 indicates. In addition and in contrast, however, it has to take two levels into account, since we are dealing here with state formulas together with path formulas, so that formulas may hold in states or on paths — this situation is familiar from model checking where one has this dichotomy as well.

Definition 2.4.12. *The models $\mathcal{M}$ and $\mathcal{N}$ are called* logically equivalent *iff both*

$$\{Th_{\mathcal{M}}(s) \mid s \in S\} = \{Th_{\mathcal{N}}(s') \mid s' \in S'\}$$

and

$$\{Th_{\mathcal{M}}(\sigma) \mid \sigma \in (S \times \mathbb{R}_+)^\infty\} = \{Th_{\mathcal{N}}(\sigma') \mid \sigma' \in (S' \times \mathbb{R}_+)^\infty\}$$

hold.

Note that we take both states and infinite paths into consideration. Thus the models are logically equivalent iff these conditions are satisfied:

1. Given a state $s \in S$, there exists a state $s' \in S'$ such that $[\mathcal{M}, s \models \phi \Leftrightarrow \mathcal{N}, s' \models \phi]$ holds for all state formulas ϕ, and vice versa.
2. Given a path $\sigma \in (S \times \mathbb{R}_+)^\infty$, there exists a path $\sigma' \in (S' \times \mathbb{R}_+)^\infty$ such that $[\mathcal{M}, \sigma \models \psi \Leftrightarrow \mathcal{N}, \sigma' \models \psi]$ holds for all path formulas ψ, and vice versa.

As usual, the existence of a morphism between models entails their logical equivalence.

Proposition 2.4.13. *Let $\Phi : \mathcal{M} \to \mathcal{N}$ be a morphism. Then $\mathcal{M}$ and $\mathcal{N}$ are logically equivalent.* □

We will show that logically equivalent models are bisimilar. Bisimilarity is again introduced as a span of morphisms, and behavioral equivalence through a cospan.

Definition 2.4.14. *Let $\mathcal{M}$ and $\mathcal{N}$ be nondegenerate models for μ***CSL***.*

a. $\mathcal{M}$ and $\mathcal{N}$ are said to be bisimilar *iff there exists a model $\mathcal{Q}$ for μ***CSL** *and morphisms*

$$\mathcal{M} \xleftarrow{\Phi} \mathcal{Q} \xrightarrow{\Psi} \mathcal{N}.$$

b. $\mathcal{M}$ and $\mathcal{N}$ are said to be behaviorally equivalent *iff there exists a model $\mathcal{Q}$ for μ***CSL** *and morphisms*

$$\mathcal{M} \xrightarrow{\Phi} \mathcal{Q} \xleftarrow{\Psi} \mathcal{N}.$$

It is clear that bisimilar models are logically equivalent, because this notion of equivalence is transitive; see Proposition 2.4.7. Now suppose model $\mathcal{M} = (S, M, \mathcal{I}, V)$ is bisimilar to model $\mathcal{N} = (S', N, \mathcal{J}, W)$ with mediating model $\mathcal{Q}$ over the state space S'' and the morphisms according to Definition 2.4.14. Then the condition on bisimilarity implies in the present scenario that

1. $M(\Phi(s''))(I \times B) = N(\Psi(s''))(I \times B')$ for every state $s'' \in S''$, every rational interval I, and all common events $B \in \mathcal{B}(S), B' \in \mathcal{B}(S')$ (thus every pair of events B, B' such that $\Phi^{-1}[B] = \Psi^{-1}[B']$, as the discussion following Definition 2.2.1 indicates). Consequently, the probability for $\mathcal{M}$ changing the state during interval I and entering a state in B from state $\Phi(s'')$ equals the probability of $\mathcal{N}$ changing the state during time interval I and entering a state in B' from state $\Psi(s'')$. This illustrates again the mediating work done through model $\mathcal{Q}$.
2. For a state $s'' \in S''$, $\Phi(s'')$ is a member of the valuation for a state variable Z in model $\mathcal{M}$ iff $\Psi(s'')$ is a member for this variable in model $\mathcal{N}$, and similarly for path variables, and for atomic propositions.

The situation is a bit different with behavioral equivalence. By implication, the model which constitutes the range of the cospan is based on a Polish space. Otherwise we could not always conclude that behaviorally equivalent models are logically equivalent as well. This is so since computing the set of all states in which a given formula is valid requires the knowledge of a projective limit, and we did establish the existence of such a limit only for the case of Polish spaces, not for general analytic ones. On the other hand, these cospans are constructed usually through factoring (see, e.g., Propositions 2.2.4 and 2.4.22 below), and the factor space of a Polish space is not always a Polish one. So we need to exercise some care. Specifically, the behavioral equivalence of the models means that, if $\Phi(s) = \Psi(s')$ for states $s \in S$, $s' \in S'$ then

1. $M(s)(I \times \Phi^{-1}[B'']) = N(s')(I \times \Psi^{-1}[B''])$ whenever $B'' \in \mathcal{B}(S'')$ is a Borel set in S'', and I is a rational interval. Consequently, the probability of $\mathcal{M}$ changing the state during interval I and entering a state $s^\bullet \in S$ with $\Phi(s^\bullet) \in B''$ from state s equals the probability of $\mathcal{N}$ changing the state during time interval I and entering a state $s^\star \in S'$ with $\Psi(s^\star) \in B''$ from state s'.
2. s is a member for the valuation of a state variable Z in model $\mathcal{M}$ iff s' is a member for the valuation of a state variable Z in model $\mathcal{N}$; similarly for path variables, and for atomic propositions.

Returning to the general discussion, fix the models $\mathcal{M} = (S, M, \mathcal{I}, V)$ and $\mathcal{N} = (S', N, \mathcal{J}, W)$ such that $\mathcal{M}$ and $\mathcal{N}$ are logically equivalent; hence both models are logically equivalent. Each model has the equivalence relations $\zeta_\mathcal{M}$ and $\omega_\mathcal{M}$ or $\zeta_\mathcal{N}$ and $\omega_\mathcal{N}$ associated with it, as defined in Section 2.4.4.

The stochastic relations M_∞ and N_∞ will be investigated with respect to bisimilarity first, and it will be shown first that they are bisimilar *as stochastic relations on Polish spaces.*

Lemma 2.4.15. *$\zeta_\mathcal{M}$ and $\zeta_\mathcal{N}$ spawn each other; so do $\omega_\mathcal{M}$ and $\omega_\mathcal{N}$.*

Proof (Sketch) 0. We will show only that $\zeta_\mathcal{M}$ spawns $\zeta_\mathcal{N}$; interchanging the rôles of $\mathcal{M}$ and $\mathcal{N}$ will show that $\zeta_\mathcal{N}$ spawns $\zeta_\mathcal{M}$. The argumentation for $\omega_\mathcal{M}$ and $\omega_\mathcal{N}$ is nearly verbatim the same, so the reader is invited to fill in the details.

1. Define for the state $s \in S$ the map $\Upsilon([s]_{\zeta_\mathcal{M}}) := [s']_{\zeta_\mathcal{N}}$, whenever $Th_\mathcal{M}(s) = Th_\mathcal{N}(s')$. Because $s_1 \; \zeta_\mathcal{M} \; s_2$ iff $Th_\mathcal{M}(s_1) = Th_\mathcal{M}(s_2)$, and similarly for $\mathcal{N}$, the map is well defined. For the state formula ϕ its class $[\![\phi]\!]_\mathcal{M}$ can be represented as

$$\bigcup\{[s]_{\zeta_\mathcal{M}} \mid \mathcal{M}, s \models \phi\};$$

thus it is readily verified that $\Upsilon_{[\![\phi]\!]_\mathcal{M}} = [\![\phi]\!]_\mathcal{N}$. Consequently,

$$\{\Upsilon_{[\![\phi]\!]_\mathcal{M}} \mid \phi \text{ is a state formula}\}$$

is a generator of $\mathbf{\Sigma}(\mathcal{B}(S'), \zeta_\mathcal{N})$. This generator is closed under intersections, since the conjunction of two state formulas is again one. □

We know that both $(\zeta_\mathcal{M}, \Delta_{\mathbb{R}_+} \times \omega_\mathcal{M})$ and $(\zeta_\mathcal{N}, \Delta_{\mathbb{R}_+} \times \omega_\mathcal{N})$ are congruences for the stochastic relations M_∞ resp. N_∞. We will show now that they are simulation equivalent, so that the situation is here very similar to that prevailing for logically equivalent modal logics in Section 2.3.2.

Proposition 2.4.16. *Let $\mathcal{M}$ and $\mathcal{N}$ be logically equivalent models. Then the congruences $\mathsf{c}_\mathcal{M} := (\zeta_\mathcal{M}, \Delta_{\mathbb{R}_+} \times \omega_\mathcal{M})$ and $\mathsf{c}_\mathcal{N} := (\zeta_\mathcal{N}, \Delta_{\mathbb{R}_+} \times \omega_\mathcal{N})$ are simulation equivalent.*

Proof (Sketch) 1. We know that $\zeta_\mathcal{M}$ and $\zeta_\mathcal{N}$ are in a mutually spawning relationship; so are $\omega_\mathcal{M}$ and $\omega_\mathcal{N}$. Consequently, $\Delta_{\mathbb{R}_+} \times \omega_\mathcal{M}$ and $\Delta_{\mathbb{R}_+} \times \omega_\mathcal{N}$ are related through spawning as well, where

$$\{I \times [\![\psi]\!]_{\mathcal{M}} \mid I \text{ is a rational interval, } \psi \text{ is a path formula}\}$$

and

$$\{I \times [\![\psi]\!]_{\mathcal{N}} \mid I \text{ is a rational interval, } \psi \text{ is a path formula}\}$$

are the generators that relate to each other.

2. Using the map $\Upsilon : S/\zeta_{\mathcal{M}} \to S'/\zeta_{\mathcal{N}}$ defined in the proof of Lemma 2.4.15, we show that

$$M_\infty(s)(I \times [\![\psi]\!]_{\mathcal{M}}) = N_\infty(s')(I \times [\![\psi]\!]_{\mathcal{N}})$$

for each $s \in S$, $s' \in \Upsilon([s]_{\zeta_{\mathcal{M}}})$, and for each rational interval I and each path formula ψ. Because $s' \in \Upsilon([s]_{\zeta_{\mathcal{M}}})$ means $Th_{\mathcal{M}}(s) = Th_{\mathcal{N}}(s')$, we obtain for an arbitrary rational number p

$$\begin{aligned} M_\infty(s)(I \times [\![\psi]\!]_{\mathcal{M}}) \le p &\Leftrightarrow \mathcal{M}, s \models \mathcal{P}_{\le p}(\mathcal{X}^I\ \psi) \\ &\Leftrightarrow \mathcal{N}, s' \models \mathcal{P}_{\le p}(\mathcal{X}^I\ \psi) \\ &\Leftrightarrow N_\infty(s')(I \times [\![\psi]\!]_{\mathcal{M}}) \le p; \end{aligned}$$

consequently, both probabilities are identical. This implies that $(\zeta_{\mathcal{M}}, \Delta_{\mathbb{R}_+} \times \omega_{\mathcal{M}})$ simulates $(\zeta_{\mathcal{N}}, \Delta_{\mathbb{R}_+} \times \omega_{\mathcal{N}})$. Interchanging the rôles of $\mathcal{M}$ and $\mathcal{N}$ gives the result now. □

This yields the properties we are interested in for the associated stochastic relations.

Proposition 2.4.17. *Let $\mathcal{M}$ and $\mathcal{N}$ be logically equivalent models. Then the associated stochastic relations $M_\infty : S \rightsquigarrow (\mathbb{R}_+ \times S)^\infty$ and $N_\infty : S' \rightsquigarrow (\mathbb{R}_+ \times S')^\infty$ are bisimilar and behaviorally equivalent.* □

Tuning the Mediator

This result is quite welcome when being looked at from the point of view of stochastic relations: Given two models for μ**CSL** that are logically equivalent, we can show that the associated stochastic relations are bisimilar. It does not give us, however, in this present and preliminary form a *model* that mediates between $\mathcal{M}$ and $\mathcal{N}$ (a similar situation has been encountered already with stochastic Kripke models in Section 2.3.2). An analysis of the construction leading to the mediating relation will again provide information for the construction of a model $\mathcal{L}$ and the desired morphisms $\mathcal{L} \to \mathcal{M}$ and $\mathcal{L} \to \mathcal{N}$. The construction leading to Proposition 2.4.17 is again based on a semi-pullback construction.

Lemma 2.4.18. *Let $\mathcal{M}$ and $\mathcal{N}$ be logically equivalent models for μ**CSL**. Define*

$$A := \{\langle s, s'\rangle \in S \times S' \mid Th_{\mathcal{M}}(s) = Th_{\mathcal{N}}(s')\},$$
$$B := \{\langle\langle t, \sigma\rangle, \langle t, \sigma'\rangle\rangle \in (\mathbb{R}_+ \times (S \times \mathbb{R}_+)^\infty) \times (\mathbb{R}_+ \times (S' \times \mathbb{R}_+)^\infty) \mid Th_{\mathcal{M}}(\sigma) = Th_{\mathcal{N}}(\sigma')\}.$$

Then A and B are Standard Borel, and there exists a stochastic relation $L_0 : A \rightsquigarrow B$ that mediates between M_∞ and N_∞. The morphisms are composed from the corresponding projections. □

We know from Proposition 2.4.9 that $\omega_{\mathcal{M}} = (\zeta_{\mathcal{M}} \times \Delta_{\mathbb{R}_+})^\infty$; similarly for $\zeta_{\mathcal{N}}$ and $\omega_{\mathcal{N}}$. Thus B is essentially the set of all paths over A, extended by timing information.

Corollary 2.4.19. *Define A and B according to Lemma 2.4.18. There exists a bijection $\Lambda : B \to (A \times \mathbb{R}_+)^\infty$ that is also a Borel isomorphism.* □

Define for this bijection Λ the map $L' := \mathfrak{P}(\Lambda) \circ L_0$; then this is a stochastic relation $L' : A \rightsquigarrow \mathbb{R}_+ \times (A \times \mathbb{R}_+)^\infty$ that mediates between M_∞ and N_∞. But, still, this is not enough, because we cannot ascertain that L' is actually generated from a model, since we do not know whether or not L' is actually a projective limit of some sort. However, the semi-pullback is a rather flexible construction, and we will show now that we may construct from L' a mediator L_0 with the desired shape, viz., $L_0 = L_\infty$ for some stochastic relation $L : A \rightsquigarrow \mathbb{R}_+ \times A$.

In fact, put for $\langle s, s'\rangle \in A$ and for $E \in \mathcal{B}(\mathbb{R}_+ \times A)$

$$L(s, s')(E) := L'(s, s')\big(E \times \prod_{j>1} (\mathbb{R}_+ \times A)\big).$$

Thus the semi-pullback is restricted to its first component, yielding a stochastic relation $L : A \rightsquigarrow \mathbb{R}_+ \times A$, for which the projective limit can be constructed. This is what we will have a closer look at now.

Define for $n \in \mathbb{N}$ the map $\ell_n : (\mathbb{R}_+ \times A)^n \to (\mathbb{R}_+ \times S)^n$ through

$$\ell_n(t_1, s_1, s_1', \ldots, t_n, s_n, s_n') := \langle t_1, s_1, \ldots, t_n, s_n\rangle;$$

the map $r_n : (\mathbb{R}_+ \times A)^n \to (\mathbb{R}_+ \times S')^n$ is defined analogously.

Lemma 2.4.20. *Define $L_n : A \rightsquigarrow (\mathbb{R}_+ \times A)^n$ inductively from L in the same way as M_n is defined from M for the statement of Proposition 2.4.3, and let π_i be the ith projection. Then the diagram*

$$\begin{array}{ccccc}
S & \xleftarrow{\pi_1} & A & \xrightarrow{\pi_2} & S' \\
\downarrow{\scriptstyle M_n} & & \downarrow{\scriptstyle L_n} & & \downarrow{\scriptstyle N_n} \\
\mathfrak{P}((\mathbb{R}_+ \times S)^n) & \xleftarrow[\mathfrak{P}(\ell_n)]{} & \mathfrak{P}((\mathbb{R}_+ \times A)^n) & \xrightarrow[\mathfrak{P}(r_n)]{} & \mathfrak{P}((\mathbb{R}_+ \times S')^n)
\end{array}$$

commutes for every $n \in \mathbb{N}$.

Proof (Sketch) 1. The proof proceeds by induction on n. For $n = 1$ there is not much to show: By construction, L' mediates between M_∞ and N_∞, and the latter relations are projective limits, so that for $\langle s, s'\rangle \in A$ and $E \in \mathcal{B}(\mathbb{R}_+ \times S)$

$$\begin{aligned} M_1(s)(E) &= M_\infty(s)\big(E \times \prod_{j>1}(\mathbb{R}_+ \times S)\big) \\ &= L'(s,s')\big(\ell_1^{-1}[E] \times \prod_{j>1}(\mathbb{R}_+ \times A)\big) \\ &= L_1(s,s')(\ell_1^{-1}[E]). \end{aligned}$$

Similarly, the right hand side of the diagram above is shown to commute for $n = 1$.

2. Now assume the assertion is established for n; then we get from the induction hypothesis together with the Change of Variables formula (see Lemma 1.6.20) for $g : (\mathbb{R}_+ \times S)^n \to \mathbb{R}$ measurable and bounded, and for $\langle s, s'\rangle \in A$ the equality

$$\int_{(\mathbb{R}_+ \times S)^n} g(\mathbf{v})\ M_n(s)(d\mathbf{v}) = \int_{(\mathbb{R}_+ \times A)^n} (g \circ \ell_n)(\mathbf{w})\ L_n(s,s')(d\mathbf{w}).$$

This is shown first for $g = \chi_D$ for $D \in \mathcal{B}((\mathbb{R}_+ \times S)^n)$, whence it is equivalent to the induction hypothesis; then it is shown for step functions by the linearity of the integral, subsequently for nonnegative measurable and bounded g by the Monotone Convergence Theorem (Proposition 1.6.1), and finally for general g by decomposing the map into a positive and a negative part.

3. But now we can perform the induction step: Let $\langle s_0, s_0'\rangle \in A$ and $F \in \mathcal{B}((\mathbb{R}_+ \times S)^{n+1})$ be a Borel set; then

$$\begin{aligned} &M_{n+1}(s_0)(F) = \\ &\qquad \int_{(\mathbb{R}_+ \times S)^n} M(\hbar_S(\mathbf{v}))(\{\langle t, s\rangle \mid \langle \mathbf{v}, t, s\rangle \in F\})\ M_n(s_0)(d\mathbf{v}) = \\ &\int_{(\mathbb{R}_+ \times A)^n} M(\pi_1(\hbar_A(\mathbf{w})))(\{\langle t, s\rangle \mid \langle \mathbf{w}, t, s, s'\rangle \in \ell_{n+1}^{-1}[F]\}) L_n(s_0, s_0')(d\mathbf{w}) = \\ &\int_{(\mathbb{R}_+ \times A)^n} L(\hbar_A(\mathbf{w}))(\{\langle t, s, s'\rangle \mid \langle \mathbf{w}, t, s, s'\rangle \in \ell_{n+1}^{-1}[F]\}) L_n(s_0, s_0')(d\mathbf{w}) = \\ &\qquad\qquad L_{n+1}(s_0, s_0')(\ell_{n+1}^{-1}[F]). \end{aligned}$$

□

Now extend ℓ_n and r_n to the corresponding infinite products, yielding maps ℓ_∞ and r_∞.

Proposition 2.4.21. *Assume that $\mathcal{M}$ and $\mathcal{N}$ are Hennessy-Milner equivalent; construct the Standard Borel space A and the stochastic relation $L : A \rightsquigarrow \mathbb{R}_+ \times A$ as above. Then*

i. $(\pi_1, \ell_\infty) : L_\infty \to M_\infty$ *and* $(\pi_2, r_\infty) : L_\infty \to N_\infty$ *are morphisms.*
ii. M_∞ *and* N_∞ *are bisimilar with* L_∞ *as a mediator.* □

We are nearly ready for the main result, which will be stated after dealing with behavioral equivalence.

Proposition 2.4.22. *Let $\mathcal{M}$ and $\mathcal{N}$ be logically equivalent models for μ**CSL**. If $S/\zeta_{\mathcal{M}}$ is a Standard Borel space, then $\mathcal{M}$ and $\mathcal{N}$ are behaviorally equivalent.* □

This, now, is the main result:

Theorem 2.4.23. *Let $\mathcal{M}$ and $\mathcal{N}$ be nontrivial models for μ**CSL**. Consider these statements:*

a. $\mathcal{M}$ *and* $\mathcal{N}$ *are behaviorally equivalent.*
b. $\mathcal{M}$ *and* $\mathcal{N}$ *are logically equivalent.*
c. $\mathcal{M}$ *and* $\mathcal{N}$ *are bisimilar.*

Then $a \Rightarrow b \Leftrightarrow c$, and if $S/\zeta_{\mathcal{M}}$ is a Standard Borel space, then all three statements are equivalent.

Proof (Sketch) 1. The implications $c \Rightarrow b$ and $a \Rightarrow b$ both follow from Proposition 2.4.13. If $S/\zeta_{\mathcal{M}}$ is Standard Borel, then $b \Rightarrow a$ follows from Proposition 2.4.22, so that we have to take care of $b \Rightarrow c$.

2. Construct the Standard Borel space A and the stochastic relation $L : A \rightsquigarrow \mathbb{R}_+ \times (A \times \mathbb{R}_+)^\infty$ together with the maps ℓ_∞ and r_∞ as in Proposition 2.4.21. Assume that the interpretation $\mathcal{J}$ for model $\mathcal{N}$ is $\mathcal{J} = (\Sigma', \Pi')$, and define

$$\mathcal{L} := (L, A, (\Sigma^*, \Pi^*), V^*)$$

with

1. $V^* := (V(a) \times W(a)) \cap A$ for the atomic propositions $a \in \mathsf{AP}$,
2. $\Sigma^*(Z) := (\Sigma(Z) \times \Sigma'(Z)) \cap A$ for the state variable $Z \in \mathsf{SV}$,
3. $\Pi^*(P) := \{\rho \in (A \times \mathbb{R}_+)^\infty \mid \ell_\infty(\rho) \in \Pi(P), r_\infty(\rho) \in \Pi'(P)\}$ for the path variable $P \in \mathsf{PV}$.

Then both $\ell_\infty : \mathcal{L} \to \mathcal{M}$ and $r_\infty : \mathcal{L} \to \mathcal{N}$ are morphisms. □

2.5 Bibliographic Notes

The exposition follows essentially the discussion in [20], in particular, proofs which have been omitted in the present discussion can be found there. Nevertheless, some notes are in order.

Modal Logics

We follow essentially the exposition in the monograph [8], with an occasional look at Rutten's overview of coalgebras [72]. It may be interesting for the reader to look also at the massive collection [9], and in particular at the chapters on proof theory in [35]. An early collection of mathematical questions pertaining to modal logic can be found in [40]. The treatment of nondeterministic Kripke models is fairly standard, stochastic Kripke models and their morphisms were inspired by the work [58] of Larsen and Skou on testing. The paper [22] proposes stochastic Kripke models for general modal logics; the results on bisimilarity are from there.

Continuous Time Stochastic Logics

The paper [3] introduces and studies a logic called **CSL**, *continuous time stochastic logics*, with applications to model checking; after all, **CSL** is fashioned after the popular logics **CTL** studied extensively in model checking [16]. Some mathematical questions for **CSL** were discussed in [20], in particular the use of projective limits. There the fixed-point operator is introduced into this scenario, evolving **CSL** into μ**CSL**. The investigation of the relation of logical equivalence, bisimilarity, and behavioral equivalence found there seems to originate the investigation of continuous stochastic models for logics outside the direct realm of modal logics. It is also shown how previous *ad hoc* approaches to probabilistic modelling fit into the general model.

2.6 Appendix: Behavioral and Logical Equivalence Reconsidered

We have shown that logical equivalence, bisimilarity, and behavioral equivalence are the same for Kripke models that are based on analytic spaces. This appendix is intended to address the question of behavioral and logical equivalence without topological assumptions; thus we will work in general measurable spaces, and we will show that both notions are equivalent as well. We did not include bisimilarity in this discussion. If we want to show that two behavioral equivalent models are bisimilar, we are requested to construct a mediating model, and it is currently not clear how this can be done without constructing a semi-pullback which in turn requires at least analytic base spaces. For simplicity, the constructions will be carried out for the negation free Hennessy-Milner logic $\mathfrak{L} = \mathfrak{L}(\mathsf{Act}, [0,1])$, the formulas of which are given through the grammar

$$\top \mid \phi_1 \wedge \phi_2 \mid \langle a \rangle_r \phi.$$

Here $a \in \mathsf{Act}$ is an action, and the threshold r is a real number from the unit interval (see Example 2.3.3). This logic plays an important rôle in other places in this book as well, but we do not require here the set Act of possible actions to be countable, and we do not restrict ourselves to rationals for the values of thresholds. Because the logic is so simple we can keep the interpreting Kripke models simple, too.

2.6.1 Discussing the Strategy

Let us briefly reconsider the strategy for the analytic case, where we have a countable number of actions, and where the thresholds are taken from the rational numbers. Given a Kripke model over an analytical space, the logic defines an equivalence relation $\mathsf{r}_{\mathcal{K}}$ which is smooth; this is so since we only have a countable number of formulas at our disposal (having all real numbers in the unit interval rather than only the rationals is not essential, since the rationals are dense; see Section 3.4.1, in particular Lemma 3.4.1, for a discussion). This equivalence relation is used for factoring, and since it is smooth, we obtain an analytic space again, when adopting the final σ-algebra as the σ-algebra on the factor space. This σ-algebra has a fairly rich structure: it constitutes the Borel sets of an analytic space, thus making in particular Souslin's Theorem available; its generators can be computed directly through the logic, and it is in direct correspondence with the $\mathsf{r}_{\mathcal{K}}$-invariant Borel sets. All constructions remain within the realm of analytic sets, so that in particular the amalgamated sum leads to an analytic space which in turn can be made the state space of a Kripke model through standard constructions. The logic influences these discussions only through the corresponding equivalence relations, witnessed by the observation that the general criterion for bisimilarity enters the discussion, this criterion being formulated in terms of general smooth equivalence relations.

We show in this appendix that it is also possible to construct a cospan of Kripke models without having to use the machinery of Polish and analytic spaces. So we start from general measurable spaces, investigating the equivalence relation which is induced by the logic on the state space. Since analycity is not available, we will not be able to observe such a convenient interplay of the measurable structures induced by the logic on the state space and on the factor space; specifically we are no longer able to observe that the $\mathsf{r}_{\mathcal{K}}$-invariant measurable sets are exactly the inverse images of the elements of the final σ-algebra with respect to the factor map $\eta_{\mathsf{r}_{\mathcal{K}}}$. Thus we need to construct explicitly an σ-algebra on the factor space which is closely adapted to the logic, and to derive a Kripke model from it which plays the rôle of the factor model. Similarly, the amalgamation of the equivalences on the individual models needs to be investigated more closely, since the interesting properties are no longer being made available through analycity and smoothness. A technical observation notes that the sum of two Kripke models is no longer helpful, so that another avenue has to be considered. We solve this obstacle by first constructing a σ-algebra on the factor space for the amalgamation, and then construct a relation on this space; finding the σ-algebra with the right properties appears as the key point. The leading idea is based on the observation that the equivalence classes induced by the logic for logically equivalent Kripke models are in a one-to-one correspondence. This basically sketches the strategy for this excursion.

2.6.2 The Equivalence Relation Induced by the Logic $\mathfrak{L}$

Fix a Kripke model $\mathcal{K} = \big((S, \mathcal{S}), (k_a)_{a \in \mathsf{Act}}\big)$ with a measurable state space $(S, \mathcal{S})$; thus $k_a : (S, \mathcal{S}) \rightsquigarrow (S, \mathcal{S})$ is a stochastic relation for each action a. A *morfism* $f : \mathcal{K} \to \big((T, \mathcal{T}), (\ell_a)_{a \in \mathsf{Act}}\big)$ between Kripke models is an $\mathcal{S}$-$\mathcal{T}$-measurable map $f : S \to T$ such that

$$\forall a \in \mathsf{Act} : \ell_a \circ f = \mathfrak{S}(f) \circ k_a$$

holds. Note that we do not require f to be onto which otherwise is assumed nearly everywhere in this treatise (for emphasizing this, we have dubbed these maps *morfisms* rather than *mor**ph**isms*). Just for the record:

Lemma 2.6.1. *Let $f : \mathcal{K} \to \mathcal{L}$ be a morfism and ϕ a formula in $\mathfrak{L}$, then*

a. $\mathcal{K}, s \models \phi \Leftrightarrow \mathcal{L}, f(s) \models \phi$ holds for each state s of K,
b. $f^{-1}\left[[\![\phi]\!]_{\mathcal{L}}\right] = [\![\phi]\!]_{\mathcal{K}}$. $\square$

The equivalence relation $\mathsf{r}_{\mathcal{K}}$ induced by $\mathfrak{L}$ on S is defined as above through

$$s \ \mathsf{r}_{\mathcal{K}} \ s' \text{ iff } \forall \phi : \mathcal{K}, s \models \phi \Leftrightarrow \mathcal{K}, s' \models \phi.$$

Define the set $\mathcal{E}_{\mathcal{K}}$ of all extensions for formulas through

$$\mathcal{E}_{\mathcal{K}} := \{[\![\phi]\!]_{\mathcal{K}} \mid \phi \text{ is a formula}\},$$

and define as σ-algebra $\mathcal{S}^{\ddagger}_{\mathsf{r}_{\mathcal{K}}}$ on the factor space $S/\mathsf{r}_{\mathcal{K}}$ as the smallest σ-algebra which contains all the sets the inverse image of which lies in $\mathcal{E}_{\mathcal{K}}$:

$$\mathcal{S}^{\ddagger}_{\mathsf{r}_{\mathcal{K}}} := \sigma\big(\{A \subseteq S/\mathsf{r}_{\mathcal{K}} \mid \eta_{\mathsf{r}_{\mathcal{K}}}^{-1}[A] \in \mathcal{E}_{\mathcal{K}}\}\big).$$

We analyze this construction, and then we enter a discussion of behavioral and logical equivalence.

Lemma 2.6.2. *The set $\mathcal{A} := \{\eta_{\mathsf{r}_{\mathcal{K}}}\left[[\![\phi]\!]_{\mathcal{K}}\right] \mid \phi$ is a formula$\}$ is a generator of $\mathcal{S}^{\ddagger}_{\mathsf{r}_{\mathcal{K}}}$, which is closed under finite intersections.*

Proof Because each extension $[\![\phi]\!]_{\mathcal{K}}$ is $\eta_{\mathsf{r}_{\mathcal{K}}}$-invariant, and because the logic $\mathfrak{L}$ is closed under conjunction, $\mathcal{A}$ is closed under finite intersections. The factor map is onto; thus we have $[\![\phi]\!]_{\mathcal{K}} = \eta_{\mathsf{r}_{\mathcal{K}}}^{-1}\left[\eta_{\mathsf{r}_{\mathcal{K}}}\left[[\![\phi]\!]_{\mathcal{K}}\right]\right]$. Consequently, if ϕ is a formula, then $\eta_{\mathsf{r}_{\mathcal{K}}}\left[[\![\phi]\!]_{\mathcal{K}}\right] \in \mathcal{S}^{\ddagger}_{\mathsf{r}_{\mathcal{K}}}$, thus $\sigma(\mathcal{A}) \subseteq \mathcal{S}^{\ddagger}_{\mathsf{r}_{\mathcal{K}}}$. On the other hand, if $\eta_{\mathsf{r}_{\mathcal{K}}}^{-1}[A] \in \mathcal{E}_{\mathcal{K}}$ for some $A \subseteq S/\mathsf{r}_{\mathcal{K}}$, then there exists a formula ϕ such that $[\![\phi]\!]_{\mathcal{K}} = \eta_{\mathsf{r}_{\mathcal{K}}}^{-1}[A]$; hence $A = \eta_{\mathsf{r}_{\mathcal{K}}}\left[[\![\phi]\!]_{\mathcal{K}}\right]$, since $\eta_{\mathsf{r}_{\mathcal{K}}}$ is onto. This yields

$$\{A \subseteq S/\mathsf{r}_{\mathcal{K}} \mid \eta_{\mathsf{r}_{\mathcal{K}}}^{-1}[A] \in \mathcal{E}_{\mathcal{K}}\} \subseteq \mathcal{A},$$

establishing the equality. □

This has the following as an immediate consequence.

Corollary 2.6.3. *The factor map $\eta_{\mathsf{r}_{\mathcal{K}}} : S \to S/\mathsf{r}_{\mathcal{K}}$ is $\mathcal{S}$-$\mathcal{S}^{\ddagger}_{\mathsf{r}_{\mathcal{K}}}$-measurable.*

Proof Consider

$$\mathcal{D} := \{A \in \mathcal{S}^{\ddagger}_{\mathsf{r}_{\mathcal{K}}} \mid \eta_{\mathsf{r}_{\mathcal{K}}}^{-1}[A] \in \mathcal{S}\}.$$

Because $\mathcal{S} \ni [\![\phi]\!]_{\mathcal{K}} = \eta_{\mathsf{r}_{\mathcal{K}}}^{-1}\left[\eta_{\mathsf{r}_{\mathcal{K}}}\left[[\![\phi]\!]_{\mathcal{K}}\right]\right]$, we may conclude that $\mathcal{A} \subseteq \mathcal{D}$, where the generator $\mathcal{A}$ is defined as in Lemma 2.6.2. Because $\mathcal{D}$ is a σ-algebra, we obtain $\sigma(\mathcal{A}) \subseteq \mathcal{D}$. Thus the assertion follows from Lemma 2.6.2. □

Because the final σ-algebra with respect to a map is the largest σ-algebra on the codomain rendering this map measurable, we obtain the following as an immediate consequence.

Corollary 2.6.4. *Let $\mathcal{S}/\mathsf{r}_{\mathcal{K}}$ be the final σ-algebra with respect to $\eta_{\mathsf{r}_{\mathcal{K}}} : S \to S/\mathsf{r}_{\mathcal{K}}$ and $\mathcal{S}$. Then $\mathcal{S}^{\ddagger}_{\mathsf{r}_{\mathcal{K}}} \subseteq \mathcal{S}/\mathsf{r}_{\mathcal{K}}$.* □

A closer analysis reveals that we can say even more, viz., that $\mathcal{S}^{\ddagger}_{\mathsf{r}_{\mathcal{K}}}$ coincides with the factor algebra, provided the $\eta_{\mathsf{r}_{\mathcal{K}}}$-invariant sets are generated from the formulas. Recall that $\mathbf{\Sigma}(\mathcal{S}, \mathsf{r}_{\mathcal{K}})$ is σ-algebra of $\eta_{\mathsf{r}_{\mathcal{K}}}$-invariant measurable sets.

Corollary 2.6.5. *These statements are equivalent:*

a. $\sigma\big(\{[\![\phi]\!]_{\mathcal{K}} \mid \phi \text{ is a formula}\}\big) = \mathbf{\Sigma}(\mathcal{S}, \mathsf{r}_{\mathcal{K}})$.
b. $\mathcal{S}^{\ddagger}_{\mathsf{r}_{\mathcal{K}}} = \mathcal{S}/\mathsf{r}_{\mathcal{K}}$.

Proof 1. For establishing $a \Rightarrow b$ we show first that

$$\mathcal{S}^{\ddagger}_{\mathsf{r}_{\mathcal{K}}} = \{A \subseteq \mathcal{S}/\mathsf{r}_{\mathcal{K}} \mid \eta_{\mathsf{r}_{\mathcal{K}}}^{-1}[A] \in \sigma(\mathcal{E}_{\mathcal{K}})\}.$$

The construction of $\mathcal{S}^{\ddagger}_{\mathsf{r}_{\mathcal{K}}}$ implies that it is contained in the latter σ-algebra. For establishing the reverse inclusion, we argue as follows. Because each element of $\sigma(\mathcal{E}_{\mathcal{K}})$ is $\eta_{\mathsf{r}_{\mathcal{K}}}$-invariant, the π-λ-Theorem 1.3.1 is used to show that $\eta_{\mathsf{r}_{\mathcal{K}}}[B] \in \mathcal{S}^{\ddagger}_{\mathsf{r}_{\mathcal{K}}}$ for each $B \in \sigma(\mathcal{E}_{\mathcal{K}})$. Because $\eta_{\mathsf{r}_{\mathcal{K}}}$ is onto, we know $\eta_{\mathsf{r}_{\mathcal{K}}}\left[\eta_{\mathsf{r}_{\mathcal{K}}}^{-1}[A]\right] = A$ for $A \subseteq \mathcal{S}/\mathsf{r}_{\mathcal{K}}$; hence the inclusion follows.

The assertion follows now from the observation that the factor σ-algebra $\mathcal{S}/\mathsf{r}_{\mathcal{K}}$ can be written as $\mathcal{S}/\mathsf{r}_{\mathcal{K}} = \sigma(\{\eta_{\mathsf{r}_{\mathcal{K}}}[A] \mid A \in \mathcal{S} \text{ is } \eta_{\mathsf{r}_{\mathcal{K}}}\text{-invariant}\})$.

2. In order to prove $b \Rightarrow a$ we note first that the sets $[\![\phi]\!]_{\mathcal{K}}$ are $\mathsf{r}_{\mathcal{K}}$-invariant, so that $\sigma(\{[\![\phi]\!]_{\mathcal{K}} \mid \phi \text{ is a formula}\}) \subseteq \mathbf{\Sigma}(\mathcal{S}, \mathsf{r}_{\mathcal{K}})$. On the other hand, Lemma 2.6.2 yields

$$\eta_{\mathsf{r}_{\mathcal{K}}}^{-1}\left[\mathcal{S}_{\mathsf{r}_{\mathcal{K}}}^{\ddagger}\right] = \sigma(\{[\![\phi]\!]_{\mathcal{K}} \mid \phi \text{ is a formula}\}).$$

Now take $A \in \mathbf{\Sigma}(\mathcal{S}, \mathsf{r}_{\mathcal{K}})$. Then $A = \eta_{\mathsf{r}_{\mathcal{K}}}^{-1}\left[\eta_{\mathsf{r}_{\mathcal{K}}}[A]\right]$ with $\eta_{\mathsf{r}_{\mathcal{K}}}[A] \in \mathcal{S}/\mathsf{r}_{\mathcal{K}}$. Consequently the assumption implies $A \in \eta_{\mathsf{r}_{\mathcal{K}}}^{-1}\left[\mathcal{S}_{\mathsf{r}_{\mathcal{K}}}^{\ddagger}\right]$, from which the reverse inclusion follows. □

Comparing the construction for the general case with the one for analytic spaces, it follows that we can determine the crucial σ-algebra $\mathcal{S}_{\mathsf{r}_{\mathcal{K}}}^{\ddagger}$ through the factor map. This is actually a straightforward consequence of Corollary 1.7.13 and Corollary 2.6.5.

Corollary 2.6.6. *Assume that S is an analytic space with $\mathcal{S} = \mathcal{B}(S)$, and that* Act *is countable. Then $\mathcal{S}_{\mathsf{r}_{\mathcal{K}}}^{\ddagger} = \mathcal{B}(S/\mathsf{r}_{\mathcal{K}})$.* □

If $\mathcal{S}_{\mathsf{r}_{\mathcal{K}}}^{\ddagger}$ is a proper sub-σ-algebra of $\mathcal{S}/\mathsf{r}_{\mathcal{K}}$, then we conclude with Proposition 1.7.21 that the equivalence relation induced by the logic constitutes no longer a congruence for the Kripke model. This is a fairly peculiar situation which indicates that analytic state spaces play a somewhat special rôle (and invites further investigations for the general case).

Example 2.6.7. Put $S := \{x, y, z\}$ as the state space, $\mathcal{S} := \{\emptyset, S, \{x\}, \{y, z\}\}$, as the σ-algebra over S, and fix $\mu \in \mathfrak{S}(S, \mathcal{S})$. Let $\mathsf{Act} := \{*\}$ be a singleton set of actions, and put $k_*(s) := \mu$ for all $s \in S$. Then we know for the Kripke model $\mathcal{K} := \left((S, \mathcal{S}), (k_a)_{a \in \mathsf{Act}}\right)$ that $\{x\}$ is $\mathsf{r}_{\mathcal{K}}$-invariant; hence

$$\mathbf{\Sigma}(\mathcal{S}, \mathsf{r}_{\mathcal{K}}) = \mathcal{S}.$$

Furthermore we establish by induction that $\mathcal{K}, s \models \phi \Leftrightarrow \mathcal{K}, s' \models \phi$ for any states $s, s' \in S$, so that $[\![\phi]\!]_{\mathcal{K}} \neq \emptyset$ implies $[\![\phi]\!]_{\mathcal{K}} = S$. Consequently,

$$\sigma\left(\{[\![\phi]\!]_{\mathcal{K}} \mid \phi \text{ is a formula}\}\right) = \{\emptyset, S\}.$$

This implies by Corollary 2.6.5 that $\mathcal{S}_{\mathsf{r}_{\mathcal{K}}}^{\ddagger} \neq \mathcal{S}/\mathsf{r}_{\mathcal{K}}$. Thus we cannot dispose of the assumption that the state space is analytic. ⊣

Now consider the Kripke model $\mathcal{K}$. Let $a \in \mathsf{Act}$ be an action; the factor relation $k_{a,\mathsf{r}_{\mathcal{K}}}$ is defined through

$$k_{a,\mathsf{r}_{\mathcal{K}}}\left([s]_{\mathsf{r}_{\mathcal{K}}}\right)(A) := k_a(s)(\eta_{\mathsf{r}_{\mathcal{K}}}^{-1}[A])$$

whenever $A \in \mathcal{S}_{\mathsf{r}_{\mathcal{K}}}^{\ddagger}$ (compare the definition of the factor relation in the general setting on page 53). This definition is possible since we know from Corollary 2.6.4 that $\eta_{\mathsf{r}_{\mathcal{K}}}^{-1}[A] \in \mathcal{S}$ for $A \in \mathcal{S}_{\mathsf{r}_{\mathcal{K}}}^{\ddagger}$; it determines in fact a stochastic relation.

Proposition 2.6.8. $k_{a,\mathsf{r}_\mathcal{K}} : (S/\mathsf{r}_\mathcal{K}, \mathcal{S}^\ddagger_{\mathsf{r}_\mathcal{K}}) \rightsquigarrow (S/\mathsf{r}_\mathcal{K}, \mathcal{S}^\ddagger_{\mathsf{r}_\mathcal{K}})$ *is a stochastic relation.*

Proof 1. Because $k_{a,\mathsf{r}_\mathcal{K}}\big([s]_{\mathsf{r}_\mathcal{K}}\big) = \big(\mathfrak{S}(\eta_{\mathsf{r}_\mathcal{K}}) \circ k_a\big)(s)$, it follows immediately that $k_{a,\mathsf{r}_\mathcal{K}}\big([s]_{\mathsf{r}_\mathcal{K}}\big)$ is a subprobability on $\mathcal{S}^\ddagger_{\mathsf{r}_\mathcal{K}}$ for each $s \in S$.

2. Fix $A \in \mathcal{S}^\ddagger_{\mathsf{r}_\mathcal{K}}$; then the $\mathcal{S}^\ddagger_{\mathsf{r}_\mathcal{K}}$-measurability of $v \mapsto k_{a,\mathsf{r}_\mathcal{K}}(v)(A)$ has to be established. Let for this

$$\mathcal{D} := \{A \in \mathcal{S}^\ddagger_{\mathsf{r}_\mathcal{K}} \mid v \mapsto k_{a,\mathsf{r}_\mathcal{K}}(v)(A) \text{ is } \mathcal{S}^\ddagger_{\mathsf{r}_\mathcal{K}}\text{-measurable}\}.$$

We observe these properties

(i) $\mathcal{D}$ is closed under complementation.
(ii) $\mathcal{D}$ is closed under disjoint countable unions.
(iii) $\mathcal{A} \subseteq \mathcal{D}$, where $\mathcal{A}$ is the generator defined in Lemma 2.6.2.

The first and the second property follow from the usual properties of measurable functions; so only the last property needs to be verified. Because

$$\mathcal{K}, s \models \langle a \rangle_r \phi \Leftrightarrow k_a(s)\big([\![\phi]\!]_\mathcal{K}\big) \geq r,$$

we infer

$$\begin{aligned}\{[s]_{\mathsf{r}_\mathcal{K}} \mid k_{a,\mathsf{r}_\mathcal{K}}\big([s]_{\mathsf{r}_\mathcal{K}}\big)\big(\eta_{\mathsf{r}_\mathcal{K}}\left[[\![\phi]\!]_\mathcal{K}\right]\big) \geq r\} &= \eta_{\mathsf{r}_\mathcal{K}}\left[\{s \in S \mid k_a(s)\big([\![\phi]\!]_\mathcal{K}\big) \geq r\}\right] \\ &= \eta_{\mathsf{r}_\mathcal{K}}\left[[\![\langle a \rangle_r \phi]\!]_\mathcal{K}\right],\end{aligned}$$

and the latter set is a member of $\mathcal{S}^\ddagger_{\mathsf{r}_\mathcal{K}}$. Consequently, $\eta_{\mathsf{r}_\mathcal{K}}\left[[\![\phi]\!]_\mathcal{K}\right] \in \mathcal{D}$ for each formula ϕ. Using the π-λ-Theorem 1.3.1 and Lemma 2.6.2, we conclude now that $\mathcal{D} = \mathcal{S}^\ddagger_{\mathsf{r}_\mathcal{K}}$ holds. □

The factor map defines a morfism between the stochastic relations k_a and $k_{a,\mathsf{r}_\mathcal{K}}$, as we will see now. In fact, define the Kripke model

$$\mathcal{K}_\mathfrak{L} := \big((S/\mathsf{r}_\mathcal{K}, \mathcal{S}^\ddagger_{\mathsf{r}_\mathcal{K}}), (k_{a,\mathsf{r}_\mathcal{K}})_{a \in \mathsf{Act}}\big);$$

then we make this observation, which will be useful for the investigations of behavioral and logical equivalence:

Corollary 2.6.9. $\eta_{\mathsf{r}_\mathcal{K}} : \mathcal{K} \to \mathcal{K}_\mathfrak{L}$ *is a morfism.*

Proof Because we have

$$k_{a,\mathsf{r}_\mathcal{K}}\big([s]_{\mathsf{r}_\mathcal{K}}\big)(A) = k_a(s)\big(\eta^{-1}_{\mathsf{r}_\mathcal{K}}[A]\big) = \big(\mathfrak{S}(\eta_{\mathsf{r}_\mathcal{K}}) \circ k_a\big)(s)(A)$$

for all actions $a \in \mathsf{Act}$, for all states $s \in S$, and for all sets $A \in \mathcal{S}^\ddagger_{\mathsf{r}_\mathcal{K}}$, the claim is easily established. □

2.6.3 Logical Equivalence

Now let $\mathcal{L} = \big((T, \mathcal{T}), (\ell_a)_{a \in \mathsf{Act}}\big)$ be another Kripke model. Denote the equivalence relation defined by the logic $\mathfrak{L}$ on T by $\mathsf{r}_{\mathcal{L}}$. All constructions with the σ-algebra $\mathcal{S}$ and the equivalence relation $\mathsf{r}_{\mathcal{K}}$ are carried out with $\mathcal{T}$ and $\mathsf{r}_{\mathcal{L}}$, so we may construct a σ-algebra $\mathcal{T}^{\ddagger}_{\mathsf{r}_{\mathcal{L}}}$ on $T/\mathsf{r}_{\mathcal{L}}$, and we obtain a new Kripke model $\mathcal{L}_{\mathfrak{L}} = \big((T/\mathsf{r}_{\mathcal{L}}, \mathcal{T}^{\ddagger}_{\mathsf{r}_{\mathcal{L}}}), (\ell_{a,\mathsf{r}_{\mathcal{K}}})_{a \in \mathsf{Act}}\big)$ together with the morfism $\eta_{\mathsf{r}_{\mathcal{L}}} : \mathcal{L} \to \mathcal{L}_{\mathfrak{L}}$.

Define the relation

$$\mathfrak{R} := \{\langle s, t\rangle \in S \times T \mid Th_{\mathcal{K}}(s) = Th_{\mathcal{L}}(t)\}.$$

Consequently, $\langle s, t\rangle \in \mathfrak{R}$ iff s and t satisfy exactly the same formulas. In particular, we know that then $k_a(s)([\![\phi]\!]_{\mathcal{K}}) = \ell_a([\![\phi]\!]_{\mathcal{L}})$ holds for all formulas ϕ. This is so because $k_a(s)\big([\![\phi]\!]_{\mathcal{K}}\big) \geq r \Leftrightarrow \ell_a(t)\big([\![\phi]\!]_{\mathcal{L}}\big) \geq r$ for each r in the unit interval (it would suffice to restrict ourselves to rational numbers r).

In addition to relation $\mathfrak{R}$ a relation $\mathfrak{R}_0$ derived from it on the Cartesian product of the factor spaces is defined:

$$\mathfrak{R}_0 := \{\langle [s]_{\mathsf{r}_{\mathcal{K}}}, [t]_{\mathsf{r}_{\mathcal{L}}}\rangle \mid \langle s, t\rangle \in \mathfrak{R}\}.$$

Call models $\mathcal{K}$ and $\mathcal{L}$ *behaviorally equivalent* iff there exists a cospan

$$\mathcal{K} \xrightarrow{f} \mathcal{M} \xleftarrow{g} \mathcal{L}$$

of morfisms, as above. Logical equivalence may be defined through the relation $\mathfrak{R}$: $\mathcal{K}$ and $\mathcal{L}$ are said to be *logically equivalent* iff the relation $\mathfrak{R}$ is both right and left total. This is but a simple reformulation of the usual definition of logical equivalence; see Definition 2.3.6. Assume for the rest of this section that the Kripke models $\mathcal{K}$ and $\mathcal{L}$ are logically equivalent.

Lemma 2.6.10. *$\mathfrak{R}_0$ is the graph of a bijection $\tau : S/\mathsf{r}_{\mathcal{K}} \to T/\mathsf{r}_{\mathcal{L}}$; τ is $\mathcal{S}^{\ddagger}_{\mathsf{r}_{\mathcal{K}}}$-$\mathcal{T}^{\ddagger}_{\mathsf{r}_{\mathcal{L}}}$-measurable. Similarly, $\mathfrak{R}_0^{-1}$ is the graph of a bijection $\theta : S/\mathsf{r}_{\mathcal{K}} \to T/\mathsf{r}_{\mathcal{L}}$, which is $\mathcal{T}^{\ddagger}_{\mathsf{r}_{\mathcal{L}}}$-$\mathcal{S}^{\ddagger}_{\mathsf{r}_{\mathcal{K}}}$-measurable. τ and θ are inverse to each other.*

Proof 1. Define $\tau([s]_{\mathsf{r}_{\mathcal{K}}}) := [t]_{\mathsf{r}_{\mathcal{L}}}$ iff $\langle [s]_{\mathsf{r}_{\mathcal{K}}}, [t]_{\mathsf{r}_{\mathcal{L}}}\rangle \in \mathfrak{R}_0$. Then $\tau : S/\mathsf{r}_{\mathcal{K}} \to T/\mathsf{r}_{\mathcal{L}}$ is obviously well defined and injective. Because $\mathfrak{R}$ is a right total relation, τ is surjective as well.

2. Consider $\mathcal{D} := \{B \in \mathcal{T}^{\ddagger}_{\mathsf{r}_{\mathcal{L}}} \mid \tau^{-1}[B] \in \mathcal{S}^{\ddagger}_{\mathsf{r}_{\mathcal{K}}}\}$. Then $\mathcal{D}$ is closed under complementation and countable disjoint unions. Let ϕ be a formula; then $\tau^{-1}[\eta_{\mathsf{r}_{\mathcal{L}}}[[\![\phi]\!]_{\mathcal{L}}]] = \eta_{\mathsf{r}_{\mathcal{K}}}[[\![\phi]\!]_{\mathcal{K}}]$. Consequently, $\eta_{\mathsf{r}_{\mathcal{L}}}[[\![\phi]\!]_{\mathcal{L}}] \in \mathcal{D}$, so $\mathcal{D}$ contains a generator which is closed under finite intersections by Lemma 2.6.2. Thus $\mathcal{D} = \mathcal{T}^{\ddagger}_{\mathsf{r}_{\mathcal{L}}}$ by the π-λ-Theorem 1.3.1. Thus τ is $\mathcal{S}^{\ddagger}_{\mathsf{r}_{\mathcal{K}}}$-$\mathcal{T}^{\ddagger}_{\mathsf{r}_{\mathcal{L}}}$-measurable.

3. The argumentation for θ is verbatim the same, after interchanging the rôles of $\mathcal{K}$ and $\mathcal{L}$. It is also obvious that τ and θ are inverse to each other. □

If $\Re$ would not be a left total relation, the map τ would only be partially defined; if $\Re$ would not be right total, τ would not be surjective. Consequently, this construction works only with logically equivalent Kripke models.

But τ and θ are even richer in structure.

Lemma 2.6.11. *Define τ and θ as in Lemma 2.6.10. Then $\tau : \mathcal{K}_{\mathfrak{L}} \to \mathcal{L}_{\mathfrak{L}}$ and $\theta : \mathcal{L}_{\mathfrak{L}} \to \mathcal{K}_{\mathfrak{L}}$ are morfisms.*

Proof 1. Assume that $\tau([s]_{\mathsf{r}_\mathcal{K}}) = [t]_{\mathsf{r}_\mathcal{L}}$; then, since $\eta_{\mathsf{r}_\mathcal{K}}^{-1}[\eta_{\mathsf{r}_\mathcal{K}}[[\![\phi]\!]_\mathcal{K}]] = [\![\phi]\!]_\mathcal{K}$, and similarly for $[\![\phi]\!]_\mathcal{L}$,

$$\begin{aligned} k_{a,\mathsf{r}_\mathcal{K}}([s]_{\mathsf{r}_\mathcal{K}})(\eta_{\mathsf{r}_\mathcal{K}}[[\![\phi]\!]_\mathcal{K}]) &= k_a(s)([\![\phi]\!]_\mathcal{K}) \\ &= \ell_a(t)([\![\phi]\!]_\mathcal{L}) \\ &= \ell_{a,\mathsf{r}_\mathcal{L}}([t]_{\mathsf{r}_\mathcal{L}})(\eta_{\mathsf{r}_\mathcal{L}}[[\![\phi]\!]_\mathcal{L}]). \end{aligned}$$

2. We want to show that

$$\ell_{a,\mathsf{r}_\mathcal{L}}(\tau([s]_{\mathsf{r}_\mathcal{K}}))(A) = k_{a,\mathsf{r}_\mathcal{K}}([s]_{\mathsf{r}_\mathcal{K}})(\tau^{-1}[A])$$

holds for each $s \in S$ and each measurable set $A \in \mathcal{T}^{\ddagger}_{\mathsf{r}_\mathcal{L}}$. Consider for fixed $a \in \mathsf{Act}$ and $s \in S$ the set

$$\mathcal{D} := \{A \in \mathcal{T}^{\ddagger}_{\mathsf{r}_\mathcal{L}} \mid \ell_{a,\mathsf{r}_\mathcal{L}}(\tau([s]_{\mathsf{r}_\mathcal{K}}))(A) = k_{a,\mathsf{r}_\mathcal{K}}([s]_{\mathsf{r}_\mathcal{K}})(\tau^{-1}[A])\}.$$

Then $\mathcal{D}$ is closed under complementation and countable disjoint unions, and part 1 shows that $\eta_{\mathsf{r}_\mathcal{L}}[[\![\phi]\!]_\mathcal{L}] \in \mathcal{D}$ for each formula. Thus the π-λ-Theorem 1.3.1 together with Lemma 2.6.2 implies that $D = \mathcal{T}^{\ddagger}_{\mathsf{r}_\mathcal{L}}$. Consequently, τ constitutes a morfism; the same argumentation shows that θ is a morfism as well. □

2.6.4 Logical vs. Behavioral Equivalence

Form the sum $S + T$ of the state space with injections $i_S : S \to S + T$ and $i_T : T \to S + T$, respectively. Let $\mathsf{r}_\mathcal{K} \diamond \mathsf{r}_\mathcal{L}$ be the amalgamation of $\mathsf{r}_\mathcal{K}$ and $\mathsf{r}_\mathcal{L}$, so that we have for $v, v' \in S + T$ in the present scenario

$$v \ \mathsf{r}_\mathcal{K} \diamond \mathsf{r}_\mathcal{L} \ v' \iff \begin{cases} s \ \mathsf{r}_\mathcal{K} \ s', & v = i_S(s), v' = i_S(s'), \\ t \ \mathsf{r}_\mathcal{L} \ t', & v = i_T(t), v' = i_T(t'), \\ \langle s, t \rangle \in \Re, & v = i_S(s), v' = i_T(t), \\ \langle t, s \rangle \in \Re^{-1}, & v = i_T(t), v' = i_S(s) \end{cases}$$

(see page 52). This construction entails also that

$$(*) \quad \begin{cases} [i_S(s)]_{\mathsf{r}_\mathcal{K} \diamond \mathsf{r}_\mathcal{L}} &= i_S\big[[s]_{\mathsf{r}_\mathcal{K}}\big] \cup i_T\big[\tau\big([s]_{\mathsf{r}_\mathcal{K}}\big)\big], \\ [i_T(t)]_{\mathsf{r}_\mathcal{K} \diamond \mathsf{r}_\mathcal{L}} &= i_T\big[[t]_{\mathsf{r}_\mathcal{L}}\big] \cup i_S\big[\theta\big([t]_{\mathsf{r}_\mathcal{L}}\big)\big]. \end{cases}$$

Thus the equivalence class of an element of $S + T$ has both a non-void component from S and from T, and these components are linked through τ and θ, respectively. Now define maps $I_S : S/\mathsf{r}_\mathcal{K} \to (S+T)/\mathsf{r}_\mathcal{K} \diamond \mathsf{r}_\mathcal{L}$ and $I_T : T/\mathsf{r}_\mathcal{L} \to (S+T)/\mathsf{r}_\mathcal{K} \diamond \mathsf{r}_\mathcal{L}$ through

$$I_S\big([s]_{\mathsf{r}_\mathcal{K}}\big) := [i_S(s)]_{\mathsf{r}_\mathcal{K} \diamond \mathsf{r}_\mathcal{L}},$$
$$I_S\big([t]_{\mathsf{r}_\mathcal{L}}\big) := [i_T(t)]_{\mathsf{r}_\mathcal{K} \diamond \mathsf{r}_\mathcal{L}}.$$

Hence we assign to each class in the participating spaces the class of its representative in the sum; it is clear from the characterization in $(*)$ that both maps are well defined.

Lemma 2.6.12. *$I_S\left[\eta_{\mathsf{r}_\mathcal{K}}\left[[\![\phi]\!]_\mathcal{K}\right]\right] = I_T\left[\eta_{\mathsf{r}_\mathcal{L}}\left[[\![\phi]\!]_\mathcal{L}\right]\right]$ holds for each formula ϕ, and the set $\{I_S\left[\eta_{\mathsf{r}_\mathcal{K}}\left[[\![\phi]\!]_\mathcal{K}\right]\right] \mid \phi \text{ is a formula}\}$ is closed under finite intersections.*

Proof The first claim is established by this direct computation:

$$\begin{aligned} I_S\left[\eta_{\mathsf{r}_\mathcal{K}}\left[[\![\phi]\!]_\mathcal{K}\right]\right] &= \{I_S(b) \mid b \in \eta_{\mathsf{r}_\mathcal{K}}\left[[\![\phi]\!]_\mathcal{K}\right]\} \\ &= \{[i_S(s)]_{\mathsf{r}_\mathcal{K} \diamond \mathsf{r}_\mathcal{L}} \mid s \in [\![\phi]\!]_\mathcal{K}\} \\ &\overset{(\natural)}{=} \{[i_T(t)]_{\mathsf{r}_\mathcal{K} \diamond \mathsf{r}_\mathcal{L}} \mid t \in [\![\phi]\!]_\mathcal{L}\} \\ &= \{I_T(c) \mid c \in \eta_{\mathsf{r}_\mathcal{L}}\left[[\![\phi]\!]_\mathcal{L}\right]\} \\ &= I_T\left[\eta_{\mathsf{r}_\mathcal{L}}\left[[\![\phi]\!]_\mathcal{L}\right]\right] \end{aligned}$$

The crucial equality $(\natural)$ follows immediately from $[i_S(s)]_{\mathsf{r}_\mathcal{K} \diamond \mathsf{r}_\mathcal{L}} = [\tau(i_S(s))]_{\mathsf{r}_\mathcal{K} \diamond \mathsf{r}_\mathcal{L}}$ and $[i_T(t)]_{\mathsf{r}_\mathcal{K} \diamond \mathsf{r}_\mathcal{L}} = [\theta(i_T(t))]_{\mathsf{r}_\mathcal{K} \diamond \mathsf{r}_\mathcal{L}}$ together with $\theta = \tau^{-1}$(Lemma 2.6.10). Because $\eta_{\mathsf{r}_\mathcal{K}}\left[[\![\phi]\!]_\mathcal{K}\right]$ is an I_S-invariant set, it follows from Lemma 1.4.30 that $\{I_S\left[\eta_{\mathsf{r}_\mathcal{K}}\left[[\![\phi]\!]_\mathcal{K}\right]\right] \mid \phi \text{ is a formula}\}$ is closed under finite intersections. □

Define the σ-algebra $\mathcal{W}$ on $(S+T)/\mathsf{r}_\mathcal{K} \diamond \mathsf{r}_\mathcal{L}$ as

$$\mathcal{W} := \sigma\big(\{I_S\left[\eta_{\mathsf{r}_\mathcal{K}}\left[[\![\phi]\!]_\mathcal{K}\right]\right] \mid \phi \text{ is a formula}\}\big);$$

then $\mathcal{W} = \sigma\big(\{I_T\left[\eta_{\mathsf{r}_\mathcal{L}}\left[[\![\phi]\!]_\mathcal{L}\right]\right] \mid \phi \text{ is a formula}\}\big)$ follows from Lemma 2.6.12, and the maps I_S and I_T turn out to be measurable.

Lemma 2.6.13. *$I_S : S \to (S+T)/\mathsf{r}_\mathcal{K} \diamond \mathsf{r}_\mathcal{L}$ is $\mathcal{S}^\ddagger_{\mathsf{r}_\mathcal{K}}$-$\mathcal{W}$-measurable, and $I_T : T \to (S+T)/\mathsf{r}_\mathcal{K} \diamond \mathsf{r}_\mathcal{L}$ is $\mathcal{T}^\ddagger_{\mathsf{r}_\mathcal{L}}$-$\mathcal{W}$-measurable.*

Proof We focus on the map I_S and apply essentially the π-λ-Theorem 1.3.1 again. Let

$$\mathcal{D} := \{C \in \mathcal{W} \mid I_S^{-1}[C] \in \mathcal{S}^\ddagger_{\mathsf{r}_\mathcal{K}}\}.$$

Because $\eta_{\mathsf{r}_\mathcal{K}}\left[\llbracket\phi\rrbracket_\mathcal{K}\right]$ is an I_S-invariant set, Lemma 1.4.30 tells us for a formula ϕ that $I_S^{-1}\left[I_S\left[\eta_{\mathsf{r}_\mathcal{K}}\left[\llbracket\phi\rrbracket_\mathcal{K}\right]\right]\right] = \eta_{\mathsf{r}_\mathcal{K}}\left[\llbracket\phi\rrbracket_\mathcal{K}\right]$, and the latter is an element of $\mathcal{S}^\ddagger_{\mathsf{r}_\mathcal{K}}$. Consequently, $I_S\left[\eta_{\mathsf{r}_\mathcal{K}}\left[\llbracket\phi\rrbracket_\mathcal{K}\right]\right] \in \mathcal{D}$ for each formula ϕ. Now $\mathcal{D}$ is a σ-algebra, in particular closed under complementation and disjoint countable unions. The family $\{I_S\left[\eta_{\mathsf{r}_\mathcal{K}}\left[\llbracket\phi\rrbracket_\mathcal{K}\right]\right] \mid \phi \text{ is a formula}\}$ generates $\mathcal{W}$ and is closed under finite intersections by Lemma 2.6.12, so the π-λ-Theorem implies $\mathcal{D} = \mathcal{W}$. This establishes the assertion. □

This construction yields a measurable space $((S+T)/\mathsf{r}_\mathcal{K} \diamond \mathsf{r}_\mathcal{L}, \mathcal{W})$. It will serve as the state space for a Kripke model for which we will construct the transition law now. Before we do that, we need to make sure that the transition laws $k_{a,\mathsf{r}_\mathcal{K}}$ and $\ell_{a,\mathsf{r}_\mathcal{L}}$ coincide on certain crucial sets.

Lemma 2.6.14. *Assume that $\langle s,t\rangle \in \mathfrak{R}$. Then*

$$k_{a,\mathsf{r}_\mathcal{K}}([s]_{\mathsf{r}_\mathcal{K}})(I_S^{-1}\left[C\right]) = \ell_{a,\mathsf{r}_\mathcal{L}}([t]_{\mathsf{r}_\mathcal{L}})(I_T^{-1}\left[C\right])$$

for all $C \in \mathcal{W}$.

Proof It is by the π-λ-Theorem (Lemma 1.6.30) sufficient to take the set C from the generator $\{I_S\left[\eta_{\mathsf{r}_\mathcal{K}}\left[\llbracket\phi\rrbracket_\mathcal{K}\right]\right] \mid \phi \text{ is a formula}\}$ of $\mathcal{W}$. Thus, let ϕ be a formula; then

$$\begin{aligned}
k_{a,\mathsf{r}_\mathcal{K}}([s]_{\mathsf{r}_\mathcal{K}})\big(I_S^{-1}\left[I_S\left[\eta_{\mathsf{r}_\mathcal{K}}\left[\llbracket\phi\rrbracket_\mathcal{K}\right]\right]\right]\big) &= k_{a,\mathsf{r}_\mathcal{K}}([s]_{\mathsf{r}_\mathcal{K}})\big(\eta_{\mathsf{r}_\mathcal{K}}\left[\llbracket\phi\rrbracket_\mathcal{K}\right]\big)\\
&= k_a(s)(\eta_{\mathsf{r}_\mathcal{K}}^{-1}\left[\eta_{\mathsf{r}_\mathcal{K}}\left[\llbracket\phi\rrbracket_\mathcal{K}\right]\right])\\
&= k_a(s)(\llbracket\phi\rrbracket_\mathcal{K})\\
&\stackrel{(\natural)}{=} \ell_a(t)(\llbracket\phi\rrbracket_\mathcal{L})\\
&= \ell_{a,\mathsf{r}_\mathcal{L}}([t]_{\mathsf{r}_\mathcal{L}})\big(\eta_{\mathsf{r}_\mathcal{L}}\left[\llbracket\phi\rrbracket_\mathcal{L}\right]\big)\\
&= \ell_{a,\mathsf{r}_\mathcal{L}}([t]_{\mathsf{r}_\mathcal{L}})\big(I_T^{-1}\left[I_T\left[\eta_{\mathsf{r}_\mathcal{L}}\left[\llbracket\phi\rrbracket_\mathcal{L}\right]\right]\right]\big).
\end{aligned}$$

The equation $(\natural)$ permits the comparisons between different Kripke models; it follows from the assumption that $\langle s,t\rangle \in \mathfrak{R}$. □

Now we are poised to define the transition law on the compound factor space. We put

$$m_a([i_S(s)]_{\mathsf{r}_\mathcal{K}\diamond\mathsf{r}_\mathcal{L}})(C) := k_{a,\mathsf{r}_\mathcal{K}}([s]_{\mathsf{r}_\mathcal{K}})(I_S^{-1}\left[C\right])$$

for $C \in \mathcal{W}$. It follows from Lemma 2.6.14 that also

$$m_a([i_T(t)]_{\mathsf{r}_\mathcal{K}\diamond\mathsf{r}_\mathcal{L}})(C) = \ell_{a,\mathsf{r}_\mathcal{L}}([t]_{\mathsf{r}_\mathcal{L}})(I_T^{-1}\left[C\right])$$

holds, provided $[i_S(s)]_{\mathsf{r}_\mathcal{K}\diamond\mathsf{r}_\mathcal{L}} = [i_T(t)]_{\mathsf{r}_\mathcal{K}\diamond\mathsf{r}_\mathcal{L}}$. This is so because the latter condition is equivalent to $\langle [s]_{\mathsf{r}_\mathcal{K}}, [t]_{\mathsf{r}_\mathcal{L}}\rangle \in \mathfrak{R}_0$ which in turn is equivalent to $\langle s,t\rangle \in \mathfrak{R}$.

This defines for each action $a \in \mathsf{Act}$ a map

$$m_a : (S+T)/\mathsf{r}_\mathcal{K} \diamond \mathsf{r}_\mathcal{L} \to \mathfrak{S}\left((S+T)/\mathsf{r}_\mathcal{K} \diamond \mathsf{r}_\mathcal{L}, \mathcal{W}\right),$$

and we have to make sure that it defines a Kripke model, i.e., that it is a stochastic relation. Thus we have to establish measurability.

Lemma 2.6.15. $m_a : ((S+T)/\mathsf{r}_{\mathcal{K}} \diamond \mathsf{r}_{\mathcal{L}}, \mathcal{W}) \rightsquigarrow ((S+T)/\mathsf{r}_{\mathcal{K}} \diamond \mathsf{r}_{\mathcal{L}}, \mathcal{W})$ *is a stochastic relation for each action* $a \in \mathsf{Act}$.

Proof 1. We have to show that $w \mapsto m_a(w)(G)$ constitutes a $\mathcal{W}$-measurable map on $(S+T)/\mathsf{r}_{\mathcal{K}} \diamond \mathsf{r}_{\mathcal{L}}$ for each set $G \in \mathcal{W}$. This is established through the π-λ-Theorem 1.3.1 by investigating the set of all sets for which the assertion holds.

2. In fact, $\mathcal{D} := \{G \in \mathcal{W} \mid w \in m_a(w)(G) \text{ is } \mathcal{W}\text{-measurable}\}$ is closed under complementation and under countable disjoint unions. Now assume that $G = I_S\left[\eta_{\mathsf{r}_{\mathcal{K}}}\left[[\![\phi]\!]_{\mathcal{K}}\right]\right]$ is an element of the generators of $\mathcal{W}$, then

$$\begin{aligned} m_a\big([s]_{\mathsf{r}_{\mathcal{K}} \diamond \mathsf{r}_{\mathcal{L}}}\big)(G) \geq r &\Leftrightarrow k_{a,\mathsf{r}_{\mathcal{K}}}\big([s]_{\mathsf{r}_{\mathcal{K}}}\big)\big(\eta_{\mathsf{r}_{\mathcal{K}}}\left[[\![\phi]\!]_{\mathcal{K}}\right]\big) \geq r \\ &\Leftrightarrow [s]_{\mathsf{r}_{\mathcal{K}}} \in [\![\langle a \rangle_r \phi]\!]_{\mathcal{K}}. \end{aligned}$$

Thus

$$\{w \in (S+T)/\mathsf{r}_{\mathcal{K}} \diamond \mathsf{r}_{\mathcal{L}} \mid m_a(w)\big(I_S\left[\eta_{\mathsf{r}_{\mathcal{K}}}\left[[\![\phi]\!]_{\mathcal{K}}\right]\right]\big) \geq r\} = I_S\left[\eta_{\mathsf{r}_{\mathcal{K}}}\left[[\![\langle a \rangle_r \phi]\!]_{\mathcal{K}}\right]\right],$$

which is a generator itself.

Hence $\mathcal{D}$ contains the generator of $\mathcal{W}$ which is closed under finite intersections by Lemma 2.6.12, so that $\mathcal{D} = \mathcal{W}$, which establishes measurability. □

Thus we may use m_a as the transition law for a Kripke model.

Corollary 2.6.16. $\mathcal{K}_{\mathfrak{L}} \xrightarrow{I_S} \mathcal{M} \xleftarrow{I_T} \mathcal{L}_{\mathfrak{L}}$ *is a cospan of morfisms.* □

This is the main result.

Proposition 2.6.17. *Logical and behavioral equivalence are the same for Kripke models over general measurable spaces for the negation-free Hennessy-Milner logic* $\mathfrak{L}$.

Proof 1. Since morfisms preserve and reflect validity, behaviorally equivalent Kripke models are logically equivalent.

2. Let $\mathcal{K} = \big((S, \mathcal{S}), (k_a)_{a \in \mathsf{Act}}\big)$ and $\mathcal{L} = \big((T, \mathcal{T}), (\ell_a)_{a \in \mathsf{Act}}\big)$ be the Kripke models under consideration. Construct the factor space $(S+T)/\mathsf{r}_{\mathcal{K}} \diamond \mathsf{r}_{\mathcal{L}}$ for the amalgamation $\mathsf{r}_{\mathcal{K}} \diamond \mathsf{r}_{\mathcal{L}}$ of the equivalence relations $\mathsf{r}_{\mathcal{K}}$ and $\mathsf{r}_{\mathcal{L}}$ which are constructed from logic $\mathfrak{L}$ over S respectively T together with the maps $I_S : S/\mathsf{r}_{\mathcal{K}} \to (S+T)/\mathsf{r}_{\mathcal{K}} \diamond \mathsf{r}_{\mathcal{L}}$ and $I_T : T/\mathsf{r}_{\mathcal{L}} \to (S+T)/\mathsf{r}_{\mathcal{K}} \diamond \mathsf{r}_{\mathcal{L}}$. Construct the σ-algebra $\mathcal{W}$ from these data, and define the stochastic relation $m_a : \big((S+T)/\mathsf{r}_{\mathcal{K}} \diamond \mathsf{r}_{\mathcal{L}}, \mathcal{W}\big) \rightsquigarrow \big((S+T)/\mathsf{r}_{\mathcal{K}} \diamond \mathsf{r}_{\mathcal{L}}, \mathcal{W}\big)$. Put

$$\mathcal{M} := \Big(\big((S+T)/\mathsf{r}_{\mathcal{K}} \diamond \mathsf{r}_{\mathcal{L}}, \mathcal{W}\big), (m_a)_{a \in \mathsf{Act}}\Big).$$

Then this diagram gives the desired cospan of morfisms.

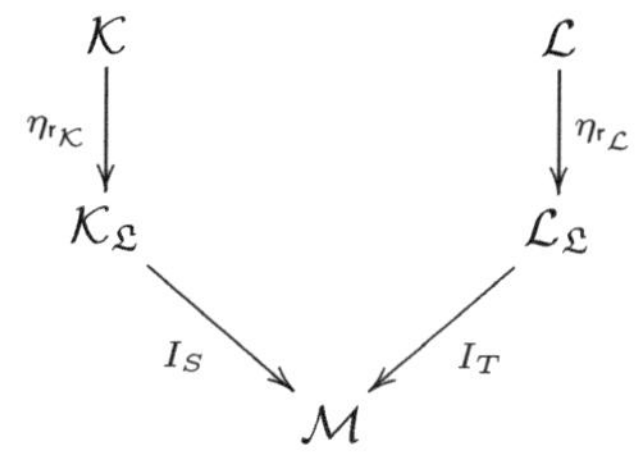

The factor maps $\eta_{\mathsf{r}_\mathcal{K}}$ and $\eta_{\mathsf{r}_\mathcal{L}}$ are morfisms by Corollary 2.6.9; I_S and I_T are morfisms by Corollary 2.6.16. □

One may wonder why we went through this slightly complicated construction process in order to obtain the σ-algebra on the target space. Taking the sum $S + T$ of the measurable spaces $(S, \mathcal{S})$ and $(T, \mathcal{T})$ with the sum-σ-algebra $\mathcal{S} + \mathcal{T}$, factoring this space through the amalgamated equivalence relation $\mathsf{r}_\mathcal{K} \diamond \mathsf{r}_\mathcal{L}$ and constructing the σ-algebra $(\mathcal{S} + \mathcal{T})^{\ddagger}_{\mathsf{r}_\mathcal{K} \diamond \mathsf{r}_\mathcal{L}}$ is not possible. Alas, this approach requires an underlying Kripke model which this process is just intended to construct.

Chapter 3
The Giry Monad: Randomized Morphisms

3.1 Introduction

Morphisms relate stochastic relations in a way that preserves the probabilistic structure. We did define a morphism $\mathsf{f} : \mathsf{K} \to \mathsf{L}$ between stochastic relations $\mathsf{K} = \big((X, \mathcal{A}), (Y, \mathcal{B}), K\big)$ and $\mathsf{L} = \big((A, \mathcal{D}), (B, \mathcal{E}), L\big)$ as a pair (f, g) of surjective measurable maps so that $\mathfrak{S}\,(g) \circ K = L \circ f$. This means that for each measurable subset F of B, and for each $x \in X$ the equality $K(x)\big(g^{-1}\left[F\right]\big) = L(f(x))(F)$ holds. Stochastic relations form a category with these morphisms, and the kernels of morphisms are exactly the congruences; see Section 1.7.3.

The subprobability functor is the functorial part of the Giry monad, analogous to the power set functor which is the functorial part of the Manes monad. Each monad carries its own special category with it — the Kleisli category — so that pure mathematical curiosity demands asking the question what can be said about morphisms in this monad. To be more specific, let K and L be stochastic relations again; then we say that $\mathsf{F} : \mathsf{K} \curvearrowright \mathsf{L}$ is a randomized morphism iff $\mathsf{F} = (\Phi, \Psi)$ is a pair of Kleisli morphisms $\Phi : (X, \mathcal{A}) \rightsquigarrow (A, \mathcal{D})$ and $\Psi : (Y, \mathcal{B}) \rightsquigarrow (B, \mathcal{E})$ such that we have $L * \Phi = \Psi * K$ for their Kleisli product, where the Kleisli product is of course the composition that is defined in the Kleisli category for the Giry monad. We will see that moving from morphisms to randomized morphisms is akin to moving from equivalence relations on a space to this kind of relations on the space of subprobabilities, and we will see that we need to impose additional conditions for this new notion of morphisms to be useful for us.

It might be noted that a similar process has been undertaken in stochastic dynamic programming when discussing policies: a policy assigns a decision maker (a.k.a. player) a rule about what to do next. A randomized policy permits the decision maker to randomize this decision. Hence instead of assigning an action to a state, a randomized policy assigns a probability distribution

E.-E. Doberkat, *Stochastic Coalgebraic Logic*, Monographs in Theoretical Computer Science. An EATCS Series, DOI 10.1007/978-3-642-02995-0_3,

over the actions to a state. This is discussed at length in Hinderer's seminal work [47] on stochastic dynamic programming.

We introduce in Section 3.2 the algebraic framework for this discussion. First we briefly discuss the notion of a monad for an endofunctor and move quickly to the subprobability functor as the endofunctor on the category of measurable spaces (and on some important subcategories). The Kleisli composition is introduced, and we discuss the Kleisli extension of a stochastic relation, which is essentially a measure transformation, and some of its elementary properties. One of the interesting questions arising immediately is factoring: we know that congruences are essentially the kernels of morphisms, and we want to know whether or not this is the case for the randomized morphisms as well. This question is investigated in Section 3.3 for the restricted class of near-grounded morphisms; this seems to be the most sensible class of randomized morphisms in which this question can be investigated at all — the more general setting appears to be way too general. The main tool in these investigations is derived again from smooth equivalence relations and their invariant sets, in particular in the way invariance is reflected on various spaces of subprobabilities and their factors.

This is then applied to a simple Hennessy-Milner type logic which is intentionally being held quite simple again: formulas are closed under conjunctions and under a modal operator which in turn is indexed by a degree of satisfaction. We investigate this logic first with respect to randomized morphisms and find that the factor model has some interesting properties. Using this logic, we investigate questions of logical and behavioral equivalence, but this time through randomized morphisms for comparing Kripke models. Rather than comparing states proper of a Kripke model, we compare distributions over states. The appropriate morphisms are randomized ones. The problem of the existence of spans or cospans of morphisms is posed in this context for the randomized variety.

This kind of question arises naturally when modelling stochastic phenomena through a Kripke model for a suitable logic when one does not want to give too much attention to the behavior of individual states but rather to their distribution. This is of interest, e.g.,

1. in Mathematical Economics. Here the members of a measurable space are agents who influence the behavior of a large economy, the σ-algebra being coalitions of these agents. Individual agents usually cannot influence the market, so one pays attention to the weight which is assigned to the coalitions through subprobabilities; see the introductory section of [80] for a concise discussion. When modelling events in an economy through a modal logic where actions may bring forth changes in the economy's state, it appears sensible to compare distributions rather than individuals with each other.
2. in Systems Biology. When modelling biological systems, the behavior of the individual units (like cells) may for some questions be not as important as their collective behavior; this indicates that stochastic modelling,

i.e., modelling on the level of distributions, is for many purposes more effective than discussing individual units; see [84, Section 1.3] or [50, Section 3.5, p. 108].

The randomized variants of logical and behavioral equivalence or bisimulation are defined through randomized morphisms, and we relate these notions to each other and to their cousins which are defined through strong morphisms. Technically, a special rôle is played by ergodic morphisms; i.e., those randomized morphisms which are binary-valued and are adapted to the logic. Factorizations and the structure theorems which we explored for randomized congruences turn out to be helpful as well.

3.2 The Giry Monad

This section briefly reviews the notion of a monad, i.e., a triplet consisting of an endofunctor and two natural transformations, the unit and the multiplication. We show in which way unit and multiplication are defined for the subprobability functor, giving rise to the Giry monad. Associated with each monad is the Kleisli construction. It assigns to the monad a category the objects of which are the same as in the underlying category, the morphisms of which, however, are composed through the monad's component. This construction is given here for the Giry monad as well, the corresponding composition being well known in Probability Theory as the convolution of transition probabilities; see, e.g., [5, §64] or [77, Chapter II.6].

Let us briefly recall the definition of a monad.

Definition 3.2.1. *A* monad $\langle\mathfrak{T}, \mathfrak{e}, \mathfrak{m}\rangle$ *over the category* **C** *is an endofunctor* $\mathfrak{T} : \mathbf{C} \to \mathbf{C}$ *together with the natural transformations* $\mathfrak{e} : 1\!\!1_{\mathbf{C}} \stackrel{\bullet}{\to} \mathfrak{T}$ *(the unit) and* $\mathfrak{m} : \mathfrak{T}^2 \stackrel{\bullet}{\to} \mathfrak{T}$ *(the multiplication) such that these diagrams commute:*

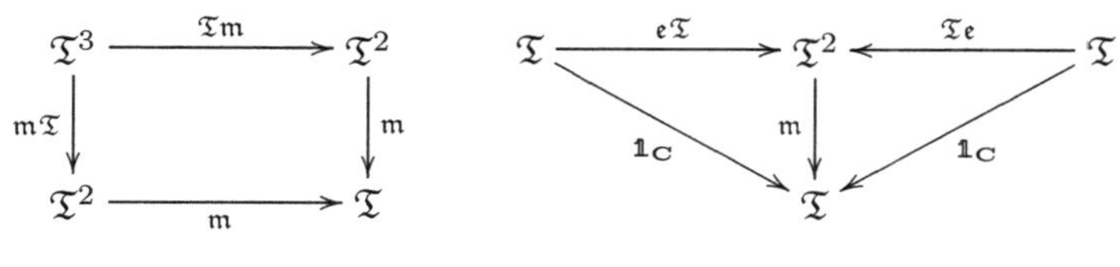

The commutativity of the leftmost diagram is expressed for an object x of **C** through

$$\mathfrak{m}_c \circ \mathfrak{T}(\mathfrak{m}_c) = \mathfrak{m}_c \circ \mathfrak{m}_{\mathfrak{T}(x)},$$

while the commutativity of the rightmost diagram is written down as

$$\mathfrak{m}_c \circ \mathfrak{e}_{\mathfrak{T}(x)} = id_c = \mathfrak{m}_c \circ \mathfrak{T}(\mathfrak{e}_c).$$

These expressions are sometimes easier to handle than the functorial notation in the diagrams above.

We will deal mostly with the *Giry monad* over the category **Meas** of measurable spaces or one of its subcategories. The endofunctor underlying this monad is the subprobability functor $\mathfrak{S}$, the unit $\mathfrak{e} : \mathbb{1}_{\mathbf{Meas}} \stackrel{\bullet}{\to} \mathfrak{S}$ assigns to each point x in a measurable space the Dirac measure δ_x defined on that point. The multiplication $\mathfrak{m} : \mathfrak{S}^2 \stackrel{\bullet}{\to} \mathfrak{S}$ is defined through

$$\mathfrak{m}_X(M)(A) := \int_X \tau(A)\ M(d\tau)$$

whenever $(X, \mathcal{A})$ is a measurable space, $A \in \mathcal{A}$ is measurable, and $M \in \mathfrak{S}^2(X, \mathcal{A})$. Multiplication is well behaved in a topological context, as the statement below shows. Before stating and proving this result, integration with respect to the image of monad multiplication is characterized.

Lemma 3.2.2. *Let $(X, \mathcal{A})$ be a measurable space. Then*

$$(*) \int_X f\ d\mathfrak{m}_X(M) = \int_{\mathfrak{S}(X,\mathcal{A})} \big(\int_X f(x)\ \tau(dx)\big) M(d\tau)$$

whenever $M \in \mathfrak{S}^2(X, \mathcal{A})$, and $f : X \to \mathbb{R}$ is bounded and measurable.

Proof If $f = \chi_A$ is the indicator function of a measurable set $A \in \mathcal{A}$, then Equation $(*)$ is just the definition of $\mathfrak{m}_X(M)$. Thus the equation holds for step functions as well by the integral's linearity. If $f \geq 0$, then f may be approximated through a sequence of step functions by Proposition 1.3.5; so the assertion follows from the Monotone Convergence Theorem (Proposition 1.6.1). In the general case, decompose f into a positive and a negative part and apply the equation separately. □

This technical result helps in establishing continuity of the monad's multiplication on Polish spaces.

Proposition 3.2.3. *If X is a Polish space, then $\mathfrak{m}_X : \mathfrak{S}^2(X) \to \mathfrak{S}(X)$ is continuous in the respective topologies of weak convergence.*

Proof Let $M_n \rightharpoonup_w M$ for a sequence $(M_n)_{n\in\mathbb{N}}$ in $\mathfrak{S}^2(X)$. If $f : X \to \mathbb{R}$ is continuous and bounded, then by the definition of the topology of weak convergence, $\mu \mapsto \int_X f\ d\mu$ is a continuous map on $\mathfrak{S}(X)$; thus by Lemma 3.2.2

$$\begin{aligned}\int_X f(x)\ \mathfrak{m}_X(M_n)(dx) &= \int_{\mathfrak{S}(X)} \big(\int_X f(x)\ \tau(dx)\big)\ M_n(d\tau)\\ &\to \int_{\mathfrak{S}(X)} \big(\int_X f(x)\ \tau(dx)\big)\ M(d\tau)\\ &= \int_X f(x)\ \mathfrak{m}_X(M)(dx).\end{aligned}$$

Consequently, $\mathfrak{m}_X(M_n) \rightharpoonup_w \mathfrak{m}_X(M)$. □

Let $\langle \mathfrak{T}, \mathfrak{e}, \mathfrak{m} \rangle$ be a monad over category $\mathbf{C}$, then this monad defines a new category, the *Kleisli category* for this monad: the objects of this category are the objects of $\mathbf{C}$, and a morphism $a \to b$ in the new category is a morphism $a \to \mathfrak{T}(b)$ in $\mathbf{C}$. Let $f : a \to b$ and $g : b \to c$ be morphisms in the Kleisli category; then their composition $g * f$ is defined through

$$g * f := \mathfrak{m}_c \circ \mathfrak{T}(g) \circ f.$$

The laws for the monad make sure that the Kleisli composition has all the required properties for rendering this construction a category.

Applying this construction to the Giry monad $\langle \mathfrak{S}, \mathfrak{e}, \mathfrak{m} \rangle$, we obtain from Proposition 1.6.18 that

- the Kleisli morphisms are just the stochastic relations,
- the Kleisli product $L * K$ of two stochastic relations $K : (X, \mathcal{A}) \rightsquigarrow (Y, \mathcal{B})$ and $L : (Y, \mathcal{B}) \rightsquigarrow (Z, \mathcal{C})$ is just what is called the convolution of the corresponding transition subprobabilities in Probability Theory; see [5, §64]:

$$\begin{aligned}(L * K)(x)(E) &= \mathfrak{m}_Z(\mathfrak{S}(L) \circ K)(x)(E) \\ &= \int_{\mathfrak{S}(Z)} \tau(E)\ \mathfrak{S}(L)\big(K(x)\big)(d\tau) \\ &= \int_Y L(y)(E)\ K(x)(dy)\end{aligned}$$

by the definition of the Kleisli composition and by the Change of Variables formula (Lemma 1.6.20).

These observations will be fairly helpful for the discussion that follows now.

3.3 Randomized Congruences and Morphisms

All preparations are now in place for a definition of randomized congruences and randomized morphisms based on the Kleisli category for the Giry monad. In order to define congruences, we will first extend the stochastic relation $K : X \rightsquigarrow Y$, i.e., the map $K : X \to \mathfrak{S}(Y)$ to a map $\overline{K} : \mathfrak{S}(X) \to \mathfrak{S}(Y)$ (its Kleisli extension) in the canonical way (in fact, [60, 48] use this extension as a starting point). The extension renders a map for which a congruence can be defined in a way that is customary in Universal Algebra [13, 42].

We will, however, want to take care that we interlock the base space and the space of all subprobabilities on it properly, for otherwise the definition of a congruence would be too general to be of any use here. We will define randomized morphisms as those stochastic relations for which composition yields a commutative diagram, the composition being given through the Kleisli construction. In the same spirit there is some care to be exercised in not permitting too large a gap to open between the base space and its subprobabilities. It turns out that both constructions are really the randomizations of the constructions that are well-known in their non-randomized version (so that, roughly, considering Dirac measures δ_x in the randomized world corresponds very closely to the elements proper in the non-randomized context). Moreover, each randomized morphism yields a randomized congruence in a natural way.

Define the *Kleisli extension* $\overline{K}$ of a stochastic relation $K : (X, \mathcal{A}) \rightsquigarrow (Y, \mathcal{B})$ through

$$\overline{K} := \mathfrak{m}_{(Y,\mathcal{B})} \circ \mathfrak{S}\,(K)$$

with $\mathfrak{m}_{(Y,\mathcal{B})} : \mathfrak{S}^2(Y, \mathcal{B}) \to \mathfrak{S}\,(Y, \mathcal{B})$ as the multiplication in the Giry monad. We observe the following properties, which in particular relate the composition of this extension to the Kleisli extension of the Kleisli product.

Lemma 3.3.1. *Let $K : (X, \mathcal{A}) \rightsquigarrow (Y, \mathcal{B})$ and $L : (Y, \mathcal{B}) \rightsquigarrow (Z, \mathcal{C})$ be stochastic relations.*

a. $\overline{K}(\mu)(B) = \int_X K(x)(B)\ \mu(dx)$, *whenever $B \subseteq Y$ is a measurable set.*
b. $\overline{L} \circ \overline{K} = \overline{L * K}$.
c. If X and Y are Polish spaces endowed with their Borel sets, and K is continuous, so is $\overline{K}$.

Proof 1. Let $\mu \in \mathfrak{S}\,(X, \mathcal{A})$ be a subprobability on $(X, \mathcal{A})$, $B \in \mathcal{B}$ measurable; then

$$\overline{K}(\mu)(B) = \int_{\mathfrak{S}(X)} \tau(B)\ \mathfrak{S}\,(K)\,(\mu)(d\tau) = \int_X K(x)(B)\ \mu(dx).$$

The first equation follows from the definition of the multiplication in the Giry monad, the second one comes from the Change of Variables formula, Lemma 1.6.20. This establishes property *a*. Property *b*. is proved by a direct computation, taking into account that

$$\int_Y f\ d\overline{K}(\mu) = \int_X \int_Y f(y)\ K(x)(dy)\ \mu(dx)$$

whenever $f : Y \to \mathbb{R}$ is bounded and $\mathcal{B}$-measurable:

$$\begin{aligned}
(\overline{K} \circ \overline{L})(\mu)(E) &= \overline{L}(\overline{K})(\mu)(E)\\
&= \int_Y L(y)(E)\ \overline{K}(\mu)(dy)\\
&= \int_X \Big(\int_Y L(y)(E)\ K(x)(dy)\Big)\mu(dx)\\
&= \overline{(L * K)}(\mu)(E).
\end{aligned}$$

2. If $K : X \to \mathfrak{S}(Y)$ is continuous, so is $\mathfrak{S}(K) : \mathfrak{S}(X) \to \mathfrak{S}^2(Y)$. In fact, one notes first that

$$\int_{\mathfrak{S}(X)} F \, d\mathfrak{S}(K)(\mu) = \int_X F \circ K \, d\mu$$

whenever $F : \mathfrak{S}(X) \to \mathbb{R}$ is measurable and bounded, by the Change of Variables formula from Lemma 1.6.20. Consequently, if F is continuous, then $F \circ K$ is, so that for $\mu_n \rightharpoonup_w \mu$

$$\begin{aligned}\int_{\mathfrak{S}(X)} F \, d\mathfrak{S}(K)(\mu_n) &= \int_X F \circ K \, d\mu_n \\ &\to \int_X F \circ K \, d\mu \\ &= \int_{\mathfrak{S}(X)} F \, d\mathfrak{S}(K)(\mu).\end{aligned}$$

Consequently, assertion c follows from the observation in Proposition 3.2.3 that $\mathfrak{m}_Y$ is continuous in the weak topology whenever Y is Polish. □

An alternative proof of part b. takes only properties of the monad's multiplication into account:

$$\begin{aligned}\overline{L} \circ \overline{K} &= \mathfrak{m}_{(Z,\mathcal{C})} \circ \mathfrak{S}(L) \circ \mathfrak{m}_{(Y,\mathcal{B})} \circ \mathfrak{S}(K) \\ &\overset{(*)}{=} \mathfrak{m}_{(Z,\mathcal{C})} \circ \mathfrak{m}_{\mathfrak{S}(Z,\mathcal{C})} \circ \mathfrak{S}^2(L) \circ \mathfrak{S}(K) \\ &\overset{(\dagger)}{=} \mathfrak{m}_{(Z,\mathcal{C})} \circ \mathfrak{S}\left(\mathfrak{m}_{(Z,\mathcal{C})}\right) \circ \mathfrak{S}^2(L) \circ \mathfrak{S}(K) \\ &\overset{(\ddagger)}{=} \mathfrak{m}_{(Z,\mathcal{C})} \circ \mathfrak{S}\left(\mathfrak{m}_{(Z,\mathcal{C})} \circ \mathfrak{S}(L) \circ K\right) \\ &= \mathfrak{m}_{(Z,\mathcal{C})} \circ \mathfrak{S}(L * K) \\ &= \overline{L * K},\end{aligned}$$

because $\mathfrak{m}$ is a natural transformation $(*)$ and constitutes the multiplication in the monad $(\dagger)$, and because $\mathfrak{S}$ is a functor $(\ddagger)$.

We obtain the following from the integral representation of the Kleisli extension $\overline{K}$ in Lemma 3.3.1 through Proposition 1.6.15 as an immediate consequence:

Corollary 3.3.2. *Whenever $K : (X, \mathcal{A}) \rightsquigarrow (Y, \mathcal{B})$ is a stochastic relation, $\overline{K} : \mathfrak{S}(X, \mathcal{A}) \to \mathfrak{S}(Y, \mathcal{B})$ is a $\mathcal{A}^\bullet$-$\mathcal{B}^\bullet$- measurable map.* □

3.3.1 Randomizing Morphisms

We define randomized morphisms as those morphisms that correspond to the Kleisli category for the Giry monad through stochastic relations. Let $\mathsf{K} = (X, Y, K)$ and $\mathsf{L} = (A, B, L)$ be stochastic relations, and assume that $\Phi : X \rightsquigarrow A$ and $\Psi : Y \rightsquigarrow B$.

Suppose we interpret X and Y as alphabets for input and output, respectively, of a system, and that we observe inputs and outputs not directly but rather assign distributions to them through Φ or Ψ. Thus input $x \in X$ is being observed with probability $\Phi(x)$, and output y with probability $\Psi(y)$. Let $E \subseteq B$ be a measurable set; then

$$\int_A L(a)(E)\ \Phi(x)(da)$$

is the expected probability of hitting set E when the input to L is being averaged over through Φ; similarly,

$$\int_Y \Psi(y)(E)\ K(x)(dy)$$

is the probability of hitting this set subject to the output of K being averaged over through Ψ. The observations are compatible whenever both probabilities coincide.

This consideration leads to the definition of a randomized morphism between stochastic relations.

Definition 3.3.3. *Assume that* $\mathsf{K} = (X, Y, K)$ *and* $\mathsf{L} = (A, B, L)$ *are stochastic relations;* $\mathsf{F} = (\Phi, \Psi)$ *is called a* randomized morphism *between* K *and* L *(in symbols:* $\mathsf{F} : \mathsf{K} \curvearrowright \mathsf{L}$*) iff*

a. $\Phi : X \rightsquigarrow A$ *and* $\Psi : Y \rightsquigarrow B$ *are stochastic relations,*
b. $\overline{\Phi}$ *and* $\overline{\Psi}$ *are onto,*
c. $\Psi * K = L * \Phi$ *holds.*

Consequently, a randomized morphism makes the diagram on the left-hand side below commutative, products taken in the Kleisli category associated with the Giry monad; morphisms for the Kleisli product are indicated through a double shafted arrow. For comparison we display also the diagram for $\mathsf{f} = (f, g) : \mathsf{K} \to \mathsf{L}$ to be a non-randomized morphism. Thus $f : X \to A$ and $g : Y \to B$ are surjective Borel maps, and the diagram is commutative using ordinary map composition.

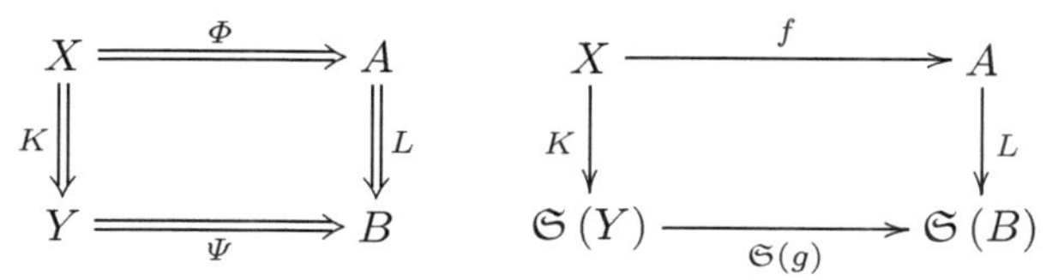

Spelling the condition out for a randomized morphism in terms of integrals, we see that

$$\big(\Psi * K\big)(x)(D) = \int_Y \Psi(y)(D)\ K(x)(dy) =$$

$$\int_A L(a)(D)\ \Phi(x)(da) = \big(L * \Phi\big)(x)(D)$$

should hold for $x \in X$ and the Borel set $D \in \mathcal{B}(B)$. Recall that expanding the diagram on the right hand side for the non-randomized version entails

$$L(f(x))(D) = \big(L \circ f\big)(x)(D) = \big(\mathfrak{S}\,(g) \circ K)\big)(x)(D) = K(x)\big(g^{-1}\,[D]\big);$$

thus the probability of hitting an element of D after input $f(x)$ in L equals the probability of hitting an element of the inverse image $g^{-1}\,[D]$ after input $x \in X$. Comparing both definitions shows that the randomized version entertains an additional level of averaging.

Surjectivity of the maps underlying the morphism f is introduced in order to make sure that each element in the range originates from some element in the domain. Similarly, surjectivity of the Kleisli transforms of the stochastic relations underlying a randomized morphism assures that each distribution on the range can be accounted for.

Randomized morphisms are related to their non-randomized cousins in a similar way to that in which randomized policies are related to policies in stochastic dynamic programming: whereas policies assign each state an action to take, a randomized policy assigns each state a probability distribution over actions.

Returning to the general discussion, let $\mathsf{F} = (\Phi, \Psi) : \mathsf{K} \curvearrowright \mathsf{L}$ and $\mathsf{G} = (\Gamma, \Theta) : \mathsf{L} \curvearrowright \mathsf{M}$ be randomized morphisms; then their composition $\mathsf{G} * \mathsf{F} : \mathsf{K} \curvearrowright \mathsf{L}$ is defined in the obvious way through $\mathsf{G} * \mathsf{F} := (\Gamma * \Phi, \Theta * \Psi)$.

Proposition 3.3.4. *Stochastic relations over general measurable spaces as objects and with randomized morphisms as morphisms form a category when the composition of morphisms is defined through the Kleisli construction. The same is true for stochastic relations over Standard Borel spaces or over analytic spaces.*

Proof The crucial property is the surjectivity of the induced maps. This property, however, follows immediately from Lemma 3.3.1, property b.□

Randomized morphisms have non-randomized morphisms as special cases. It is instructive to look at the proof in some detail.

Lemma 3.3.5. *Let* $\mathsf{K} = (X, Y, K), \mathsf{L} = (A, B, L)$ *be stochastic relations, and assume that* $\mathsf{f} = (f, g) : \mathsf{K} \to \mathsf{L}$ *is a non-randomized morphism. Put* $\Phi := \mathfrak{e}_A \circ f, \Psi := \mathfrak{e}_B \circ g$*; then* $\mathsf{F}_\mathsf{f} : (\Phi, \Psi) : \mathsf{K} \curvearrowright \mathsf{L}$ *is a randomized morphism.*

Proof 1. Note that $\Phi(x) = \delta_{f(x)}$ and $\Psi(y) = \delta_{g(y)}$. Because we know for $D \subseteq A$ measurable that $\delta_{f(x)}(D) = \chi_{f^{-1}[D]}(x)$, we obtain

$$\begin{aligned}
\overline{\Phi}(\mu)(D) &= \int_X \Phi(x)(D)\ \mu(dx) \\
&= \int_X \delta_{f(x)}(D)\ \mu(dx) \\
&= \mu(f^{-1}\left[D\right]) \\
&= \mathfrak{S}\left(f\right)(\mu)(D).
\end{aligned}$$

Thus we obtain $\overline{\Phi} = \mathfrak{S}\left(f\right)$, and similarly $\overline{\Psi} = \mathfrak{S}\left(g\right)$. Both $\overline{\Phi}$ and $\overline{\Psi}$ are onto on account of f and g being onto by Proposition 1.6.19.

2. Let $\mu \in \mathfrak{S}\left(X\right)$ and $G \in \mathcal{B}(B)$; then

$$\begin{aligned}
\overline{\Psi}(\overline{K})(\mu)(G) &= \int_Y \Psi(y)(G)\ \overline{K}(\mu)(dy) \\
&= \overline{K}(\mu)(g^{-1}\left[G\right]) \\
&= \int_X K(x)(g^{-1}\left[G\right])\ \mu(dx) \\
&= \int_X \mathfrak{S}\left(g\right)(K(x))(G)\ \mu(dx) \\
&= \int_X L(f(x))(G)\ \mu(dx).
\end{aligned}$$

The last equality follows from $L \circ f = \mathfrak{S}\left(g\right) \circ K$. Similarly,

$$\begin{aligned}
\overline{L}(\overline{\Phi}(\mu))(G) &= \int_A L(a)(G)\ \overline{\Phi}(\mu)(da) \\
&= \int_A L(a)(G)\ \mathfrak{S}\left(f\right)(\mu)(da) \\
&= \int_X L(f(x))(G)\ \mu(dx)
\end{aligned}$$

by the Change of Variables formula (Lemma 1.6.20). Thus we see that $\overline{\Psi} \circ \overline{K} = \overline{L} \circ \overline{\Phi}$ holds, from which $\overline{\Psi * K} = \overline{L * \Phi}$ is obtained through Lemma 3.3.1. But now

$$\bigl(\Psi * K\bigr)(x) = \overline{\Psi * K}(\delta_x) = \overline{L * \Phi}(\delta_x) = \bigl(L * \Phi\bigr)(x)$$

follows for each $x \in X$. □

This statement may be proven alternatively through properties of a monad. Integration, however, provides probably a bit more insight into the probabilistic mechanisms behind the map $\mathsf{f} \mapsto \mathsf{F}_\mathsf{f}$.

Corollary 3.3.6. *Map each stochastic relation* K *over analytic spaces to itself, and map a non-randomized morphism* $\mathsf{f} : \mathsf{K} \to \mathsf{L}$ *to* $\mathsf{F}_\mathsf{f} : \mathsf{K} \curvearrowright \mathsf{L}$*. Then this constitutes a functor* $\mathfrak{F}$ *from the category of stochastic relations with non-randomized morphisms over analytic spaces to the category of stochastic relations with randomized morphisms over analytic spaces.*

Proof This is essentially due to a favorable property of the unit in a monad. Let X, Y, Z be the input spaces for the three relations, and $f : X \to Y, g : Y \to Z$ be Borel maps; then

$$\begin{aligned}\big((\mathfrak{e}_Z \circ g) * (\mathfrak{e}_Y \circ f)\big)(x)(E) &= \int_Y (\mathfrak{e}_Z \circ g)(y)(E)\ (\mathfrak{e}_Y \circ f)(x)(dy)\\ &= \int_Y \delta_{g(y)}(E)\ \delta_{f(x)}(dy)\\ &= \delta_{(g\circ f)(x)}(E)\\ &= \big(\mathfrak{e}_Z \circ (g \circ f)\big)(x)(E).\end{aligned}$$

Consequently,

$$\mathfrak{e}_Z \circ (g \circ f) = (\mathfrak{e}_Z \circ g) * (\mathfrak{e}_Y \circ f)$$

is inferred. From this,

$$\mathfrak{F}(\mathsf{g} \circ \mathsf{f}) = \mathfrak{F}(\mathsf{g}) * \mathfrak{F}(\mathsf{f})$$

is derived. All other properties of a functor are trivial. □

The functor $\mathfrak{F}$ forgets the deterministic nature of a non-randomized morphism and interprets deterministic behavior as randomized, albeit with a somewhat limited set of possibilities to choose from.

3.3.2 Relating Randomized Morphisms and Congruences

Recall from Section 1.7.3 that the pair (α, β) of smooth equivalence relations on X respectively Y is a *congruence* for $\mathsf{K} = (X, Y, K)$ iff $K(x)(B) = K(x')(B)$ holds whenever $x\ \alpha\ x'$, and $B \in \boldsymbol{\Sigma}\big(\mathcal{B}(Y), \beta\big)$. In contrast, a randomized congruence for K is based on a pair of smooth equivalence relations (ρ, τ) on $\mathfrak{S}(X)$ respectively $\mathfrak{S}(Y)$. We require both equivalences to be neargrounded, so that we can exercise some control from the base space over these relations. In particular we are able to relate the relation to the invariant subsets on the base space, which is not only intuitively appealing but provides the desired bracket between these spaces.

Before defining a randomized congruence, let us see what happens when we lift a congruence for a stochastic relation.

Lemma 3.3.7. *Let* $\mathsf{K} = (X, Y, K)$ *be a stochastic relation over analytic spaces, and assume that* $\mathsf{c} := (\alpha, \beta)$ *is a pair of smooth equivalence relations on* X *respectively* Y. *Put* $\overline{\mathsf{c}} := (\overline{\alpha}, \overline{\beta})$. *These conditions are equivalent for* K*:*

a. c *is a congruence.*
b. $\overline{\mathsf{c}}$ *has these properties*

a. $\overline{K}(\mu)\ \overline{\beta}\ \overline{K}(\mu')$ *whenever* $\mu\ \overline{\alpha}\ \mu'$,
b. $\overline{\alpha}$ *and* $\overline{\beta}$ *are near-grounded.*

Proof 1. $a \Rightarrow b$: The map $x \mapsto K(x)(B)$ is $\mathbf{\Sigma}(\mathcal{B}(X), \alpha)$-$\mathcal{B}(\mathbb{R})$-measurable whenever $B \in \mathbf{\Sigma}(\mathcal{B}(Y), \beta)$ is a β-invariant Borel set of Y, by Proposition 1.7.20. Taking this into account, an application of Lemma 1.8.2 now yields for the Kleisli extensions

$$\overline{K}(\mu)(B) = \int_X K(x)(B)\ \mu(dx) = \int_X K(x)(B)\ \mu'(dx) = \overline{K}(\mu')(B)$$

whenever $\mu\ \overline{\alpha}\ \mu'$ and $B \in \mathbf{\Sigma}(\mathcal{B}(Y), \beta)$. Thus $\overline{K}(\mu)\ \overline{\beta}\ \overline{K}(\mu)$. It is apparent from Corollary 1.8.3 that $\overline{\alpha}$ and $\overline{\beta}$ are near-grounded.

2. $b \Rightarrow a$: Observe that $\overline{K}(\delta_x)\ \overline{\beta}\ \overline{K}(\delta_{x'})$, provided $x\ \alpha\ x'$, because the latter implies $\delta_x\ \overline{\alpha}\ \delta_{x'}$ by Corollary 1.8.3. But then

$$K(x)(B) = \overline{K}(\delta_x)(B) = \overline{K}(\delta_{x'})(B) = K(x')(B)$$

whenever $B \in \mathbf{\Sigma}(\mathcal{B}(Y), \beta)$ is a β-invariant Borel set in Y. Thus (α, β) is a congruence for K. □

This suggests the following definition of a randomized congruence.

Definition 3.3.8. *Let* $\mathsf{C} := (\rho, \tau)$ *be a pair of smooth equivalence relations on* $\mathfrak{S}(X)$ *respectively* $\mathfrak{S}(Y)$ *with* X *and* Y *analytic.* C *is said to be a* randomized congruence *for the stochastic relation* $\mathsf{K} = (X, Y, K)$ *iff*

a. $\mu\ \rho\ \mu'$ *implies* $\overline{K}(\mu)\ \tau\ \overline{K}(\mu')$ *for all* $\mu, \mu' \in \mathfrak{S}(X)$,
b. both ρ *and* τ *are near-grounded.*

Congruence C *is said to be* grounded *iff both* ρ *and* τ *are grounded.*

Apart from near-groundedness, a congruence has the usual properties which are required for a relation that is supposed to reflect the structure in a coalgebraic setting [42]. A stochastic relation can be considered as a coalgebra for functor $\mathfrak{S}$; hence a coalgebraic view of congruences is interesting. Rutten [72] argues that congruences correspond in universal coalgebra to the largest bisimulation relation. This point of view is vital in the discussion of coalgebras for the powerset functor. It needs to be adapted to the situation at hand, however, given that bisimulations for stochastic relations have a somewhat more complicated structure than one would expect, and given that $\mathfrak{S}$ does not enjoy some properties (such as not dealing decently with weak pullbacks) that are usually taken for granted when investigating coalgebras [22].

We see from Lemma 3.3.7 that lifting a congruence to the space of subprobabilities yields a randomized congruence, as one would reasonably expect. A converse property is available as well: take a randomized congruence; then its trace on the base space will define a congruence.

Corollary 3.3.9. *If* $\mathsf{C} = (\rho, \tau)$ *is a randomized congruence for the analytic stochastic relation* $\mathsf{K} = (X, Y, K)$*, then* $\lfloor \mathsf{C} \rfloor := (\lfloor \rho \rfloor, \lfloor \tau \rfloor)$ *is a congruence for* K.

Proof Let $x, x' \in X$ with $x \lfloor \rho \rfloor x'$, and thus with $\delta_x \; \rho \; \delta_{x'}$. Consequently,

$$\big(K(x) = \big)\overline{K}(\delta_x) \; \tau \; \overline{K}(\delta_{x'})\big(= K(x')\big);$$

thus, because τ is near-grounded, $K(x) \; \overline{\lfloor \tau \rfloor} \; K(x')$. This means that $K(x)(B)$ equals $K(x')(B)$ for $B \in \mathbf{\Sigma}\big(\mathcal{B}(Y), \lfloor \tau \rfloor\big)$. □

If $\mathsf{F} := (\Phi, \Psi)$ is a randomized morphism $\mathsf{K} \curvearrowright \mathsf{L}$, we define as usual its kernel $\mathsf{ker}(\mathsf{F})$ through

$$\mathsf{ker}(\mathsf{F}) := (\mathsf{ker}(\Phi), \mathsf{ker}(\Psi)).$$

Similarly, the kernel

$$\mathsf{ker}\big(\overline{\mathsf{F}}\big) := \big(\mathsf{ker}\big(\overline{\Phi}\big), \mathsf{ker}\big(\overline{\Psi}\big)\big)$$

is defined for the Kleisli version. It is clear that both kernels are based on smooth equivalence relations, since Φ and Ψ as well as $\overline{\Phi}$ and $\overline{\Psi}$ are Borel maps between analytic spaces.

Randomized congruences are used for defining near-grounded morphisms. Before doing this, we have a look at the kernel of a stochastic relation.

Lemma 3.3.10. *Let* $\Phi : X \rightsquigarrow A$ *be a stochastic relation between the analytic spaces* X *and* A*. Then* $\overline{\mathsf{ker}(\Phi)} \subseteq \mathsf{ker}\big(\overline{\Phi}\big)$.

Proof Let $\langle \mu, \mu' \rangle \in \overline{\mathsf{ker}(\Phi)}$, so that $\mu(B) = \mu'(B)$ whenever $B \subseteq X$ is $\mathsf{ker}(\Phi)$-invariant and Borel. Thus $\int_X f \, d\mu = \int_X f \, d\mu'$ whenever $f : X \to \mathbb{R}$ is bounded and $\mathbf{\Sigma}\big(\mathcal{B}(X), \mathsf{ker}(\Phi)\big)$-measurable by Lemma 1.8.2. It is plain that $x \mapsto \Phi(x)(D)$ is $\mathbf{\Sigma}\big(\mathcal{B}(X), \mathsf{ker}(\Phi)\big)$-measurable for each Borel set $D \subseteq A$. Consequently, $\overline{\Phi}(\mu) = \overline{\Phi}(\mu')$, so that $\langle \mu, \mu' \rangle \in \mathsf{ker}\big(\overline{\Phi}\big)$. □

Example 3.3.11. Let $f : [0,1]^2 \to [0,1]$ so that for each $x \in [0,1]$ the map $f(x, \cdot)$ is continuous and strictly increasing; we call $f(x, \cdot)$ the *orbit* of f at x. Put

$$\Phi(x)(A) := \int_A f(x,y)\ dy,$$

thus $\Phi : [0,1] \rightsquigarrow [0,1]$. Let μ be a distribution on $[0,1]$, and $A \subseteq [0,1]$ be a Borel set; then

$$\begin{aligned}\overline{\Phi}(\mu)(A) &= \int_0^1 \Phi(x)(A)\ \mu(dx)\\ &= \int_A \int_0^1 f(x,y)\ \mu(dx)\ dy;\end{aligned}$$

interchanging the order of integration being admissible by Fubini's Theorem. Since

$$\mathsf{ker}\,(\Phi) = \{\langle x, x'\rangle \mid f(x,\cdot) = f(x',\cdot)\},$$

the $\mathsf{ker}\,(\Phi)$-classes are just the orbits of f. Now let μ and μ' be distributions that coincide on the orbits of f; then

$$\int_0^1 f(x,y)\ \mu(dx) = \int_0^1 f(x,y)\ \mu'(dx)$$

for almost all $y \in [0,1]$. $\dashv$

Definition 3.3.12. *The randomized morphism* $\mathsf{F} = (\Phi, \Psi) : \mathsf{K} \curvearrowright \mathsf{L}$ *is called* near-grounded *iff both* $\mathsf{ker}\left(\overline{\Phi}\right) = \overline{\mathsf{ker}\,(\Phi)}$ *and* $\mathsf{ker}\left(\overline{\Psi}\right) = \overline{\mathsf{ker}\,(\Psi)}$.

Let $(\Phi, \Psi) : (X, Y, K) \curvearrowright (A, B, L)$ be a near-grounded morphism; then this property is witnessed through both $\mathsf{ker}\left(\overline{\Phi}\right) \subseteq \overline{\mathsf{ker}\,(\Phi)}$ and $\mathsf{ker}\left(\overline{\Psi}\right) \subseteq \overline{\mathsf{ker}\,(\Psi)}$, since the other inclusion is always true by Lemma 3.3.10. Having a look at $\overline{\Phi}$, one sees that the behavior of $\overline{\Phi}$ is uniquely determined by the subprobabilities on its kernel, since

$$\overline{\Phi}(\mu) = \overline{\Phi}(\mu') \Leftrightarrow \forall B \in \mathbf{\Sigma}\big(\mathcal{B}(X), \mathsf{ker}\,(\Phi)\big) : \mu(B) = \mu'(B).$$

We will investigate randomized morphisms and their interplay with randomized congruences now a bit further.

Lemma 3.3.13. *Let* $\mathsf{F} : \mathsf{K} \curvearrowright \mathsf{L}$ *be a near-grounded randomized morphism with* K *analytic. Then* $\mathsf{ker}\left(\overline{\mathsf{F}}\right)$ *is a randomized congruence for* K.

Proof Assume $\overline{\Phi}(\mu) = \overline{\Phi}(\mu')$; then Lemma 3.3.1 yields $\overline{\Psi}(\overline{K}(\mu)) = \overline{\Psi}(\overline{K}(\mu'))$. Because F is near-grounded, both $\mathsf{ker}\left(\overline{\Phi}\right)$ and $\mathsf{ker}\left(\overline{\Psi}\right)$ are near-grounded. This is so since $\left\lfloor \mathsf{ker}\left(\overline{\Phi}\right)\right\rfloor = \mathsf{ker}\,(\Phi)$ holds, for

$$x \left\lfloor \mathsf{ker}\left(\overline{\Phi}\right)\right\rfloor x' \Leftrightarrow \overline{\Phi}(\delta_x) = \overline{\Phi}(\delta_{x'}) \Leftrightarrow \Phi(x) = \Phi(x') \Leftrightarrow x\ \mathsf{ker}\,(\Phi)\ x'.$$

Thus $\mathbf{\Sigma}\big(\mathcal{B}(X), \mathsf{ker}\,(\Phi)\big) = \mathbf{\Sigma}\big(\mathcal{B}(X), \left\lfloor \mathsf{ker}\left(\overline{\Phi}\right)\right\rfloor\big)$. Since F is near-grounded, we know $\mathsf{ker}\left(\overline{\Phi}\right) \subseteq \overline{\mathsf{ker}\,(\Phi)}$; consequently we have

$$\overline{\Phi}(\mu) = \overline{\Phi}(\mu') \Rightarrow \forall B \in \mathbf{\Sigma}\big(\mathcal{B}(X), \lfloor \mathsf{ker}\,(\overline{\Phi}) \rfloor\big) : \mu(B) = \mu'(B)$$
$$\Leftrightarrow \mu \,\lfloor \mathsf{ker}\,(\overline{\Phi}) \rfloor\, \mu',$$

yielding $\mathsf{ker}\,(\overline{\Phi}) \subseteq \overline{\lfloor \mathsf{ker}\,(\overline{\Phi}) \rfloor}$; similarly for Ψ. Thus both $\mathsf{ker}\,(\overline{\Phi})$ and $\mathsf{ker}\,(\overline{\Psi})$ are near-grounded. □

This implies that the kernel of a randomized morphism for a stochastic relation is a congruence for that relation as well. Notice the difference: while being a randomized congruence for K addresses the map $\overline{K}$, being a congruence for K entails that testing K on invariant Borel sets is necessary.

We did postulate near-groundedness above as a sensible condition to interlock the behavior on X and on $\mathfrak{S}\,(X)$ in a suitable manner. The question arises about whether groundedness would also be a suitable condition. It turns out that this condition is way too strong, since it only reflects the well known phenomenon of non-randomized morphisms suitably dressed up.

Lemma 3.3.14. *Given Standard Borel spaces X and Y, let $\Phi : X \rightsquigarrow Y$ be a stochastic relation such that $\overline{\Phi}$ is onto. Then $\mathsf{ker}\,(\overline{\Phi})$ is grounded iff there exists a surjective Borel map $f : X \to Y$ with $\Phi = \mathfrak{e}_Y \circ f$.*

Proof 0. Because of Proposition 1.4.19 we may and do assume that Φ and f are continuous.

1. Proposition 1.8.17 implies that $\mathfrak{e}_Y \circ f$ defines a grounded equivalence relation on $\mathfrak{S}\,(X)$ whenever f is continuous and onto.

2. Now let Φ be continuous with surjective Kleisli extension $\overline{\Phi}$ so that $\mathsf{ker}\,(\overline{\Phi})$ is grounded. Then Proposition 1.8.17 shows that $\overline{\Phi}$ maps Dirac measures to Dirac measures again. Thus $\Phi(x) = \overline{\Phi}(\delta_x) = \delta_y$ for some $y \in Y$; moreover, since Φ is continuous, $\delta_{f(x)} := \Phi(x)$ defines a continuous map $f : X \to Y$ such that $\overline{\Phi}(\mu) = \mathfrak{S}\,(f)\,(\mu)$. Hence surjectivity of f remains to be shown.

3. $\mathfrak{S}\,(f)$ is surjective; so given $y \in Y$ there exists $\mu \in \mathfrak{S}\,(X)$ with $\mathfrak{S}\,(f)\,(\mu) = \delta_y$. Hence we have $\delta_y(B) = \mu(f^{-1}[B])$ for each Borel set $B \in \mathcal{B}(Y)$. In particular, $\mu(f^{-1}[\{y\}]) = 1$, so that $f^{-1}[\{y\}] \neq \emptyset$. Thus there exists $x \in X$ with $f(x) = y$. □

Proposition 3.3.15. *Let $\mathsf{F} : \mathsf{K} \curvearrowright \mathsf{L}$ be a randomized morphism. Then $\mathsf{ker}\,(\overline{\mathsf{F}})$ is grounded iff there exists a morphism $\mathsf{f} : \mathsf{K} \to \mathsf{L}$ such that $\mathsf{F} = \mathsf{F}_{\mathsf{f}}$, i.e., F is deterministic.*

Proof This follows immediately from Lemma 3.3.14 in conjunction with Lemma 3.3.5. □

There is, however, a useful interconnection between non-randomized congruences and kernels of near-randomized morphisms. This is quite easy to establish and will be rather helpful below.

Proposition 3.3.16. *Let $\mathsf{F} : \mathsf{K} \curvearrowright \mathsf{L}$ be a near-grounded morphism. Then $\mathsf{ker}\,(\mathsf{F})$ is a congruence for K, and $\lfloor \mathsf{ker}\,(\overline{\mathsf{F}}) \rfloor = \mathsf{ker}\,(\mathsf{F})$.*

Proof From Lemma 3.3.13 we infer that $\mathsf{ker}\left(\overline{\mathsf{F}}\right)$ is a randomized congruence. Let $\mathsf{F} = (\Phi, \Psi)$. A simple calculation shows that $\left\lfloor \mathsf{ker}\left(\overline{\Phi}\right)\right\rfloor = \mathsf{ker}\left(\Phi\right)$ and $\left\lfloor \mathsf{ker}\left(\overline{\Psi}\right)\right\rfloor = \mathsf{ker}\left(\Psi\right)$ both hold. □

All preparations are in place now to investigate factoring through a randomized congruence.

3.3.3 Factoring Through a Randomized Congruence

When showing that congruences are really the kernels of morphisms, one readily proceeds to discuss factoring. A morphism in universal (co-)algebra can be factored uniquely through the factor space associated with its kernel.

We will look into this problem by investigating first the factor of the Kleisli extension of a stochastic relation with respect to a general congruence. Alas, this is not automatically the Kleisli extension of a factored relation, but if the relation is grounded, the isomorphism from Section 1.8.1 becomes helpful and renders this map a factor, at least up to a Borel isomorphism. In general, we show that there is a one-to-one correspondence of near-grounded randomized congruences and the kernels of near-grounded randomized morphisms. This relationship is investigated now.

Let $\mathsf{K} = (X, Y, K)$ be a stochastic relation over the analytic spaces X and Y, and assume that $\mathsf{D} = (\rho, \tau)$ is a randomized congruence for K. Define

$$\overline{K}/\mathsf{D} : \begin{cases} \mathfrak{S}(X)/\rho & \to \mathfrak{S}(Y)/\tau \\ [\mu]_\rho & \mapsto \left[\overline{K}(\mu)\right]_\tau . \end{cases}$$

It is clear from Lemma 1.8.9 that the map $\overline{K}/\mathsf{D}$ is well defined, and that it constitutes a measurable map. $\overline{K}/\mathsf{D}$ is defined as the canonical map on factors that renders

$$\begin{array}{ccc} \mathfrak{S}(X) & \xrightarrow{\overline{K}} & \mathfrak{S}(Y) \\ {\scriptstyle \eta_\rho}\downarrow & & \downarrow{\scriptstyle \eta_\tau} \\ \mathfrak{S}(X)/\rho & \xrightarrow[\overline{K}/\mathsf{D}]{} & \mathfrak{S}(Y)/\tau \end{array}$$

commutative. This diagram is a diagram of maps between sets, and it is not obvious whether or not it is the Kleisli extension of a stochastic relation at all. Consider on the other hand the stochastic relation

$$\mathsf{K}/\lfloor \mathsf{D} \rfloor := (X/\lfloor \rho \rfloor, Y/\lfloor \tau \rfloor, K/\lfloor \mathsf{D} \rfloor).$$

Then the diagram below is obtained. Note that ∂_ρ (which is defined on page 59) is a Borel map by Lemma 1.8.9, because D is near-grounded.

$$
\begin{array}{ccc}
(\bigstar)\quad \mathfrak{S}(X)/\rho & \xrightarrow{\ \partial_\rho\ } & \mathfrak{S}(X/\lfloor\rho\rfloor) \\
\big\downarrow{\scriptstyle \overline{K}/\mathsf{D}} & & \big\downarrow{\scriptstyle \overline{K/\lfloor\mathsf{D}\rfloor}} \\
\mathfrak{S}(Y)/\tau & \xrightarrow[\ \partial_\tau\ ]{} & \mathfrak{S}(Y/\lfloor\tau\rfloor)
\end{array}
$$

The diagram commutes: take $\mu \in \mathfrak{S}(X)$ and a Borel set $B \in \mathcal{B}(Y/\lfloor\tau\rfloor)$; then

$$
\begin{aligned}
\overline{K/\lfloor\mathsf{D}\rfloor}\left(\partial_\rho([\mu]_\rho)\right)(B) &\overset{(\sharp)}{=} \int_{X/\lfloor\rho\rfloor} (K/\lfloor\mathsf{D}\rfloor)(\zeta)(B)\ \partial_\rho([\mu]_\rho)(d\zeta) \\
&\overset{(\flat)}{=} \int_{X/\lfloor\rho\rfloor} (K/\lfloor\mathsf{D}\rfloor)(\zeta)(B)\ \mathfrak{S}\left(\eta_{\lfloor\rho\rfloor}\right)(\mu)(d\zeta) \\
&\overset{(\natural)}{=} \int_X (K/\lfloor\mathsf{D}\rfloor)([x]_{\lfloor\rho\rfloor})(B)\ \mu(dx) \\
&= \int_X K(x)(\eta_{\lfloor\tau\rfloor}^{-1}[B])\ \mu(dx) \\
&= \overline{K}(\mu)(\eta_{\lfloor\tau\rfloor}^{-1}[B]) \\
&= \partial_\tau\left(\left[\overline{K}(\mu)\right]_\tau\right)(B) \\
&= \partial_\tau\left((\overline{K}/\mathsf{D})([\mu]_\rho)\right)(B).
\end{aligned}
$$

Equation ($\sharp$) is the definition of $K/\lfloor\mathsf{D}\rfloor \mapsto \overline{K/\lfloor\mathsf{D}\rfloor}$, equation ($\flat$) expands the definition of map ∂_ρ, the next equation ($\natural$) applies the Change of Variables formula. The last group of equations uses the definition of map ∂_τ and the construction of the factor map $\overline{K}/\mathsf{D}$ again.

Associate with the randomized congruence $\mathsf{D} = (\rho, \tau)$ the maps

$$
\begin{aligned}
E_\rho &: X \ni x \mapsto \mathfrak{e}_{X/\lfloor\rho\rfloor}([x]_{\lfloor\rho\rfloor}) \in \mathfrak{S}(X/\lfloor\rho\rfloor), \\
E_\tau &: Y \ni y \mapsto \mathfrak{e}_{Y/\lfloor\tau\rfloor}([y]_{\lfloor\tau\rfloor}) \in \mathfrak{S}(Y/\lfloor\tau\rfloor).
\end{aligned}
$$

Thus $E_\rho = \mathfrak{e}_{X/\lfloor\rho\rfloor} \circ \eta_{\lfloor\rho\rfloor}, E_\tau = \mathfrak{e}_{Y/\lfloor\tau\rfloor} \circ \eta_{\lfloor\tau\rfloor}$. Define $\mathsf{E}_\mathsf{D} := (E_\rho, E_\tau)$.

If D is grounded, both components (ρ, τ) are, so that in this case both ∂_ρ and ∂_τ are Borel isomorphisms by Proposition 1.8.12. Thus $\overline{K}/\mathsf{D}$ is the Kleisli map of a stochastic relation up to a Borel isomorphism, provided D is a grounded congruence.

Proposition 3.3.17. *Let* D *be a congruence for the stochastic relation* K*. Then*

a. $\mathsf{E}_\mathsf{D} : \mathsf{K} \curvearrowright \mathsf{K}/\lfloor\mathsf{D}\rfloor$ *is a randomized morphism with* $\mathsf{ker}(\mathsf{E}_\mathsf{D}) = \lfloor\mathsf{D}\rfloor$.
b. If D *is grounded, the Kleisli extension* $\overline{\mathsf{K}/\lfloor\mathsf{D}\rfloor}$ *of* $\mathsf{K}/\lfloor\mathsf{D}\rfloor$ *is Borel isomorphic to the factor* $\overline{\mathsf{K}}/\mathsf{D}$ *of the Kleisli extension of* K *with respect to* D.

Proof 1. We infer from Corollary 3.3.9 that $\lfloor \mathsf{D} \rfloor$ is a congruence for K. This implies that $(\eta_{\lfloor\rho\rfloor}, \eta_{\lfloor\tau\rfloor}) : \mathsf{K} \to \mathsf{K}/\lfloor \mathsf{D} \rfloor$ is a morphism. Lemma 3.3.5 shows now that E_D is a randomized morphism.

2. If D is grounded, diagram ($\bigstar$) above gives $\overline{\mathsf{K}/\lfloor \mathsf{D} \rfloor} = \partial_\rho^{-1} \circ \overline{\mathsf{K}}/\mathsf{D} \circ \partial_\tau$ for the Borel isomorphisms ∂_ρ and ∂_τ. □

Now let $\mathsf{F} = (\Phi, \Psi) : \mathsf{K} \curvearrowright \mathsf{L}$ be a near-grounded randomized morphism; then we conclude from Lemma 3.3.13 and from Proposition 3.3.16 that $\mathsf{ker}\left(\overline{\mathsf{F}}\right)$ is a randomized congruence for K, and we know that $\left\lfloor \mathsf{ker}\left(\overline{\mathsf{F}}\right) \right\rfloor = \mathsf{ker}\left(\mathsf{F}\right)$. Thus we obtain a randomized morphism

$$\mathsf{E}_{\mathsf{ker}\left(\overline{\mathsf{F}}\right)} : \mathsf{K} \curvearrowright \mathsf{K}/\mathsf{ker}\left(\mathsf{F}\right)$$

from the construction in Proposition 3.3.17. We can say even a wee bit more.

Proposition 3.3.18. *Let* $\mathsf{F} : \mathsf{K} \curvearrowright \mathsf{L}$ *be a near-grounded randomized morphism. Then there exists a unique morphism* $\mathsf{G} : \mathsf{K}/\mathsf{ker}\left(\mathsf{F}\right) \curvearrowright \mathsf{L}$ *that makes the diagram*

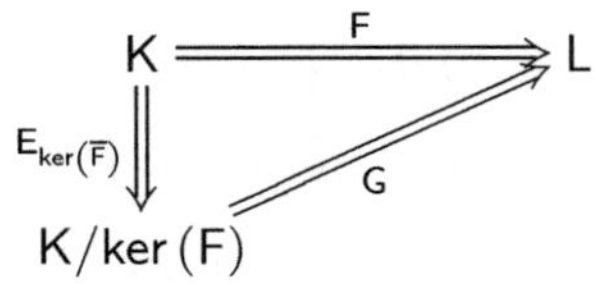

commutative.

Proof 1. Assume $\mathsf{F} = (\Phi, \Psi)$ with $\Phi : X \rightsquigarrow A$ and $\Psi : Y \rightsquigarrow B$. Put

$$\begin{aligned} \Gamma([x]_{\mathsf{ker}(\Phi)}) &:= \Phi(x), \\ \Theta([y]_{\mathsf{ker}(\Psi)}) &:= \Psi(y). \end{aligned}$$

Then evidently $\Gamma : X/\mathsf{ker}\left(\Phi\right) \rightsquigarrow A$ and $\Theta : Y/\mathsf{ker}\left(\Psi\right) \rightsquigarrow B$, and it is apparent that both $\overline{\Gamma}$ and $\overline{\Theta}$ are onto. Thus we have to show that $\mathsf{G} := (\Gamma, \Theta)$ is a randomized morphism $\mathsf{G} : \mathsf{K}/\mathsf{ker}\left(\mathsf{F}\right) \curvearrowright \mathsf{L}$ that makes the diagram above commute.

2. G is a randomized morphism: let $[x]_{\mathsf{ker}(\Phi)} \in X/\mathsf{ker}\left(\Phi\right)$ and $D \in \mathcal{B}(B)$; then we obtain on account of $\mathsf{F} = (\Phi, \Psi)$ being a randomized morphism

$$\begin{aligned} (L * \Gamma)\left([x]_{\mathsf{ker}(\Phi)}\right)(D) &= \int_A L(a)(D)\ \Gamma([x]_{\mathsf{ker}(\Phi)})(da) \\ &= \int_A L(a)(D)\ \Phi(x)(da) \\ &= \int_Y \Psi(y)(D)\ K(x)(dy) \\ &\overset{(\dagger)}{=} \int_{Y/\mathsf{ker}(\Psi)} \Theta(t)(D)\ \ (\mathsf{K}/\mathsf{ker}\left(\mathsf{F}\right))\left([x]_{\mathsf{ker}(\Phi)}\right)(dt) \\ &= \left(\Theta * \mathsf{K}/\mathsf{ker}\left(\mathsf{F}\right)\right)\left([x]_{\mathsf{ker}(\Phi)}\right)(D). \end{aligned}$$

Equation (†) follows from the Change of Variables formula. Thus we have shown that $L * \Gamma = \Theta * \mathsf{K}/\mathsf{ker}(\mathsf{F})$ holds, which implies that G is in fact a randomized morphism.

3. Now let again $D \in \mathcal{B}(A)$ be a Borel set, and $x \in X$. Then

$$\left(\Gamma * E_{\mathsf{ker}(\Phi)}\right)(x)(D) = \int_{X/\mathsf{ker}(\Phi)} \Gamma(s)(D)\ E_{\mathsf{ker}(\Phi)}(x)(ds) = \Gamma([x]_{\mathsf{ker}(\Phi)})(D) = \Phi(x)(D).$$

Thus $\Phi = \Gamma * E_{\mathsf{ker}(\Phi)}$, and Γ is uniquely determined. Similarly one shows that $\Psi = \Theta * E_{\mathsf{ker}(\Psi)}$, and that Θ is unique. But this means $\mathsf{F} = \mathsf{G} * \mathsf{E}_{\mathsf{ker}(\overline{\mathsf{F}})}$, as claimed. □

Now we are in a position to relate the kernels of near-grounded morphisms to near-grounded congruences.

Corollary 3.3.19. *These conditions are equivalent for a congruence* E *for stochastic relation* K

a. $\mathsf{E} = \mathsf{ker}(\mathsf{F})$ *for a near-grounded morphism* $\mathsf{F} : \mathsf{K} \curvearrowright \mathsf{L}$.
b. $\mathsf{E} = \lfloor \mathsf{D} \rfloor$ *for a randomized congruence* D *for* K.

Proof The implication $a \Rightarrow b$ follows from Proposition 3.3.16; $b \Rightarrow a$ is an immediate consequence of Proposition 3.3.17, part *a.* □

A stronger observation has been made in the non-randomized case. Here it could be shown that the kernels of morphisms are exactly the congruences; see Section 1.7.3. The case of randomized congruences and morphisms turns out to be considerably more involved and colorful.

3.4 Randomized Morphisms for Modal Logic

Randomized morphisms will be instrumental for the investigation of models in which not the individual states but rather their distribution is of primary interest; as indicated in the introduction, examples for systems like that may be drawn from economics or biology[1]. But then the comparison of systems can be based as well on a comparison of the respective distributions, complementing the comparison of states proper. Consider — for the sake of the argument — a logic $\widetilde{\mathfrak{L}}$ for specifying a system, and two Kripke models $\mathcal{K}$ and $\mathcal{L}$ for the logic. A comparison of the models on the level of distributions would mean that, given a distribution μ for $\mathcal{K}$, there exists a distribution ν for $\mathcal{L}$ such that $\mu(\llbracket\phi\rrbracket_{\mathcal{K}}) = \nu(\llbracket\phi\rrbracket_{\mathcal{L}})$ holds for each formula ϕ of $\widetilde{\mathfrak{L}}$, and vice versa.

[1] As a member of a German university the present author increasingly lives under the impression that the value of research is measured exclusively in terms of external money which is generated from it and on its immediate economic impact: certainly another example for this phenomenon. They probably know what they are doing.

This is conceived of in the same way as saying $\mathcal{K}, s \models \phi$ iff $\mathcal{L}, t \models \phi$ (and vice versa) when comparing state s of $\mathcal{K}$ to state t of $\mathcal{L}$, but on a different level. Stressing the analogy further, we look at morphisms: for the investigation of the case of states proper, strong morphisms turned out to be a suitable instrument, because they preserve validity. Since randomized morphisms can be made to transform distributions, we will look at this variant of morphisms as a tool for investigating the relationship of Kripke models to each other on the level of distributions. In this way, behavioral equivalence and bisimilarity are available through the existence of cospans respectively of spans of randomized morphisms.

This section will investigate these problems in greater detail. The problem of relating different ways of describing the behavior of a probabilistic system is interesting for a number of reasons: First, it provides further insight into the inner workings of the Giry monad. Second, it will show that logical equivalence and distributional equivalence are unexpectedly closely related, and so are both forms of behavioral equivalence. This permits a better understanding of the underlying logic and shows that the description of the system's behavior through probability distributions is essentially no more powerful than a description that relates states directly to each other. Finally, the results will show that bisimilarity is a concept that, when discussed through distributions, does not yield results that are directly comparable to the results obtained for the familiar strong case.

WHAT DOES THEE EXPECT, GENTLE READER? Well, we will define a simple Hennessy-Milner logic as the underlying object of investigations. This logic is again deliberately simple: it is negation-free and only contains conjunction and modal operators. We will see that conjunction is important and — remarkably enough — that it cannot be replaced by disjunction as the only Boolean connective. The logic gives us a smooth equivalence relation on the state space of a Kripke model; this relation is investigated with a look towards its lifting to the subprobability measures on the state space. Then the interplay of the logic and randomized morphisms is investigated, and here we use the Kleisli extension of the transitions for the Kripke model as a tool. The relations between models are defined formally, and we relate them to each other using the structural properties induced by the lifting of the logic's equivalence relation to the subprobabilities; here the isomorphisms which have been studied in Section 1.8 come in fairly handy. The main result, Theorem 3.4.34, relates different kinds of behaviorally comparing Kripke models to each other; it is summarized in Figure 3.1.

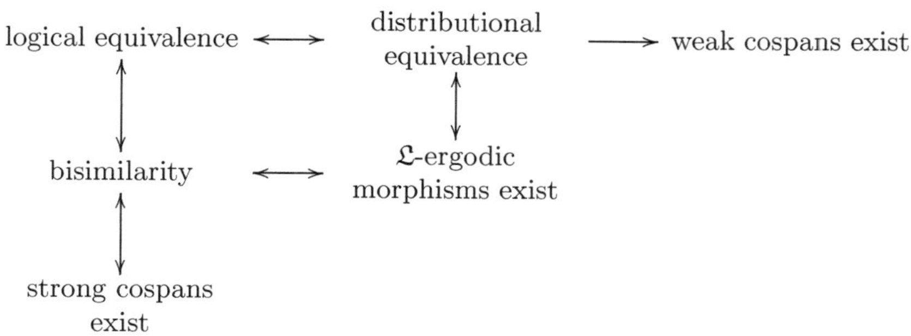

Fig. 3.1 Relations between stochastic Kripke models

3.4.1 A Simple Hennessy-Milner Logic, Again

We discuss as an illustration of the concepts of randomized congruences and randomized morphisms the by now well known logic and its interpretation through stochastic Kripke models.

At the very core of these investigations are two observations:

1. The equivalence relation that makes states equivalent iff they satisfy exactly the same formulas is smooth; see Section 2.3.2.
2. If f is a morphism between models, then a state s satisfies a formula iff $f(s)$ satisfies the same formula; see Lemma 2.3.9.

The first observation permits us to suitably factor the state spaces, building up a probabilistic structure on the factor space; the second one permits defining a bisimulation in a suitable way over the sum of certain factor spaces. This technique has been developed and exercised in Chapter 2.

We discuss these properties in this section in a randomized setting for a very simple Hennessy-Milner type logic $\mathfrak{L}$, showing that a randomized congruence is associated with the logic (which is not too much of a surprise, taking Lemma 3.3.7 into account), and discuss the way in which randomized morphisms act on these congruences. This requires specializing the notion of a randomized morphism to stochastic Kripke models, and to adapting a morphism to the logic.

The syntax of the variant of Hennessy-Milner logic $\mathfrak{L} = \mathfrak{L}(\mathsf{Act}, [0, 1])$ discussed in the present section is given by

$$\phi ::= \top \mid \phi_1 \wedge \phi_2 \mid \langle a \rangle_r \phi.$$

Here $a \in \mathsf{Act}$ is an action, and r is a real number from the unit interval; the set Act of labels is assumed to be at most countably infinite.

The semantics of $\langle a\rangle_r\phi$ says — as above — that we can make an a-move in a state s to a state that satisfies ϕ with probability at least r. Thus r acts as a threshold, and in fact as a lower bound for the probability that a state satisfies formula ϕ after action $a \in \mathsf{Act}$; see the discussion in Section 2.3.

Let $\mathcal{K} := (S, (k_a)_{a\in\mathsf{Act}})$ be a stochastic Kripke model with a Standard Borel state space S so that there is a stochastic relation $k_a : S \rightsquigarrow S$ for each action a. The interpretation of k_a reads again that $k_a(s)(E)$ is the probability that upon action $a \in \mathsf{Act}$ in state $s \in S$ the next state is a member of $E \in \mathcal{B}(S)$. Fix this Kripke model $\mathcal{K}$.

Satisfaction of a state s for a formula ϕ is defined inductively; $[\![\phi]\!]_{\mathcal{K}}$ is again defined as the set of all states s that satisfy formula ϕ, i.e. for which $\mathcal{K}, s \models \phi$ holds; in particular we have $\mathcal{K}, s \models \langle a\rangle_r\phi$ iff $k_a(s)([\![\phi]\!]_{\mathcal{K}}) \geq r$. This is a special case of the semantics defined in Section 2.3 and briefly discussed in Example 2.3.3.

The set of formulas $\mathbf{F}_{\mathfrak{L}}$ defines again an equivalence relation $\mathsf{r}_{\mathcal{K}}$ on the states upon

$$s \;\mathsf{r}_{\mathcal{K}}\; s' \text{ iff } \forall \phi \in \mathbf{F}_{\mathfrak{L}} : \big[\mathcal{K}, s \models \phi \Leftrightarrow \mathcal{K}, s' \models \phi\big].$$

Then $[\![\phi]\!]_{\mathcal{K}}$ is an $\mathsf{r}_{\mathcal{K}}$-invariant Borel set for each formula ϕ.

The set of all formulas is not countable, because we have for each action $a \in \mathsf{Act}$, each formula ϕ, and each real number $r \in [0, 1]$ a formula $\langle a\rangle_r\phi$. Nevertheless, relation $\mathsf{r}_{\mathcal{K}}$ is a smooth equivalence relation, since the rational numbers are dense, and since the underlying stochastic relation is based on σ-continuous measures.

Lemma 3.4.1. *The equivalence relation* $\mathsf{r}_{\mathcal{K}}$ *is smooth.*

Proof 1. Logic $\mathfrak{L}_{\mathbb{Q}}$ is defined through the grammar for $\mathfrak{L}$ with the assumption that the index q for formula $\langle a\rangle_q\phi$ is taken from $[0, 1] \cap \mathbb{Q}$. Define the equivalence relation r as above through $\mathbf{F}_{\mathfrak{L}_{\mathbb{Q}}}$. Then $\mathsf{r} = \mathsf{r}_{\mathcal{K}}$. Granted that, we may conclude that the relation $\mathsf{r}_{\mathcal{K}}$ is smooth, since r is smooth on account of $\mathbf{F}_{\mathfrak{L}_{\mathbb{Q}}}$ being countable.

2. Because $\mathbf{F}_{\mathfrak{L}_{\mathbb{Q}}} \subseteq \mathbf{F}_{\mathfrak{L}}$, the inclusion $\mathsf{r}_{\mathcal{K}} \subseteq \mathsf{r}$ is obvious. On the other hand, if $s \;\mathsf{r}\; s'$, then we have in particular for an arbitrary formula $\phi \in \mathbf{F}_{\mathfrak{L}}$

$$k_a(s)([\![\phi]\!]_{\mathcal{K}}) \geq q \Leftrightarrow k_a(s')([\![\phi]\!]_{\mathcal{K}}) \geq q$$

for each rational q, so that

$$k_a(x)([\![\phi]\!]_{\mathcal{K}}) = k_a(x')([\![\phi]\!]_{\mathcal{K}}),$$

because $[0, 1] \cap \mathbb{Q}$ is dense in $[0, 1]$. Since ϕ is arbitrary, we may conclude that this holds for each $\phi \in \mathbf{F}_{\mathfrak{L}}$ (strictly speaking, we should argue by induction on the structure of ϕ; but since the argument just given will be applied, we trust the reader to complete the proof). Hence we conclude that $s \;\mathsf{r}_{\mathcal{K}}\; s'$ holds. $\square$

We will need in Section 3.4.3 that $[\![\langle a\rangle_r\phi]\!]_{\mathcal{K}}$ be defined for each *real* $r \in [0,1]$. Usually one defines $\langle a\rangle_r\phi$ only for rational r. Lemma 3.4.1 shows that this will not affect the equivalence relation $\mathsf{r}_{\mathcal{K}}$, and that we may argue safely with rational thresholds. This relation is also a congruence for each $k_a : S \rightsquigarrow S$. This statement is well known; cf. Lemma 2.3.10; we give a short proof for the sake of completeness below. The proof shows that conjunction is necessary for establishing this property, for otherwise the π-λ-Theorem could not be invoked, and it would not be possible to conclude that the property in question transfers from a generator to the entire σ-algebra.

Lemma 3.4.2. *Let $\mathcal{K} = (S, (k_a)_{a\in\mathsf{Act}})$ be a stochastic Kripke model. Then the equivalence relation $\mathsf{r}_{\mathcal{K}}$ is a k_a-congruence for each action $a \in \mathsf{Act}$.*

Proof 1. Fix action $a \in \mathsf{Act}$. We show first that $x\ \mathsf{r}_{\mathcal{K}}\ x'$ implies that $k_a(x)([\![\phi]\!]_{\mathcal{K}}) = k_a(x')([\![\phi]\!]_{\mathcal{K}})$ whenever ϕ is a formula in $\mathfrak{L}$. Suppose that this is not true; then there is a shortest ϕ for which the equality above is false. Thus we can squeeze in a number q with $k_a(x)([\![\phi]\!]_{\mathcal{K}}) < q \leq k_a(x')([\![\phi]\!]_{\mathcal{K}})$ (or vice versa), so that $\mathcal{K}, x \not\models \langle a\rangle_q\phi$ but $\mathcal{K}, x' \models \langle a\rangle_q\phi$. This is a contradiction. Note that q could be chosen as a rational number.

2. Now consider for fixed x, x' with $x\ \mathsf{r}_{\mathcal{K}}\ x'$ the set

$$\mathcal{D} := \{D \in \mathbf{\Sigma}\big(\mathcal{B}(S), \mathsf{r}_{\mathcal{K}}\big) \mid k_a(x)(D) = k_a(x')(D)\}.$$

Then $[\![\phi]\!]_{\mathcal{K}} \in \mathcal{D}$ for each formula ϕ by the first part. Because the logic is closed under conjunctions, and since by Proposition 1.7.12

$$\mathbf{\Sigma}\big(\mathcal{B}(S), \mathsf{r}_{\mathcal{K}}\big) = \sigma(\{[\![\phi]\!]_{\mathcal{K}} \mid \phi \in \mathbf{F}_{\mathfrak{L}}\}),$$

this implies the claim through Lemma 1.6.30.

Consequently, $x\ \mathsf{r}_{\mathcal{K}}\ x'$ implies $k_a(x)(D) = k_a(x')(D)$ for all $\mathsf{r}_{\mathcal{K}}$-invariant Borel sets D of S. □

Now we relate two models through a randomized morphism; this is simply a randomized morphism for any relation corresponding to the same action.

Definition 3.4.3. *Let $\mathcal{K} = (S, (k_a)_{a\in\mathsf{Act}})$ and $\mathcal{L} = (T, (\ell_a)_{a\in\mathsf{Act}})$ be stochastic Kripke models. The stochastic relation $\Phi : S \rightsquigarrow T$ is called a* randomized morphism *from $\mathcal{K}$ to $\mathcal{L}$; in symbols $\Phi : \mathcal{K} \curvearrowright \mathcal{L}$ iff $\Phi : (S, k_a) \curvearrowright (T, \ell_a)$ for each $a \in \mathsf{Act}$.*

Thus given $\Phi : \mathcal{K} \curvearrowright \mathcal{L}$, we have $\ell_a * \Phi = \Phi * k_a$ for each action $a \in \mathsf{Act}$, or, equivalently, that $\overline{\ell_a} \circ \overline{\Phi} = \overline{\Phi} \circ \overline{k_a}$. Again, this is compared to the non-randomized situation. Recall from Definition 2.3.7 that $f : \mathcal{K} \to \mathcal{L}$ is a strong morphism for transition systems $\mathcal{K}$ and $\mathcal{L}$ iff $f : (S, k_a) \to (T, \ell_a)$ is a morphism for each $a \in \mathsf{Act}$. Thus $f : S \to T$ is a surjective Borel map such that $\ell_a \circ f = \mathfrak{S}(f) \circ k_a$ for each action a. Consequently, given action $a \in \mathsf{Act}$, we know for a Borel set $B \subseteq T$ and a state $s \in S$ that $\ell_a(f(s))(B) = k_a(s)(f^{-1}[B])$ holds, so that the probability of hitting a state

$t \in B$ from state $f(s)$ in $\mathcal{L}$ equals the probability of hitting a state s' with $f(s') \in B$ from state s in $\mathcal{L}$. For $\Phi : \mathcal{K} \curvearrowright \mathcal{L}$ we have $(s \in S, B \in \mathcal{B}(T))$

$$\int_T \ell_a(t)(B)\ \Phi(s)(dt) = \int_S \Phi(s')(B)\ k_a(s)(ds')$$

so that the probability of hitting from state s a new state that is a member of set B average over through $\Phi(s)$ equals the average probability of hitting this state with the morphism when averaged through $k_a(s)$. In Section 3.3.1 we had to deal with two stochastic relations, one for the domain, and the other one for the codomain of the relations involved. Since we here deal only with state spaces, domain and codomain coincide; so the situation is simplified somewhat.

The following Lemma is but a reformulation of the well-known fact that validity of a formula is respected by morphisms, and that the inverse image of $\mathsf{r}_{\mathcal{L}}$-invariant Borel sets is $\mathsf{r}_{\mathcal{K}}$-invariant.

Lemma 3.4.4. *Let $\mathcal{K}$ and $\mathcal{L}$ be stochastic Kripke models over the analytic state spaces S respectively T, and let $f : \mathcal{K} \to \mathcal{L}$ be a strong morphism. Then*

a. $\mathcal{K}, s \models \phi \Leftrightarrow \mathcal{L}, f(s) \models \phi$ holds for all $s \in S$ and for all formulas ϕ.
b. f is $\mathbf{\Sigma}(\mathcal{B}(S), \mathsf{r}_{\mathcal{K}})$-$\mathbf{\Sigma}(\mathcal{B}(T), \mathsf{r}_{\mathcal{L}})$-measurable. □

We introduce specialized randomized morphisms by taking the logic into account.

Definition 3.4.5. *Let $\mathcal{K}$ and $\mathcal{L}$ be Kripke models over the analytic state spaces S respectively T. Morphism $\Phi : \mathcal{K} \curvearrowright \mathcal{L}$ is called a* $\mathfrak{L}$-morphism *iff*

$$\Phi : \big(S, \mathbf{\Sigma}(\mathcal{B}(S), \mathsf{r}_{\mathcal{K}})\big) \rightsquigarrow \big(T, \mathbf{\Sigma}(\mathcal{B}(T), \mathsf{r}_{\mathcal{L}})\big).$$

Thus we construct $\mathfrak{L}$-morphisms in analogously to non-randomized morphisms between transition systems in the base category. It should be noted that we require $s \mapsto \Phi(s)(C)$ to be $\mathbf{\Sigma}(\mathcal{B}(S), \mathsf{r}_{\mathcal{K}})$-measurable for each $C \in \mathbf{\Sigma}(\mathcal{B}(T), \mathsf{r}_{\mathcal{L}})$ for Φ to be a $\mathfrak{L}$-morphism, so that the set

$$\{s \in S \mid \Phi(s)([\![\phi]\!]_{\mathcal{K}}) \geq q\}$$

is $\mathbf{\Sigma}(\mathcal{B}(S), \mathsf{r}_{\mathcal{K}})$-invariant for each formula ϕ; hence the elements of this set cannot be separated by the logic. This means that $\Phi(s)([\![\phi]\!]_{\mathcal{K}}) \geq q$ iff $\Phi(s')([\![\phi]\!]_{\mathcal{K}}) \geq q$, provided s and s' satisfy the same formulas in $\mathfrak{L}$. Consequently,

$$\Phi(s)([\![\phi]\!]_{\mathcal{K}}) = \Phi(s')([\![\phi]\!]_{\mathcal{K}})$$

holds for all formulas ϕ whenever $s\ \mathsf{r}_{\mathcal{K}}\ s'$, and vice versa; see the proof for Lemma 3.4.2.

To be more specific, from Proposition 1.7.20 a characterization of these morphisms in terms of congruences is obtained:

Proposition 3.4.6. *Let $\mathcal{K}$ and $\mathcal{L}$ be stochastic Kripke models over analytic spaces. Then the randomized morphism $\Phi : \mathcal{K} \curvearrowright \mathcal{L}$ is a $\mathfrak{L}$-morphism iff $(\mathsf{r}_{\mathcal{K}}, \mathsf{r}_{\mathcal{L}})$ is a congruence for Φ.* □

$\mathfrak{L}$-morphisms can be used to build up a category (which will not be done here). Just for illustrating the concept, we mention the following:

Corollary 3.4.7. *Let $\Phi : \mathcal{K} \curvearrowright \mathcal{L}$ and $\Psi : \mathcal{L} \curvearrowright \mathcal{P}$ be randomized morphisms, where $\mathcal{K}$, $\mathcal{L}$, and $\mathcal{P}$ are Kripke models over analytic spaces; then*

a. if both Φ and Ψ are $\mathfrak{L}$-morphisms, so is $\Psi * \Phi : \mathcal{K} \curvearrowright \mathcal{P}$,
b. if $(\mathsf{r}_{\mathcal{K}}, \mathsf{r}_{\mathcal{L}})$ is a congruence for Φ, and $(\mathsf{r}_{\mathcal{L}}, \mathsf{r}_{\mathcal{P}})$ is a congruence for Ψ, then $(\mathsf{r}_{\mathcal{K}}, \mathsf{r}_{\mathcal{P}})$ is a congruence for the Kleisli product $\Psi * \Phi$.

Proof Property *a.* follows from the fact that the Kleisli composition of Kleisli morphisms is again a Kleisli morphism. Property *b.* is a consequence of *a*, using Proposition 3.4.6. □

Corollary 3.4.8. *Let $\Phi : \mathcal{K} \curvearrowright \mathcal{L}$ be a $\mathfrak{L}$-morphism for models as above; then $\mu \; \overline{\mathsf{r}_{\mathcal{K}}} \; \mu'$ implies $\overline{\Phi}(\mu) \; \overline{\mathsf{r}_{\mathcal{L}}} \; \overline{\Phi}(\mu')$.*

Proof This follows directly from Lemma 3.3.7. □

Call a smooth equivalence relation τ on the state space S of the Kripke model $\mathcal{K} = (S, (k_a)_{a \in \mathsf{Act}})$ a *$\mathcal{K}$-congruence* iff τ is a congruence for each $k_a : S \rightsquigarrow S$. The factor system

$$\mathcal{K}/\tau := (S/\tau, (k_{a,\tau})_{a \in \mathsf{Act}})$$

has as a state space the analytic space S/τ of τ-equivalence classes with the transition rules $k_{a,\tau} : S/\tau \rightsquigarrow S/\tau$ for each action $a \in \mathsf{Act}$; see Section 1.7.3 for factoring stochastic relations.

Define for $\mathcal{K}$ the *$\mathfrak{L}$-reduced model $\mathcal{K}/\mathfrak{L}$* as $\mathcal{K}/\mathsf{r}_{\mathcal{K}}$, so that two different states in $\mathcal{K}/\mathfrak{L}$ can always be separated by the logic. We show that the reduction does not destroy the property of being a $\mathfrak{L}$-morphism (modulo factoring, of course).

Before doing that, we briefly investigate the reduced model with respect to non-randomized morphisms.

Proposition 3.4.9. *Let $f : \mathcal{K} \to \mathcal{L}$ be a strong morphism for models as above; then*

a. $\mathsf{ker}\,(f) \subseteq \mathsf{r}_{\mathcal{K}}$
b. there exists a unique strong morphism $\varpi_{f,\mathfrak{L}} : \mathcal{K}/\mathsf{ker}\,(f) \to \mathcal{K}/\mathfrak{L}$ with $\eta_{\mathfrak{L}} = \varpi_{f,\mathfrak{L}} \circ \eta_{\mathsf{ker}\,(f)}$.

Proof 1. Let $\langle s, s' \rangle \in \mathsf{ker}\,(f)$; then we have for each formula ϕ by Lemma 3.4.4

$$\mathcal{K}, s \models \phi \Leftrightarrow \mathcal{L}, f(s) \models \phi \Leftrightarrow \mathcal{L}, f(s') \models \phi \Leftrightarrow \mathcal{K}, s' \models \phi;$$

thus $\langle s, s' \rangle \in \mathsf{r}_{\mathcal{K}}$.

2. Because the morphism constructed in the latter part only depends on the congruences and not on the specific stochastic relations that are involved, the claim follows directly from $\mathsf{ker}(f) \subseteq \mathsf{r}_{\mathcal{K}}$ through Corollary 1.7.22. □

Returning to the discussion of $\mathfrak{L}$-morphisms, we show now that such a morphism between two models gives rise to a morphism between the reduced models.

Proposition 3.4.10. *Let $\mathcal{K}$ and $\mathcal{L}$ be stochastic Kripke models with analytic state spaces S respectively T, and assume that $\Phi : \mathcal{K} \curvearrowright \mathcal{L}$ is a $\mathfrak{L}$-morphism. Define $\Phi_{\mathfrak{L}} : S/\mathsf{r}_{\mathcal{K}} \rightsquigarrow T/\mathsf{r}_{\mathcal{L}}$ through $\Phi_{\mathfrak{L}} := \Phi_{(\mathsf{r}_{\mathcal{K}}, \mathsf{r}_{\mathcal{L}})}$, the factor relation induced by the congruence $(\mathsf{r}_{\mathcal{K}}, \mathsf{r}_{\mathcal{L}})$. Then $\Phi_{\mathfrak{L}} : \mathcal{K}/\mathfrak{L} \curvearrowright \mathcal{L}/\mathfrak{L}$ is an $\mathfrak{L}$-morphism.*

Proof 1. We show first that $\Phi_{\mathfrak{L}}$ is a randomized morphism $\mathcal{K}/\mathfrak{L} \curvearrowright \mathcal{L}/\mathfrak{L}$. Let $k_a : S \rightsquigarrow S$ and $\ell_a : T \rightsquigarrow T$ be the respective transition laws for action $a \in \mathsf{Act}$. Take $s \in S$ and $G \in \mathcal{B}(T/\mathsf{r}_{\mathcal{L}})$. Then

$$\begin{aligned}
(\ell_{a,\mathsf{r}_{\mathcal{L}}} * \Phi_{\mathfrak{L}})([s]_{\mathsf{r}_{\mathcal{K}}})(G) &= \int_{T/\mathsf{r}_{\mathcal{L}}} \ell_{a,\mathsf{r}_{\mathcal{L}}}(y)(G)\ \Phi_{\mathfrak{L}}([s]_{\mathsf{r}_{\mathcal{K}}})(dy) \\
&\overset{(\natural)}{=} \int_T \ell_a(t)(\eta_{\mathsf{r}_{\mathcal{L}}}^{-1}[G])\ \Phi(s)(dt) \\
&\overset{(\flat)}{=} \int_S \Phi(w)(\eta_{\mathsf{r}_{\mathcal{L}}}^{-1}[G])\ k_a(s)(dw) \\
&= \int_S \Phi_{\mathfrak{L}}([w]_{\mathsf{r}_{\mathcal{L}}})(G)\ k_a(s)(dw) \\
&\overset{(\natural)}{=} \int_{S/\mathsf{r}_{\mathcal{K}}} \Phi_{\mathfrak{L}}(v)(G)\ k_{a,\mathsf{r}_{\mathcal{K}}}([s]_{\mathsf{r}_{\mathcal{K}}})(dv) \\
&= (\Phi_{\mathfrak{L}} * k_{a,\mathsf{r}_{\mathcal{K}}})([s]_{\mathsf{r}_{\mathcal{K}}})(G).
\end{aligned}$$

The equalities marked ($\natural$) use Change of Variables (Lemma 1.6.20), the equality marked ($\flat$) derives from Φ being a randomized morphism.

Consequently, $\Phi_{\mathfrak{L}}$ is a randomized morphism $\mathcal{K}/\mathfrak{L} \curvearrowright \mathcal{L}/\mathfrak{L}$.

2. Because $\mathsf{r}_{\mathcal{K}/\mathfrak{L}}$ equals the identity $\Delta_{S/\mathfrak{L}}$ on $S/\mathfrak{L}$, Example 1.8.4 tells us that

$$\boldsymbol{\Sigma}\big(\mathcal{B}(S/\mathsf{r}_{\mathcal{K}}), \mathsf{r}_{\mathcal{K}/\mathfrak{L}}\big) = \mathcal{B}(S/\mathsf{r}_{\mathcal{K}}).$$

Similarly for $\mathcal{L}$. Thus $\Phi_{\mathfrak{L}} : \boldsymbol{\Sigma}\big(\mathcal{B}(S/\mathsf{r}_{\mathcal{K}}), \mathsf{r}_{\mathcal{K}/\mathfrak{L}}\big) \rightsquigarrow \boldsymbol{\Sigma}\big(\mathcal{B}(T/\mathsf{r}_{\mathcal{L}}), \mathsf{r}_{\mathcal{L}/\mathfrak{L}}\big)$ is a stochastic relation. Consequently, $\Phi_{\mathfrak{L}}$ is a $\mathfrak{L}$-morphism. □

3.4.2 Borel Isomorphisms

Let $\mathcal{K} = \big(S, (k_a)_{a \in \mathsf{Act}}\big)$ be a Kripke model for the logic $\mathfrak{L}$. The Kleisli extension $\overline{k_a}(\mu)$ can be expressed easily for sets of the type $[\![\phi]\!]_{\mathcal{K}}$. This representation in terms of a Lebesgue integral over the unit interval is quite surprising; it will be helpful later.

Lemma 3.4.11. *If $\phi \in \mathbf{F}_{\mathfrak{L}}$ is a formula, and $\mu \in \mathfrak{S}\left(S, \mathbf{\Sigma}(\mathcal{B}(S), \mathsf{r}_{\mathcal{K}})\right)$, then*

$$\overline{k_a}(\mu)([\![\phi]\!]_{\mathcal{K}}) = \sup_{n\in\mathbb{N}} \int_0^1 \mu([\![\langle a\rangle_{r+\frac{1}{n}}\phi]\!]_{\mathcal{K}})\ dr.$$

Proof Consider the Kleisli extension to k_a:

$$\begin{aligned}
\overline{k_a}(\mu)([\![\phi]\!]_{\mathcal{K}}) &= \int_S k_a(s)([\![\phi]\!]_{\mathcal{K}})\ \mu(ds) \\
&\overset{(\flat)}{=} \int_0^1 \mu(\{s \in S \mid k_a(s)([\![\phi]\!]_{\mathcal{K}}) > r\})\ dr \\
&\overset{(\dagger)}{=} \int_0^1 \sup_{n\in\mathbb{N}} \mu(\{s \in S \mid k_a(s)([\![\phi]\!]_{\mathcal{K}}) \geq r + \frac{1}{n}\})\ dr \\
&\overset{(\ddagger)}{=} \sup_{n\in\mathbb{N}} \int_0^1 \mu([\![\langle a\rangle_{r+\frac{1}{n}}\phi]\!]_{\mathcal{K}})\ dr
\end{aligned}$$

Equation ($\flat$) is the Choquet representation of the integral from Lemma 1.6.2; the equality ($\dagger$) makes use of the fact that

$$\left(\{s \in S \mid k_a(s)([\![\phi]\!]_{\mathcal{K}}) \geq r + \frac{1}{n}\}\right)_{n\in\mathbb{N}}$$

is an increasing sequence of measurable sets with

$$\bigcup_{n\in\mathbb{N}} \{s \in S \mid k_a(s)([\![\phi]\!]_{\mathcal{K}}) \geq r + \frac{1}{n}\} = \{s \in S \mid k_a(s)([\![\phi]\!]_{\mathcal{K}}) > r\},$$

capitalizing on the fact that a measure is σ-continuous. Equation ($\ddagger$) interchanges integration and forming the countable supremum, which is permissible by the Monotone Convergence Theorem, Proposition 1.6.1. □

Note that the representation above applies in particular to measures from $\mathfrak{S}(S)$ when restricted to $\mathbf{\Sigma}(\mathcal{B}(S), \mathsf{r}_{\mathcal{K}}) \subseteq \mathcal{B}(S)$. Consider the special case $\mu = \delta_s$ for some $s \in S$. Then the representation from Lemma 3.4.11 evaluates to the equation

$$\overline{k_a}(\delta_s)([\![\phi]\!]_{\mathcal{K}}) = k_a(s)([\![\phi]\!]_{\mathcal{K}}) = \lambda\left([0, k_a(x)([\![\phi]\!]_{\mathcal{K}})[\right)$$

with λ as the Lebesgue measure on the real line. This does not appear particularly helpful. The representation's strength, however, derives from the fact that it holds for arbitrary subprobabilities.

The equivalence relation

$$\mu_1\ \overline{\mathsf{r}_{\mathcal{K}}}\ \mu_2 \text{ iff } \forall A \in \mathbf{\Sigma}(\mathcal{B}(S), \mathsf{r}_{\mathcal{K}}) : \mu_1(A) = \mu_2(A)$$

is the *lifting* or *randomization* of relation $\mathsf{r}_{\mathcal{K}}$ to $\mathfrak{S}(S)$; see Definition 1.8.1. For states $s_1, s_2 \in G$ the relation $s_1\ \mathsf{r}_{\mathcal{K}}\ s_2$ is equivalent to $\delta_{s_1}\ \overline{\mathsf{r}_{\mathcal{K}}}\ \delta_{s_2}$ for the corresponding Dirac measures. We state two simple useful properties just for the record.

Lemma 3.4.12. *The equivalence relation* $\overline{\mathsf{r}_{\mathcal{K}}}$ *has these properties*

1. $\mu_1 \overline{\mathsf{r}_{\mathcal{K}}} \mu_2$ *iff* $\mu_1([\![\phi]\!]_{\mathcal{K}}) = \mu_1([\![\phi]\!]_{\mathcal{K}})$ *for every formula* ϕ,
2. $\overline{\mathsf{r}_{\mathcal{K}}}$ *is grounded.*

Proof Because $\mathbf{\Sigma}(\mathcal{B}(S), \mathsf{r}_{\mathcal{K}})$ is generated by $\{[\![\phi]\!]_{\mathcal{K}} \mid \phi \in \mathbf{F}_{\mathfrak{L}}\}$, the first assertion follows from Lemma 1.7.7. The second assertion derives directly from the definition of groundedness. □

The relationship between $\mathsf{r}_{\mathcal{K}}$ and the factor space $\mathfrak{S}(S/\mathsf{r}_{\mathcal{K}})$ will be investigated in some detail now. Define

$$\kappa_S : \begin{cases} \mathfrak{S}\left(S, \mathbf{\Sigma}(\mathcal{B}(S), \mathsf{r}_{\mathcal{K}})\right) \to \mathfrak{S}(S/\mathsf{r}_{\mathcal{K}}), \\ \mu \qquad\qquad\qquad\quad \mapsto \mathfrak{S}(\eta_{\mathsf{r}_{\mathcal{K}}})(\mu) \end{cases}$$

and $\mathfrak{z}_S : \mathfrak{S}(S/\mathsf{r}_{\mathcal{K}}) \to \mathfrak{S}(S)/\overline{\mathsf{r}_{\mathcal{K}}}$ by mapping ν to $[\nu']_{\overline{\mathsf{r}_{\mathcal{K}}}}$ iff $\nu'(C) = \nu(\eta_{\mathsf{r}_{\mathcal{K}}}^{-1}[C])$ for all $C \in \mathbf{\Sigma}(\mathcal{B}(S), \mathsf{r}_{\mathcal{K}})$.

Lemma 3.4.13. *Both* κ_S *and* $\mathfrak{z}_S$ *are Borel isomorphisms.*

Proof 1. We infer from Proposition 1.8.12 that $\mathfrak{z}_S$ is a Borel isomorphism, so that we merely have to take care of κ_S.

2. Because $A \in \mathcal{B}(S/\mathsf{r}_{\mathcal{K}})$ iff $\eta_{\mathsf{r}_{\mathcal{K}}}^{-1}[A] \in \mathbf{\Sigma}(\mathcal{B}(S), \mathsf{r}_{\mathcal{K}})$, we infer in fact that $\kappa_S(\mu) \in \mathfrak{S}(S/\mathsf{r}_{\mathcal{K}})$ whenever $\mu \in \mathfrak{S}\left(S, \mathbf{\Sigma}(\mathcal{B}(S), \mathsf{r}_{\mathcal{K}})\right)$. From Proposition 1.7.12 it follows that κ_S is injective. Since

$$\mathbf{\Sigma}(\mathcal{B}(S), \mathsf{r}_{\mathcal{K}}) \ni A \mapsto \eta_{\mathsf{r}_{\mathcal{K}}}[A] \in \mathcal{B}(S/\mathsf{r}_{\mathcal{K}})$$

is compatible with countable Boolean operations, $A \mapsto \nu(\eta_{\mathsf{r}_{\mathcal{K}}}[A])$ defines for a fixed $\nu \in \mathfrak{S}(S/\mathsf{r}_{\mathcal{K}})$ an element μ_ν of $\mathfrak{S}\left(S, \mathbf{\Sigma}(\mathcal{B}(S), \mathsf{r}_{\mathcal{K}})\right)$. Now $\eta_{\mathsf{r}_{\mathcal{K}}}^{-1}[\eta_{\mathsf{r}_{\mathcal{K}}}[A]] = A$ on $\mathbf{\Sigma}(\mathcal{B}(S), \mathsf{r}_{\mathcal{K}})$, so that $\kappa_S(\mu_\nu) = \nu$; hence κ_S is onto.

3. The weak*-σ-algebra on $\mathfrak{S}\left(S, \mathbf{\Sigma}(\mathcal{B}(S), \mathsf{r}_{\mathcal{K}})\right)$ is the smallest σ-algebra containing sets of the form

$$\Gamma_{A,q}^{\mathfrak{S}\left(S, \mathbf{\Sigma}(\mathcal{B}(S), \mathsf{r}_{\mathcal{K}})\right)} := \{\mu \in \mathfrak{S}\left(S, \mathbf{\Sigma}(\mathcal{B}(S), \mathsf{r}_{\mathcal{K}})\right) \mid \mu(A) < q\}$$

with $A \in \mathbf{\Sigma}(\mathcal{B}(S), \mathsf{r}_{\mathcal{K}}), q \in \mathbb{R}_+$; similarly, the sets $\Gamma_{C,q}^{\mathfrak{S}(S/\mathsf{r}_{\mathcal{K}})}$ define the weak-*-σ-algebra on $\mathfrak{S}(S/\mathsf{r}_{\mathcal{K}})$ for the Borel set $C \subseteq S/\mathsf{r}_{\mathcal{K}}$. Since

$$\mu \in \Gamma_{\eta_{\mathsf{r}_{\mathcal{K}}}^{-1}[C],q}^{\mathfrak{S}\left(S, \mathbf{\Sigma}(\mathcal{B}(S), \mathsf{r}_{\mathcal{K}})\right)} \text{ iff } \kappa_S(\mu) \in \Gamma_{C,q}^{\mathfrak{S}(S/\mathsf{r}_{\mathcal{K}})}$$

we infer that both κ_S and its inverse are measurable. □

Lemma 3.4.14. *Define for each action* $a \in \mathsf{Act}$ *the map*

$$\Theta_{a,\mathcal{K}}([\mu]_{\overline{\mathsf{r}_{\mathcal{K}}}}) := \left[\overline{k_a}(\mu)\right]_{\overline{\mathsf{r}_{\mathcal{K}}}}.$$

Then $\Theta_{a,\mathcal{K}} : \mathfrak{S}(S)/\overline{\mathsf{r}_{\mathcal{K}}} \to \mathfrak{S}(S)/\overline{\mathsf{r}_{\mathcal{K}}}$ *is Borel.*

Proof 1. Assume that $\mu_1 \overline{\mathsf{r}_{\mathcal{K}}} \mu_2$ for $\mu_1, \mu_2 \in \mathfrak{S}(S)$; thus $\mu_1([\![\phi]\!]_{\mathcal{K}}) = \mu_2([\![\phi]\!]_{\mathcal{K}})$ for all formulas $\phi \in \mathbf{F}_{\mathfrak{L}}$. But then Lemma 3.4.11 yields

$$\begin{aligned}\overline{k_a}(\mu_1)([\![\phi]\!]_{\mathcal{K}}) &= \sup_{n\in\mathbb{N}} \int_0^1 \mu_1([\![\langle a\rangle_{r+\frac{1}{n}}\phi]\!]_{\mathcal{K}})\, dr \\ &= \sup_{n\in\mathbb{N}} \int_0^1 \mu_2([\![\langle a\rangle_{r+\frac{1}{n}}\phi]\!]_{\mathcal{K}})\, dr \\ &= \overline{k_a}(\mu_2)([\![\phi]\!]_{\mathcal{K}}).\end{aligned}$$

Thus we obtain from Lemma 3.4.12 that $\overline{k_a}(\mu_1)\ \overline{\mathsf{r}_{\mathcal{K}}}\ \overline{k_a}(\mu_2)$. Hence $\Theta_{a,\mathcal{K}}$ is well defined.

2. This diagram commutes

$$\begin{array}{ccc} \mathfrak{S}(S) & \xrightarrow{\ \overline{k_a}\ } & \mathfrak{S}(S) \\ \downarrow{\scriptstyle \eta_{\overline{\mathsf{r}_{\mathcal{K}}}}} & & \downarrow{\scriptstyle \eta_{\overline{\mathsf{r}_{\mathcal{K}}}}} \\ \mathfrak{S}(S)/\overline{\mathsf{r}_{\mathcal{K}}} & \xrightarrow[\ \Theta_{a,\mathcal{K}}\ ]{} & \mathfrak{S}(S)/\overline{\mathsf{r}_{\mathcal{K}}} \end{array}$$

Since the Borel sets of $\mathfrak{S}(S)/\overline{\mathsf{r}_{\mathcal{K}}}$ are just the final σ-algebra for the factor map $\eta_{\overline{\mathsf{r}_{\mathcal{K}}}}$, and since $\overline{k_a}$ is Borel measurable, Borel measurability for $\Theta_{a,\mathcal{K}}$ follows from Lemma 1.3.3. □

3.4.3 Distributional Equivalence

After having introduced Hennessy–Milner logic and its interpretation through Kleisli models, the problem about the conditions under which interpretations are equivalent, in the sense that given a state in one model, there exists a matching state in the other model, has been investigated quite extensively; see Chapter 2.

The question considered here is of a more general nature: given a distribution on the states of one model, we assume for distributional equivalence that there exists a matching distribution on the state space of the other model.

Before discussing details, we will extend our repertoire of morphisms by introducing ergodic morphisms.

Definition 3.4.15. *Let* $\mathcal{K} = (S, (k_a)_{a\in\mathsf{Act}})$ *and* $\mathcal{L} = (T, (\ell_a)_{a\in\mathsf{Act}})$ *be stochastic Kleisli models over analytic state spaces. A stochastic relation*

$$\Phi : (S, \mathbf{\Sigma}(\mathcal{B}(S), \mathsf{r}_{\mathcal{K}})) \rightsquigarrow (T, \mathbf{\Sigma}(\mathcal{B}(T), \mathsf{r}_{\mathcal{L}}))$$

is called $\mathfrak{L}$-ergodic *($\mathcal{K} \curvearrowright^{\mathbf{e}} \mathcal{L}$) iff*

a. $\big(\Phi * k_a\big)(s)(E) = \big(\ell_a *\big)(s)(E)$ *whenever* $s \in S$ *and* $E \in \mathbf{\Sigma}\big(\mathcal{B}(T), \mathsf{r}_{\mathcal{L}}\big)$,
b. given $\phi \in \mathbf{F}_{\mathfrak{L}}$ *the map* $s \mapsto \Phi(s)(\llbracket\phi\rrbracket_{\mathcal{L}})$ *takes only the values* 0 *and* 1 *such that* $\Phi(s)(\llbracket\phi\rrbracket_{\mathcal{L}}) = 1$ *iff* $\mathcal{K}, s \models \phi$.

An $\mathfrak{L}$-ergodic morphism is — in addition to its being an $\mathfrak{L}$-morphism on the respective invariant sets — characterized through $\Phi(s)(\llbracket\phi\rrbracket_{\mathcal{L}}) = \chi_{\llbracket\phi\rrbracket_{\mathcal{K}}}(s)$ for all formulas $\phi \in \mathbf{F}_{\mathfrak{L}}$, χ_A denoting as usual the indicator function of the set A. Putting it less formally, we know that, whenever $\mathcal{K}, s \models \phi$, the state's counterpart t in model $\mathcal{L}$ will satisfy $\mathcal{L}, t \models \phi$ almost surely. Hence validity in model $\mathcal{K}$ is very closely tied to validity in model $\mathcal{L}$, albeit on the level of distributions. Note that we do not need to impose any assumptions on the value of $\Phi(s)(B)$ whenever B is an $\mathsf{r}_{\mathcal{L}}$-invariant Borel subset of T other than the validity set of some formula. This is so since standard arguments including the π-λ-Theorem 1.3.1 show that $\Phi(s)(B) \in \{0,1\}$ whenever $B \in \mathbf{\Sigma}\big(\mathcal{B}(T), \mathsf{r}_{\mathcal{L}}\big)$. In the common parlance of probability theory, the measure $\Phi(x)$ is ergodic with respect to the equivalence relation $\mathsf{r}_{\mathcal{L}}$ for each $x \in G$.

Turning to logical equivalence, we define

Definition 3.4.16. *Let* $\mathcal{K}$ *and* $\mathcal{L}$ *be Kripke models over the analytic spaces* S *respectively* T. $\mathcal{K}$ *and* $\mathcal{L}$ *are* distributionally equivalent *iff these conditions hold:*

a. for all $\mu \in \mathfrak{S}(S)$ *there exists* $\nu \in \mathfrak{S}(T)$ *with* $\forall\phi \in \mathbf{F}_{\mathfrak{L}} : \mu(\llbracket\phi\rrbracket_{\mathcal{K}}) = \nu(\llbracket\phi\rrbracket_{\mathcal{L}})$,
b. for all $\nu \in \mathfrak{S}(T)$ *there exists* $\mu \in \mathfrak{S}(S)$ *with* $\forall\phi \in \mathbf{F}_{\mathfrak{L}} : \nu(\llbracket\phi\rrbracket_{\mathcal{L}}) = \mu(\llbracket\phi\rrbracket_{\mathcal{K}})$.
□

We define the relation $\Re_{\mathcal{K},\mathcal{L}} \subseteq \mathfrak{S}(S) \times \mathfrak{S}(T)$ through

$$\Re_{\mathcal{K},\mathcal{L}} := \bigcap_{\phi \in \mathbf{F}_{\mathfrak{L}}} \{\langle\mu,\nu\rangle \mid \mu(\llbracket\phi\rrbracket_{\mathcal{K}}) = \nu(\llbracket\phi\rrbracket_{\mathcal{L}})\};$$

thus $\mathcal{K}$ and $\mathcal{L}$ are distributionally equivalent iff relation $\Re_{\mathcal{K},\mathcal{L}}$ is both left and right complete.

The definition for logical equivalence addresses the states of a Kripke model. It reads that given a state in one model there exists a state in the other model such that the respective theories coincide, so that given $x \in G$ there exists $y \in H$ with $\mathcal{K}, x \models \phi \Leftrightarrow \mathcal{L}, y \models \phi$ for each formula ϕ, and vice versa. Since $\mathcal{K}, x \models \phi$ is equivalent to $x \in \llbracket\phi\rrbracket_{\mathcal{K}}$ which in turn is equivalent to $\delta_x(\llbracket\phi\rrbracket_{\mathcal{K}}) = 1$ with δ_x as the Dirac measure on x, we can state that $\mathcal{K}$ and $\mathcal{L}$ are logically equivalent iff for each $x \in G$ there exists $y \in H$ such that

$$\forall\phi \in \mathbf{F}_{\mathfrak{L}} : \delta_x(\llbracket\phi\rrbracket_{\mathcal{K}}) = \delta_y(\llbracket\phi\rrbracket_{\mathcal{L}}),$$

and vice versa. Thus distributional equivalence generalizes logical equivalence.

Logically equivalent models are distributionally equivalent. This is not readily apparent, however, since distributional equivalence requires finding for a given distribution on one model a matching distribution on the other, but the strong version talks only about states. Because there exists a cospan of morphisms for logically equivalent models, however, this relationship can be established.

Proposition 3.4.17. *Let $\mathcal{K}$ and $\mathcal{L}$ be logically equivalent models. Then $\mathcal{K}$ and $\mathcal{L}$ are distributionally equivalent.*

Proof Since $\mathcal{K}$ and $\mathcal{L}$ are logically equivalent, there exists a cospan

$$\mathcal{K} \xrightarrow{\gamma} \mathcal{E} \xleftarrow{\chi} \mathcal{L}$$

for a suitable model $\mathcal{E}$ and strong morphisms γ and χ by Theorem 2.3.11. Morphisms are based on surjective Borel maps. Take μ in $\mathfrak{S}(S)$; then $\mathfrak{S}(\gamma)(\mu) \in \mathfrak{S}(E)$, since $\chi : G \to E$ is onto, so is $\mathfrak{S}(\chi) : \mathfrak{S}(S) \to \mathfrak{S}(E)$ by Proposition 1.6.19, so that there exists $\nu \in \mathfrak{S}(T)$ with $\mathfrak{S}(\chi)(\nu) = \mathfrak{S}(\gamma)(\mu)$. Now consider a formula $\phi \in \mathbf{F}_{\mathfrak{L}}$. From $[\![\phi]\!]_{\mathcal{K}} = \gamma^{-1}\left[[\![\phi]\!]_{\mathcal{E}}\right]$, and, similarly, $[\![\phi]\!]_{\mathcal{L}} = \chi^{-1}\left[[\![\phi]\!]_{\mathcal{E}}\right]$ for χ, we infer

$$\mu([\![\phi]\!]_{\mathcal{K}}) = \mathfrak{S}(\gamma)(\mu)([\![\phi]\!]_{\mathcal{E}}) = \mathfrak{S}(\chi)(\nu)([\![\phi]\!]_{\mathcal{E}}) = \nu([\![\phi]\!]_{\mathcal{L}}).$$

□

If the models $\mathcal{K}$ and $\mathcal{L}$ are logically equivalent, then the factor spaces $S/\mathsf{r}_{\mathcal{K}}$ and $T/\mathsf{r}_{\mathcal{L}}$ are Borel isomorphic by Corollary 2.2.5. A similar result is established now for the case of distributional equivalence, factoring the corresponding space of subprobabilities on the respective state spaces, rather than the state spaces proper.

Proposition 3.4.18. *Let $\mathcal{K}$ and $\mathcal{L}$ be distributionally equivalent Kripke models. Define $\tau : \mathfrak{S}(S)/\overline{\mathsf{r}_{\mathcal{K}}} \to \mathfrak{S}(L)/\overline{\mathsf{r}_{\mathcal{L}}}$ through*

$$\tau([\mu]_{\overline{\mathsf{r}_{\mathcal{K}}}}) := [\nu]_{\overline{\mathsf{r}_{\mathcal{L}}}} \text{ iff } \langle \mu, \nu \rangle \in \mathfrak{R}_{\mathcal{K},\mathcal{L}}.$$

Then τ is a Borel isomorphism.

Proof 1. Let $\mu_1 \ \overline{\mathsf{r}_{\mathcal{K}}}\ \mu_2$ and select $\nu_i \in \mathfrak{S}(T)$ with $\langle \mu_i, \nu_i \rangle \in \mathfrak{R}_{\mathcal{K},\mathcal{L}}$ for $i = 1, 2$; this is possible since the models are distributionally equivalent. Thus $\nu_1 \ \overline{\mathsf{r}_{\mathcal{L}}}\ \nu_2$, so that τ is well defined. τ is plainly injective, and from the construction it is clear that τ is also onto. Let $\widetilde{\tau}$ be the inverse of τ.

2. We show that the image $\tau[A]$ of a Borel set $A \subseteq \mathfrak{S}(S)/\overline{\mathsf{r}_{\mathcal{K}}}$ is a Borel set in $\mathfrak{S}(T)/\overline{\mathsf{r}_{\mathcal{L}}}$; the argumentation is similar to the one in the proof of Lemma 1.8.11. $\tau[A]$ is an analytic set by Proposition 1.4.26, and

$$\tau\left[\left(\mathfrak{S}(S)/\overline{\mathsf{r}_{\mathcal{K}}}\right) \setminus A\right] = \left(\mathfrak{S}(T)/\overline{\mathsf{r}_{\mathcal{L}}}\right) \setminus \tau[A].$$

In fact, if $b = [\nu]_{\mathsf{r}_{\mathcal{L}}} \in \tau\left[\left(\mathfrak{S}(S)/\overline{\mathsf{r}_{\mathcal{K}}}\right) \setminus A\right]$, then we can find $a = [\mu]_{\mathsf{r}_{\mathcal{K}}} \notin A$ with $b = \tau(a)$. If there exists $a' = [\mu']_{\mathsf{r}_{\mathcal{K}}} \in A$ with $b = \tau(a')$, then we infer for each formula ϕ that $\mu([\![\phi]\!]_{\mathcal{K}}) = \nu([\![\phi]\!]_{\mathcal{L}}) = \mu'([\![\phi]\!]_{\mathcal{K}})$, contradicting $[\mu]_{\mathsf{r}_{\mathcal{K}}} \neq [\mu']_{\mathsf{r}_{\mathcal{K}}}$. This gives one inclusion; the other is trivial. Consequently, both $\tau[A]$ and its complement are analytic sets, so they are Borel sets by Souslin's Theorem 1.4.29. Hence $\widetilde{\tau}$ is Borel measurable; interchanging the rôles of S and T shows that τ is Borel. □

Assume for the rest of this section that $\mathcal{K}$ and $\mathcal{L}$ are distributionally equivalent stochastic Kripke models.

Proposition 3.4.19. *The map* τ *makes this diagram*

$$\begin{array}{ccc} \mathfrak{S}(S)/\overline{\mathsf{r}_{\mathcal{K}}} & \xrightarrow{\Theta_{a,\mathcal{K}}} & \mathfrak{S}(S)/\overline{\mathsf{r}_{\mathcal{K}}} \\ \downarrow{\scriptstyle\tau} & & \downarrow{\scriptstyle\tau} \\ \mathfrak{S}(T)/\overline{\mathsf{r}_{\mathcal{L}}} & \xrightarrow[\Theta_{a,\mathcal{L}}]{} & \mathfrak{S}(T)/\overline{\mathsf{r}_{\mathcal{L}}} \end{array}$$

commutative for every action $a \in \mathsf{Act}$.

Proof Recall from Lemma 3.4.14 that $\Theta_{a,\mathcal{K}}([\mu]_{\overline{\mathsf{r}_{\mathcal{K}}}}) = \left[\overline{k_a}(\mu)\right]_{\overline{\mathsf{r}_{\mathcal{K}}}}$ for $\mu \in \mathfrak{S}(S)$. It is sufficient to show that $\langle \mu, \nu\rangle \in \Re_{\mathcal{K},\mathcal{L}}$ implies $\langle \overline{k_a}(\mu), \overline{\ell_a}(\nu)\rangle \in \Re_{\mathcal{K},\mathcal{L}}$.

In fact, let $\phi \in \mathbf{F}_{\mathfrak{L}}$; then

$$\begin{aligned} \overline{\ell_a}(\nu)([\![\phi]\!]_{\mathcal{L}}) &= \sup_{n\in\mathbb{N}} \int_0^1 \nu([\![\langle a\rangle_{r+\frac{1}{n}}\phi]\!]_{\mathcal{L}})\ dr \\ &= \sup_{n\in\mathbb{N}} \int_0^1 \mu([\![\langle a\rangle_{r+\frac{1}{n}}\phi]\!]_{\mathcal{K}})\ dr \\ &= \overline{k_a}(\mu)([\![\phi]\!]_{\mathcal{K}}) \end{aligned}$$

by Lemma 3.4.11. □

Now consider this diagram with $\nabla := (\mathfrak{z}_T \circ \kappa_T)^{-1} \circ \tau \circ (\mathfrak{z}_S \circ \kappa_S)$.

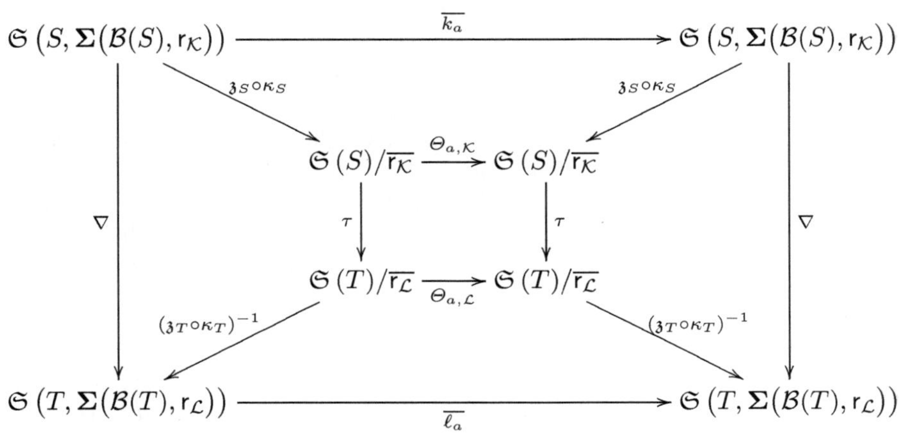

We infer from Proposition 1.6.15 that $s \mapsto k_a(s)(E)$ is $\mathbf{\Sigma}(\mathcal{B}(S), \mathsf{r}_\mathcal{K})$-measurable whenever $E \in \mathbf{\Sigma}(\mathcal{B}(S), \mathsf{r}_\mathcal{K})$. Consequently, we obtain from Corollary 3.3.2 that $\overline{k_a} : \mathfrak{S}\left(S, \mathbf{\Sigma}(\mathcal{B}(S), \mathsf{r}_\mathcal{K})\right) \to \mathfrak{S}\left(S, \mathbf{\Sigma}(\mathcal{B}(S), \mathsf{r}_\mathcal{K})\right)$, so that $\overline{k_a}$ has the proper domain and codomain; the same argument shows that $\overline{\ell_a}$ maps $\mathfrak{S}\left(T, \mathbf{\Sigma}(\mathcal{B}(T), \mathsf{r}_\mathcal{L})\right)$ to $\mathfrak{S}\left(T, \mathbf{\Sigma}(\mathcal{B}(T), \mathsf{r}_\mathcal{L})\right)$. The upper trapezoid commutes because we have for $\mu \in \mathfrak{S}\left(S, \mathbf{\Sigma}(\mathcal{B}(S), \mathsf{r}_\mathcal{K})\right)$

$$\Theta_{a,\mathcal{K}}\big(\mathfrak{z}_S \circ \kappa_S\big)(\mu) = \big[\overline{k_a}(\mu)\big]_{\overline{\mathsf{r}_\mathcal{K}}} = \big((\mathfrak{z}_S \circ \kappa_S) \circ \overline{k_a}\big)(\mu).$$

Similarly, the lower one commutes. By Proposition 3.4.19 the inner rectangle commutes, so that the outer rectangle commutes as well. We infer from this that $\nabla(\mu)([\![\phi]\!]_\mathcal{L}) = \mu([\![\phi]\!]_\mathcal{K})$ for all formulas $\phi \in \mathbf{F}_\mathfrak{L}$, and for all $\mu \in \mathfrak{S}\left(S, \mathbf{\Sigma}(\mathcal{B}(S), \mathsf{r}_\mathcal{K})\right)$. If we would be able to represent ∇ as the Kleisli extension to some stochastic relation, then we would be done. Unfortunately, the construction does not warrant such a construction. It will become apparent, however, that we can single out the images of Dirac measures under ∇. This will produce a stochastic relation with the desired properties.

Define to this end $\Lambda(x)(A) := \nabla(\delta_x)(A)$ for $x \in G$ and $A \in \mathbf{\Sigma}(\mathcal{B}(T), \mathsf{r}_\mathcal{L})$; then the construction entails that $\Lambda : (S, \mathbf{\Sigma}(\mathcal{B}(S), \mathsf{r}_\mathcal{K})) \rightsquigarrow (T, \mathbf{\Sigma}(\mathcal{B}(T), \mathsf{r}_\mathcal{L}))$ is a stochastic relation, and we infer from the construction of ∇ that

$$\Lambda(x)([\![\phi]\!]_\mathcal{L}) = \delta_x([\![\phi]\!]_\mathcal{K}).$$

Thus $x \mapsto \Lambda(x)([\![\phi]\!]_\mathcal{L})$ takes only the values 0 or 1 for a formula ϕ with

$$\Lambda(x)([\![\phi]\!]_\mathcal{L}) = 1 \Leftrightarrow x \in [\![\phi]\!]_\mathcal{K}.$$

Lemma 3.4.20. $\left(\Lambda * k_a\right)(x)(A) = \left(\ell_a * \Lambda\right)(x)(A)$ *holds for all* $x \in G, A \in \mathbf{\Sigma}(\mathcal{B}(T), \mathsf{r}_\mathcal{L})$ *and for each action* $a \in \mathsf{Act}$.

Proof 1. We compute first the probabilities for a formula ϕ. It is easy to see that

$$\left(\Lambda * k_a\right)(x)([\![\phi]\!]_\mathcal{L}) = \int_S \Lambda(y)([\![\phi]\!]_\mathcal{L})\ k_a(x)(dy) = k_a(x)([\![\phi]\!]_\mathcal{K}).$$

On the other hand we infer from Lemma 3.4.11 that

$$\begin{aligned}\left(\ell_a * \Lambda\right)(x)([\![\phi]\!]_\mathcal{L}) &= \int_T \ell_a(y)([\![\phi]\!]_\mathcal{L})\ \Lambda(x)(dy) \\ &= \sup_{n \in \mathbb{N}} \int_0^1 \Lambda(x)([\![\langle a \rangle_{r+\frac{1}{n}} \phi]\!]_\mathcal{L})\ dr\end{aligned}$$

with $\Lambda(x)([\![\langle a \rangle_{r+\frac{1}{n}} \phi]\!]_\mathcal{L}) = 1$ iff $r \leq k_a(x)([\![\phi]\!]_\mathcal{K}) - \frac{1}{n}$. Thus the latter integral evaluates to

$$\sup_{n\in\mathbb{N}}\int_0^{k_a(x)(\llbracket\phi\rrbracket_{\mathcal{K}})-\frac{1}{n}} 1\ dr = \sup_{n\in\mathbb{N}}\bigl(k_a(x)(\llbracket\phi\rrbracket_{\mathcal{K}}) - \frac{1}{n}\bigr)$$
$$= k_a(x)(\llbracket\phi\rrbracket_{\mathcal{K}}).$$

Consequently, $(\Lambda * k_a)(x)(\llbracket\phi\rrbracket_{\mathcal{L}}) = (\ell_a * \Lambda)(x)(\llbracket\phi\rrbracket_{\mathcal{L}})$ for all $x \in G$ and all formulas $\phi \in \mathbf{F}_{\mathfrak{L}}$.

2. Now consider

$$\mathcal{D} := \{A \in \mathbf{\Sigma}\bigl(\mathcal{B}(T), \mathsf{r}_{\mathcal{L}}\bigr) \mid \forall x \in G : (\Lambda * k_a)(x)(A) = (\ell_a * \Lambda)(x)(A)\}.$$

Then $\mathcal{D}$ is closed under complementation, and under countable disjoint unions. The first part of the proof shows that $\llbracket\phi\rrbracket_{\mathcal{L}} \in D$ for each formula ϕ. Since $\{\llbracket\phi\rrbracket_{\mathcal{L}} \mid \phi \in \mathbf{F}_{\mathfrak{L}}\}$ is closed under intersections, from the π-λ-Theorem 1.3.1 it is inferred that $\mathcal{D} = \sigma\bigl(\{\llbracket\phi\rrbracket_{\mathcal{L}} \mid \phi \in \mathbf{F}_{\mathfrak{L}}\}\bigr)$. □

Summarizing, we have partially established the following.

Proposition 3.4.21. *Let $\mathcal{K}$ and $\mathcal{L}$ be distributionally equivalent stochastic Kripke models over analytic spaces. Then the following statements are equivalent:*

a. $\mathcal{K}$ and $\mathcal{L}$ are distributionally equivalent.
b. There exists $\mathfrak{L}$-ergodic morphisms $\Lambda : \mathcal{K} \curvearrowright^{\mathsf{e}} \mathcal{L}$ and $\Xi : \mathcal{L} \curvearrowright^{\mathsf{e}} \mathcal{K}$.

Proof We have just given the argument and a construction for proving $a \Rightarrow b$. For establishing $b \Rightarrow a$, take $\mu \in \mathfrak{S}(S)$ and a formula ϕ. Then

$$\overline{\Lambda}(\mu)(\llbracket\phi\rrbracket_{\mathcal{L}}) = \int_S \Lambda(x)(\llbracket\phi\rrbracket_{\mathcal{L}})\ \mu(dx) = \int_{\llbracket\phi\rrbracket_{\mathcal{K}}} 1\ d\mu = \mu(\llbracket\phi\rrbracket_{\mathcal{K}}),$$

since Λ is binary. Thus $\Lambda(\mu) \in \mathfrak{S}(T)$ matches $\mu \in \mathfrak{S}(S)$ on the logic; similarly, $\Xi(\nu) \in \mathfrak{S}(S)$ matches $\nu \in \mathfrak{S}(T)$ there. □

We consider as a corollary the factor systems $\mathcal{K}/\mathfrak{L}$ and $\mathcal{L}/\mathfrak{L}$. Define for a stochastic relation $\Phi : (S, \mathbf{\Sigma}(\mathcal{B}(S), \mathsf{r}_{\mathcal{K}})) \rightsquigarrow (T, \mathbf{\Sigma}(\mathcal{B}(T), \mathsf{r}_{\mathcal{L}}))$ as in Section 3.4.1 the factor relation $\Phi_{\mathfrak{L}} : S/\mathsf{r}_{\mathcal{K}} \to \mathfrak{S}(T/\mathsf{r}_{\mathcal{L}})$ upon setting

$$\Phi_{\mathfrak{L}}([x]_{\mathsf{r}_{\mathcal{K}}})(C) := \Phi(x)(\eta_{\mathsf{r}_{\mathcal{L}}}^{-1}[C]).$$

From Proposition 3.4.21 we obtain the following as a first consequence.

Corollary 3.4.22. *Let $\mathcal{K}$ and $\mathcal{L}$ be distributionally equivalent Kripke models over analytic spaces. Then there exists weak morphisms $\widetilde{\Lambda} : \mathcal{K}/\mathfrak{L} \curvearrowright \mathcal{L}/\mathfrak{L}$ and $\widetilde{\Xi} : \mathcal{L}/\mathfrak{L} \curvearrowright \mathcal{K}/\mathfrak{L}$.*

Proof As expected, we obtain $\widetilde{\Lambda} := \Lambda_{\mathfrak{L}}$ and $\widetilde{\Xi} := \Xi_{\mathfrak{L}}$ from Λ respectively Ξ upon factoring. We will show that $\widetilde{\Lambda}$ is in fact a weak morphism: Let $x \in G$ and $E \in \mathcal{B}(T/\mathsf{r}_{\mathcal{L}})$, then

$$\begin{aligned}
(\ell_{a,\mathfrak{L}} * \Lambda_{\mathfrak{L}})\left([x]_{\mathsf{r}_{\mathcal{K}}}\right)(E) &= \int_{T/\mathsf{r}_{\mathcal{L}}} \ell_{a,\mathfrak{L}}(v)(E)\ \Lambda_{\mathfrak{L}}([x]_{\mathsf{r}_{\mathcal{K}}})(dv) \\
&\overset{(\natural)}{=} \int_{T} \ell_a(y)(\eta_{\mathsf{r}_{\mathcal{L}}}^{-1}[E])\ \Lambda(x)(dy) \\
&\overset{(£)}{=} \int_{S} \Lambda(t)(\eta_{\mathsf{r}_{\mathcal{L}}}^{-1}[E])\ k_a(x)(dt) \\
&\overset{(\natural)}{=} \int_{S/\mathsf{r}_{\mathcal{K}}} \Lambda_{\mathfrak{L}}(w)(E)\ k_{a,\mathfrak{L}}([x]_{\mathsf{r}_{\mathcal{K}}})(dw) \\
&= (\Lambda_{\mathfrak{L}} * k_{a,\mathfrak{L}})\left([x]_{\mathsf{r}_{\mathcal{K}}}\right)(E).
\end{aligned}$$

The lines marked ($\natural$) derive from the Change of Variables formula; the line marked ($£$) results from Proposition 3.4.21. It should be noted that $\eta_{\mathsf{r}_{\mathcal{L}}}^{-1}[E]$ is a member of $\mathbf{\Sigma}(\mathcal{B}(T), \mathsf{r}_{\mathcal{L}})$; hence $\Lambda(x)$ is defined for it for each $x \in G$.

Summarizing, $\widetilde{\Lambda} := \Lambda_{\mathfrak{L}}$ is the desired weak morphism $\mathcal{K}/\mathfrak{L} \curvearrowright \mathcal{L}/\mathfrak{L}$. Similarly, one shows that $\widetilde{\Xi} := \Xi_{\mathfrak{L}}$ will do the job for $\mathcal{L}/\mathfrak{L} \curvearrowright \mathcal{K}/\mathfrak{L}$. □

As a second consequence one obtains that distributional equivalence entails the existence of a cospan between these models. This is captured through the notion of behavioral equivalence.

Definition 3.4.23. *The stochastic Kripke models $\mathcal{K}$ and $\mathcal{L}$ are called* weakly behaviorally equivalent *iff there is a cospan $\mathcal{K} \curvearrowright \mathcal{M} \curvearrowleft \mathcal{L}$ of weak morphisms for some suitable stochastic Kripke model $\mathcal{M}$.*

Assume that $\mathcal{K} \xrightarrow{f} \mathcal{M} \xleftarrow{g} \mathcal{L}$ is a span of strong morphisms. Let $s \in S$, because g is onto, we can find $t \in T$ with $f(s) = g(t)$. Thus we have for each formula ϕ

$$\mathcal{K}, s \models \phi \Leftrightarrow \mathcal{K}, f(s) \models \phi \Leftrightarrow \mathcal{K}, g(t) \models \phi \Leftrightarrow \mathcal{L}, t \models \phi.$$

Consequently, given $s \in S$ there exists $t \in T$ so that both states satisfy exactly the same formulas; conversely, we can find for each state in $\mathcal{L}$ a state in $\mathcal{K}$ with this property. This observation accounts for behavioral equivalence. Weak behavioral equivalence accounts for the existence of a cospan, but we will see in Theorem 3.4.34 that the connection is deeper.

It is quite easy to see now that distributional equivalence implies weak behavioral equivalence.

Proposition 3.4.24. *Distributionally equivalent Kripke models are weakly behaviorally equivalent.*

Proof Let $\mathcal{K}$ and $\mathcal{L}$ be distributionally equivalent Kripke models, and select arbitrarily one of them, say, $\mathcal{L}$. The factor maps $\eta_{\mathsf{r}_{\mathcal{K}}} : \mathcal{K} \to \mathcal{K}/\mathfrak{L}$ respectively $\eta_{\mathsf{r}_{\mathcal{L}}} : \mathcal{L} \to \mathcal{L}/\mathfrak{L}$ are strong morphisms by Proposition 3.4.9, yielding the weak morphisms $\mathsf{F}_{\eta_{\mathsf{r}_{\mathcal{K}}}}$ respectively $\mathsf{F}_{\eta_{\mathsf{r}_{\mathcal{L}}}}$; see Lemma 3.3.5. Weak morphisms are closed under Kleisli composition, because they are morphisms in a category (see Section 3.2). Put $\mathcal{K} := \mathcal{L}/\mathfrak{L}$; then $\widetilde{\Lambda} * \mathsf{F}_{\eta_{\mathsf{r}_{\mathcal{K}}}} : \mathcal{K} \curvearrowright \mathcal{K}$ and $\mathsf{F}_{\eta_{\mathsf{r}_{\mathcal{L}}}} : \mathcal{L} \curvearrowright \mathcal{K}$ together form the desired cospan. □

An analysis of the proof for the existence of bisimulations for logically equivalent Kripke models and coalgebras reveals usually the following pattern. Constructing the bisimulation requires first constructing a cospan of morphisms. The equivalence relations corresponding to $\mathsf{r}_{\mathcal{K}}$ and $\mathsf{r}_{\mathcal{L}}$ are shown to generate Borel isomorphic factor spaces, the factor map being a morphism. These factors are used to build a sum which in turn is the target for the cospan. Examining this construction closer, it turns out that the individual factor spaces are isomorphic to the sum, and that the gist of the construction lies in the one-to-one correspondence of the equivalence classes. The upshot is that the same construction works mutatis mutandis for the weak case as well, so that probably a more general principle is at work.

3.4.4 Bisimulations

We want to relate bisimulations for strong and for weak morphisms as well. Before we are able to do this, we need some simple measure-theoretic preparations. They will help us cater to ergodic morphisms.

Lemma 3.4.25. *Let X be a metric space with a countable basis, and μ a probability measure on the Borel sets of X. Then there exists a unique closed set C_μ with these properties:*

1. $\mu(C_\mu) = 1$,
2. $x \in C_\mu$ *iff* $\mu(U) > 0$ *for each open neighborhood U of x.*

C_μ is called the support *of μ.*

Proof Let $\mathcal{G} := \{G \subseteq X \mid G \text{ is open}, \mu(G) = 0\}$; then $V := \bigcup \mathcal{G}$ is open, and $\mu(V) = 0$. This is so since the topology of X has a countable basis, so that we can find $(U_n)_{n\in\mathbb{N}} \subseteq \mathcal{G}$ with $\bigcup_{n\in\mathbb{N}} U_n = V$. Put $C_\mu := X \setminus V$; then $\mu(C_\mu) = 1$, and C_μ is closed. If $x \in C_\mu$ and U is an open neighborhood of x, then $\mu(U) = 0$ would imply $U \in \mathcal{G}$, and thus $U \subseteq V$, hence $x \notin C_\mu$, contradicting the assumption. □

This has as a consequence that binary-valued measures on an analytic space live on single points.

Lemma 3.4.26. *Let A be an analytic space. If the subprobability $\mu \in \mathfrak{S}(A)$ takes only the values 0 and 1, and if $\mu(A) = 1$, then there exists a unique $a \in A$ such that $\mu = \delta_a$.*

Proof By Corollary 1.4.35 and Lemma 1.4.36 there exists a separable metric topology for A that generates its Borel sets. So we may and do assume that A is a metric space with a countable basis. Assume that the support C_μ

contains more than one point. Let $x_1, x_2 \in C_\mu$ with $x_1 \neq x_2$; then there exists disjoint open neighborhoods V_i of x_i with $\mu(V_i) > 0 \quad (i = 1, 2)$. But one of them must have measure 0 on behalf of the assumption and their disjointness. This contradicts $x_1, x_2 \in C_\mu$; so there exists a unique $a \in A$ with $C_\mu = \{a\}$. □

This has as an interesting consequence that an ergodic morphism is given through a measurable map.

Lemma 3.4.27. *Let $\Psi : \mathcal{K} \curvearrowright^{\mathsf{e}} \mathcal{L}$ be an $\mathfrak{L}$-ergodic morphism. Then there exists a measurable map $\psi : S/\mathsf{r}_\mathcal{K} \to T/\mathsf{r}_\mathcal{L}$ such that $\Psi_\mathfrak{L}(v) = \delta_{\psi(v)}$ for all $v \in S/\mathsf{r}_\mathcal{K}$.*

Proof 1. For each $s \in S$ the measure $\Psi([s]_{\mathsf{r}_\mathcal{K}}) \in \mathfrak{S}(T/\mathsf{r}_\mathcal{L})$ is binary-valued. Thus there exists by Lemma 3.4.26 for each $s \in S$ some $t \in T$ such that $\Psi_\mathfrak{L}([s]_{\mathsf{r}_\mathcal{K}}) = \delta_{[t]_{\mathsf{r}_\mathcal{L}}}$. Since $[t]_{\mathsf{r}_\mathcal{L}}$ is uniquely determined, the map $\psi([s]_{\mathsf{r}_\mathcal{K}}) := [t]_{\mathsf{r}_\mathcal{L}}$ is well defined.

2. Because Ψ is ergodic, the map ψ is injective. In fact, assume that

$$\psi([s]_{\mathsf{r}_\mathcal{K}}) = \psi([s']_{\mathsf{r}_\mathcal{K}});$$

then we know for each formula $\phi \in \mathbf{F}_\mathfrak{L}$ that $\Psi_\mathfrak{L}([s]_{\mathsf{r}_\mathcal{K}})([\![\phi]\!]_\mathcal{L}) = 1$ iff $\Psi_\mathfrak{L}([s']_{\mathsf{r}_\mathcal{K}})([\![\phi]\!]_\mathcal{L}) = 1$. Consequently, we know for each $\phi \in \mathbf{F}_\mathfrak{L}$ that

$$\mathcal{K}, s \models \phi \Leftrightarrow \mathcal{K}, s' \models \phi.$$

But this means that $[s]_{\mathsf{r}_\mathcal{K}} = [s']_{\mathsf{r}_\mathcal{K}}$ holds.

3. The map ψ is Borel measurable. One observes first that

$$[\![\phi]\!]_\mathcal{K} = \eta_{\mathsf{r}_\mathcal{K}}^{-1}\left[\psi^{-1}\left[\eta_{\mathsf{r}_\mathcal{L}}\left[[\![\phi]\!]_\mathcal{L}\right]\right]\right]$$

for each $\phi \in \mathbf{F}_\mathfrak{L}$. Now

$$\begin{aligned} s \in [\![\phi]\!]_\mathcal{K} &\Leftrightarrow \Psi(s)([\![\phi]\!]_\mathcal{L}) = 1 \\ &\overset{(\natural)}{\Leftrightarrow} \Psi_\mathfrak{L}([s]_{\mathsf{r}_\mathcal{K}})(\eta_{\mathsf{r}_\mathcal{L}}\left[[\![\phi]\!]_\mathcal{L}\right]) = 1 \\ &\Leftrightarrow \psi([s]_{\mathsf{r}_\mathcal{K}}) \in \eta_{\mathsf{r}_\mathcal{L}}\left[[\![\phi]\!]_\mathcal{L}\right] \\ &\Leftrightarrow (\psi \circ \eta_{\mathsf{r}_\mathcal{L}})(s) \in \eta_{\mathsf{r}_\mathcal{L}}\left[[\![\phi]\!]_\mathcal{L}\right] \\ &\Leftrightarrow s \in \eta_{\mathsf{r}_\mathcal{K}}^{-1}\left[\psi^{-1}\left[\eta_{\mathsf{r}_\mathcal{L}}\left[[\![\phi]\!]_\mathcal{L}\right]\right]\right]. \end{aligned}$$

The equivalence $(\natural)$ uses that by definition $\Psi(s)(D)$ equals $\Psi_\mathfrak{L}([x]_{\mathsf{r}_\mathcal{K}})(\eta_{\mathsf{r}_\mathcal{L}}[D])$ whenever $D \in \mathbf{\Sigma}(\mathcal{B}(T), \mathsf{r}_\mathcal{L})$, and that for $\mathsf{r}_\mathcal{L}$-invariant D the equality $D = \eta_{\mathsf{r}_\mathcal{L}}^{-1}\left[\eta_{\mathsf{r}_\mathcal{L}}[D]\right]$ holds. Consequently,

$$\{R \in \mathcal{B}(T/\mathsf{r}_\mathcal{L}) \mid \psi^{-1}[R] \in \mathcal{B}(S/\mathsf{r}_\mathcal{K})\}$$

includes the set $\{\eta_{\mathsf{r}_\mathcal{L}}\left[[\![\phi]\!]_\mathcal{L}\right] \mid \phi \in \mathbf{F}_\mathfrak{L}\}$. The latter set is closed under intersections due to the logic being closed under conjunctions, and the former is closed under complements and disjoint countable unions. Thus it follows from the π-λ-Theorem Lemma 1.3.1 that $\psi^{-1}[R] \in \mathcal{B}(S/\mathsf{r}_{\mathcal{K}})$ for all $R \in \mathcal{B}(T/\mathsf{r}_{\mathcal{L}})$. □

Remark 3.4.28. Measurability of ψ may be established through another argument as well. Denote by $\mathfrak{e}$ the unit for the Giry monad. Then in general $\mathcal{M}$ is the initial σ-algebra with respect to $\mathfrak{e}_M$ and $\mathcal{M}^\bullet$ whenever $(M, \mathcal{M})$ is a measurable space by Lemma 1.3.4. Because this diagram

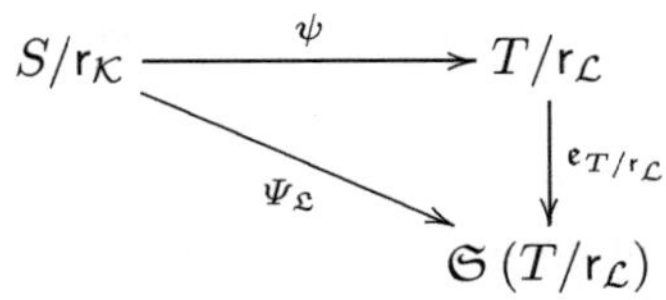

commutes, measurability of ψ follows from initiality by Lemma 1.3.4. □

The map ψ constructed in this way is nearly a morphism:

Corollary 3.4.29. *This diagram commutes, ψ being constructed according to Lemma 3.4.27:*

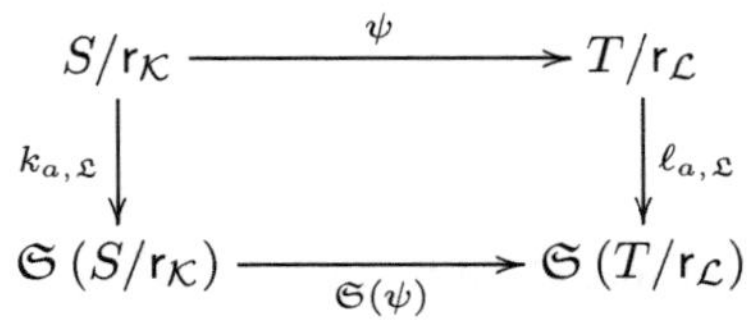

Consequently, ψ preserves and reflects validity: the equivalence

$$(*)\ \mathcal{K}/\mathfrak{L}, t \models \phi \Leftrightarrow \mathcal{L}/\mathfrak{L}, \psi(t) \models \phi$$

is true for every $t \in S/\mathsf{r}_{\mathcal{K}}$ and every formula $\phi \in \mathbf{F}_{\mathfrak{L}}$.

Proof 1. We obtain from Corollary 3.4.22 that $\Psi_{\mathfrak{L}} : \mathcal{K}/\mathfrak{L} \rightsquigarrow \mathcal{L}/\mathfrak{L}$ is a weak morphism; thus we have for $x \in G, D \in \mathcal{B}(T/\mathsf{r}_{\mathcal{L}})$

$$\begin{aligned}
(\ell_{a,\mathfrak{L}} \circ \psi)([x]_{\mathsf{r}_{\mathcal{K}}})(D) &= (\ell_{a,\mathfrak{L}} * \Psi_{\mathfrak{L}})([x]_{\mathsf{r}_{\mathcal{K}}})(D) \\
&= (\Psi_{\mathfrak{L}} * k_{a,\mathfrak{L}})([x]_{\mathsf{r}_{\mathcal{K}}})(D) \\
&= \int_{S/\mathsf{r}_{\mathcal{K}}} \delta_{\psi(z)}(D)\ k_{a,\mathfrak{L}}([x]_{\mathsf{r}_{\mathcal{K}}})(dt) \\
&= k_{a,\mathfrak{L}}([x]_{\mathsf{r}_{\mathcal{K}}})(\psi^{-1}[D]) \\
&= (\mathfrak{S}(\psi) \circ k_{a,\mathfrak{L}})([x]_{\mathsf{r}_{\mathcal{K}}})(D).
\end{aligned}$$

2. In order to establish the equivalence $(*)$, we proceed by induction on ϕ. The interesting case is $\phi = \langle a \rangle_q \phi'$, where we know that the equivalence has

already been established for formula ϕ'. In particular we know that $[\![\phi']\!]_{\mathcal{K}/\mathfrak{L}}$ equals $\psi^{-1}\left[[\![\phi']\!]_{\mathcal{L}/\mathfrak{L}}\right]$, so that

$$\begin{aligned}\mathcal{K}/\mathfrak{L}, t \models \langle a\rangle_q \phi' &\Leftrightarrow k_{a,\mathfrak{L}}(t)\big([\![\phi']\!]_{\mathcal{K}/\mathfrak{L}}\big) \geq q\\ &\Leftrightarrow \big(\mathfrak{S}(\psi) \circ k_{a,\mathfrak{L}}\big)(t)\big([\![\phi']\!]_{\mathcal{L}/\mathfrak{L}}\big) \geq q\\ &\overset{(\natural)}{\Leftrightarrow} \ell_{a,\mathfrak{L}}(\psi(t))\big([\![\phi']\!]_{\mathcal{L}/\mathfrak{L}}\big) \geq q\\ &\Leftrightarrow \mathcal{L}/\mathfrak{L}, \psi(t) \models \langle a\rangle_q \phi'.\end{aligned}$$

Equivalence $(\natural)$ derives from the commutative diagram. Hence we have established equivalence $(*)$ also for ϕ. □

We are going to show now that mutual ergodic morphisms induce an isomorphism on the factor spaces through the maps associated with them according to Lemma 3.4.27. The observation that the equivalence classes $[x]_{\mathsf{r}_\mathcal{K}}$ are exactly the atoms of the σ-algebra $\mathbf{\Sigma}\big(\mathcal{B}(S), \mathsf{r}_\mathcal{K}\big)$ will be helpful.

Lemma 3.4.30. *Let $\Psi : \mathcal{K} \curvearrowright^{\mathsf{e}} \mathcal{L}$ and $\Xi : \mathcal{L} \curvearrowright^{\mathsf{e}} \mathcal{K}$ be ergodic morphisms with associated maps $\psi : S/\mathsf{r}_\mathcal{K} \to T/\mathsf{r}_\mathcal{L}$ and $\xi : T/\mathsf{r}_\mathcal{L} \to S/\mathsf{r}_\mathcal{K}$. Then ψ and ξ are inverse to each other.*

Proof 1. Let ϕ be a formula, $s \in S$; then

$$\begin{aligned}\mathcal{K}, s \models \phi &\Leftrightarrow \Psi(s)\big([\![\phi]\!]_\mathcal{L}\big) = 1\\ &\Leftrightarrow \Psi_\mathfrak{L}\big([s]_{\mathsf{r}_\mathcal{K}}\big)\big([\![\phi]\!]_{\mathcal{L}/\mathfrak{L}}\big) = 1.\end{aligned}$$

This is implied by the construction of the factor relation in Section 1.7.3 (see p. 53) and from Proposition 3.4.10. But

$$[\![\phi]\!]_{\mathcal{L}/\mathfrak{L}} = \big\{[t]_{\mathsf{r}_\mathcal{L}} \mid \Xi_\mathfrak{L}\big([t]_{\mathsf{r}_\mathcal{L}}\big)\big([\![\phi]\!]_{\mathcal{K}/\mathfrak{L}}\big) = 1\big\}$$

for the same reasons; hence

$$\mathcal{K}, s \models \phi \Leftrightarrow \Psi_\mathfrak{L}\big([s]_{\mathsf{r}_\mathcal{K}}\big)\big(\big\{[t]_{\mathsf{r}_\mathcal{L}} \mid \Xi_\mathfrak{L}\big([t]_{\mathsf{r}_\mathcal{L}}\big)\big([\![\phi]\!]_{\mathcal{K}/\mathfrak{L}}\big) = 1\big\}\big) = 1.$$

But since $\Psi_\mathfrak{L}\big([s]_{\mathsf{r}_\mathcal{K}}\big) = \delta_{\psi([s]_{\mathsf{r}_\mathcal{K}})}$ and $\Xi_\mathfrak{L}\big([t]_{\mathsf{r}_\mathcal{L}}\big) = \delta_{\xi([t]_{\mathsf{r}_\mathcal{L}})}$, we obtain from this

$$\begin{aligned}\mathcal{K}, s \models \phi &\Leftrightarrow \xi\big(\psi([s]_{\mathsf{r}_\mathcal{K}})\big) \in [\![\phi]\!]_{\mathcal{K}/\mathfrak{L}}\\ &\Leftrightarrow \xi\big(\psi([s]_{\mathsf{r}_\mathcal{K}})\big) \subseteq \eta_{\mathsf{r}_\mathcal{K}}^{-1}\left[[\![\phi]\!]_{\mathcal{K}/\mathfrak{L}}\right] = [\![\phi]\!]_\mathcal{K}.\end{aligned}$$

Since $\xi\big(\psi([s]_{\mathsf{r}_\mathcal{K}})\big)$ is an equivalence class of $\mathsf{r}_\mathcal{K}$, and since $S \setminus [\![\phi]\!]_\mathcal{K}$ is $\mathsf{r}_\mathcal{K}$-invariant, we obtain

$$\begin{aligned}\mathcal{K}, s \not\models \phi &\Leftrightarrow \xi\big(\psi([s]_{\mathsf{r}_\mathcal{K}})\big) \notin [\![\phi]\!]_{\mathcal{K}/\mathfrak{L}}\\ &\Leftrightarrow \xi\big(\psi([s]_{\mathsf{r}_\mathcal{K}})\big) \subseteq S \setminus \eta_{\mathsf{r}_\mathcal{K}}^{-1}\left[[\![\phi]\!]_{\mathcal{K}/\mathfrak{L}}\right] = S \setminus [\![\phi]\!]_\mathcal{K}.\end{aligned}$$

2. Thus we see that

$$\xi(\psi([s]_{\mathsf{r}_\mathcal{K}})) \subseteq \bigcap\{[\![\phi]\!]_\mathcal{K} \mid \mathcal{K}, s \models \phi\} \cap \bigcap\{S \setminus [\![\phi]\!]_\mathcal{K} \mid \mathcal{K}, s \not\models \phi\} = [s]_{\mathsf{r}_\mathcal{K}},$$

the latter equality holding by Lemma 1.7.11. Since $\xi(\psi([s]_{\mathsf{r}_\mathcal{K}}))$ is an $\mathsf{r}_\mathcal{K}$-equivalence class by construction, we conclude that $\xi(\psi([s]_{\mathsf{r}_\mathcal{K}})) = [s]_{\mathsf{r}_\mathcal{K}}$, so that $\xi \circ \psi = id_{S/\mathsf{r}_\mathcal{K}}$. Similarly, $\psi \circ \xi = id_{T/\mathsf{r}_\mathcal{L}}$ is shown. □

Thus the existence of ergodic morphisms induces an isomorphism on the factor spaces of the state spaces for the Kripke models. This is now applied to distributionally equivalent transition systems.

Proposition 3.4.31. *Assume that $\mathcal{K}$ and $\mathcal{L}$ are distributionally equivalent Kripke models over analytic spaces. Then*

a. there exist strong morphisms $\mathcal{K}/\mathfrak{L} \to \mathcal{L}/\mathfrak{L}$ and $\mathcal{L}/\mathfrak{L} \to \mathcal{K}/\mathfrak{L}$,
b. the congruences $\mathsf{r}_\mathcal{K}$ and $\mathsf{r}_\mathcal{L}$ are simulation equivalent for the stochastic relations $k_a : S \rightsquigarrow S$ and $\ell_a : T \rightsquigarrow T$ for each action $a \in \mathsf{Act}$.

Proof 1. Construct the $\mathfrak{L}$-ergodic morphisms $\Psi : \mathcal{K} \curvearrowright^{\mathsf{e}} \mathcal{L}$ and $\Xi : \mathcal{L} \curvearrowright^{\mathsf{e}} \mathcal{K}$ according to Proposition 3.4.21 and let ψ and ξ be the associated point maps, as in Lemma 3.4.30. By this Lemma, ψ and ξ are inverse to each other, in particular they are onto. Thus they are strong morphisms, since the crucial diagram commutes by Corollary 3.4.29.

2. Denote by

$$\mathcal{R}_\mathcal{K} := \{[\![\phi]\!]_\mathcal{K} \mid \phi \in \mathbf{F}_\mathfrak{L}\}$$

the countable generator for the σ-algebra $\mathbf{\Sigma}(\mathcal{B}(S), \mathsf{r}_\mathcal{K})$; similarly for $\mathcal{R}_\mathcal{L}$. Both generators are closed under finite intersection, since the logic is closed under finite conjunctions. Because

$$[\![\phi]\!]_\mathcal{L} = \bigcup\{\psi([x]_{\mathsf{r}_\mathcal{K}}) \mid x \in [\![\phi]\!]_\mathcal{K}\},$$
$$[\![\phi]\!]_\mathcal{K} = \bigcup\{\xi([y]_{\mathsf{r}_\mathcal{L}}) \mid y \in [\![\phi]\!]_\mathcal{L}\},$$

we conclude that $\mathsf{r}_\mathcal{K}$ spawns $\mathsf{r}_\mathcal{L}$ via $(\psi, \mathcal{R}_\mathcal{K})$, and that $\mathsf{r}_\mathcal{L}$ spawns $\mathsf{r}_\mathcal{K}$ via $(\xi, \mathcal{R}_\mathcal{L})$. This is independent of any action $a \in \mathsf{Act}$.

3. It remains to show that $k_a(x)([\![\phi]\!]_\mathcal{K}) = \ell_a(y)([\![\phi]\!]_\mathcal{L})$ holds for all $x \in S, y \in \psi([x]_{\mathsf{r}_\mathcal{K}})$ as well as for all $y \in T, x \in \xi([y]_{\mathsf{r}_\mathcal{L}})$ whenever $\phi \in \mathbf{F}_\mathfrak{L}$ is an arbitrary formula. Discussing the first case, fix $x \in S, y \in \psi([x]_{\mathsf{r}_\mathcal{K}})$. Then

$$\begin{aligned} k_a(x)([\![\phi]\!]_\mathcal{K}) &= \int_S \Psi(x')([\![\phi]\!]_\mathcal{L})\ k_a(x)(dx') \\ &= (\Psi * k_a)(x)([\![\phi]\!]_\mathcal{L}) \\ &= (\ell_a * \Psi)(x)([\![\phi]\!]_\mathcal{L}) \\ &= \ell_{a,\mathfrak{L}}(\psi([x]_{\mathsf{r}_\mathcal{K}}))(\eta_{\mathsf{r}_\mathcal{L}}[[\![\phi]\!]_\mathcal{L}]) \\ &= \ell_a(y)([\![\phi]\!]_\mathcal{L}), \end{aligned}$$

whenever $[y]_{\mathsf{r}_{\mathcal{L}}} = \psi([x]_{\mathsf{r}_{\mathcal{K}}})$, i.e., whenever $y \in \psi([x]_{\mathsf{r}_{\mathcal{K}}})$ according to the definition of the factor relation. Thus the crucial equality from Definition 2.2.3 on page 73 is established for all $x \in S, y \in \psi([x]_{\mathsf{r}_{\mathcal{K}}})$.

4. Arguing symmetrically through the map ξ, this equality is established for all $y \in T, x \in \xi([y]_{\mathsf{r}_{\mathcal{L}}})$. This establishes part *b*. □

Turning to bisimilarity, we should make sure that the congruences we are dealing with are nontrivial. This will prevent trivial cases from creeping in, and it is technically necessary for applying the general criterion on bisimilarity. We first show that the congruences associated with distributionally equivalent Kripke models are nontrivial, provided one of the models is nondegenerate.

Lemma 3.4.32. *Assume that $\mathcal{K}$ and $\mathcal{L}$ are distributionally equivalent Kripke models. If $\mathcal{K}$ is nondegenerate, then both $\mathsf{r}_{\mathcal{K}}$ and $\mathsf{r}_{\mathcal{L}}$ are nontrivial congruences.*

Proof 0. We know that there exists a formula ϕ such that $\emptyset \neq [\![\phi]\!]_{\mathcal{K}} \neq S$. We will show that both $\mathsf{r}_{\mathcal{K}}$ and $\mathsf{r}_{\mathcal{L}}$ are not equal to the respective universal relations.

1. From the definition of the semantics it is clear that we may assume that formula ϕ can be written as $\langle a\rangle_q \phi'$ for some formula ϕ', some $a \in \mathsf{Act}$, and some number q. Thus there exists $s \in S$ with $k_a(s)\big([\![\phi']\!]_{\mathcal{K}}\big) \geq q$, and there exists $s' \in S$ with $k_a(s')\big([\![\phi']\!]_{\mathcal{K}}\big) < q$. Consequently, $\langle s, s'\rangle \notin \mathsf{r}_{\mathcal{K}}$, so that the relation $\mathsf{r}_{\mathcal{K}}$ is nontrivial.

2. Now construct $\psi : S/\mathsf{r}_{\mathcal{K}} \to T/\mathsf{r}_{\mathcal{L}}$ as in the proof of Proposition 3.4.31; then $\psi : \mathcal{K}/\mathfrak{L} \to \mathcal{L}/\mathfrak{L}$ constitutes a morphism, so that we have

$$\begin{aligned} k_a(s)\big([\![\phi']\!]_{\mathcal{K}}\big) &= k_{a,\mathfrak{L}}([s]_{\mathsf{r}_{\mathcal{K}}})\big(\eta_{\mathsf{r}_{\mathcal{K}}}[[\![\phi']\!]_{\mathcal{K}}]\big) \\ &= \ell_{a,\mathfrak{L}}(\psi([s]_{\mathsf{r}_{\mathcal{K}}}))\big(\eta_{\mathsf{r}_{\mathcal{K}}}[[\![\phi']\!]_{\mathcal{L}}]\big) \\ &= \ell_a(t)\big([\![\phi']\!]_{\mathcal{L}}\big). \end{aligned}$$

Here $t \in T$ is determined through $[t]_{\mathsf{r}_{\mathcal{K}} L} = \psi([s]_{\mathsf{r}_{\mathcal{K}}})$. Thus $\ell_a(t)([\![\phi']\!]_{\mathcal{L}}) \geq q$. Similarly, we find $t' \in T$ with $\ell_a(t')([\![\phi']\!]_{\mathcal{L}}) < q$. Thus $\langle t, t'\rangle \notin \mathsf{r}_{\mathcal{L}}$. Thus relation $\mathsf{r}_{\mathcal{L}}$ is nontrivial as well. □

Proposition 3.4.33. *Assume that $\mathcal{K}$ and $\mathcal{L}$ are distributionally equivalent and nondegenerate Kripke models. Then they are bisimilar.*

Proof 1. Since $\mathcal{K}$ and $\mathcal{L}$ are distributionally equivalent, the congruences $\mathsf{r}_{\mathcal{K}}$ and $\mathsf{r}_{\mathcal{L}}$ are simulation equivalent for the stochastic relations $k_a : S \rightsquigarrow S$ and $\ell_a : T \rightsquigarrow T$ for each $a \in \mathsf{Act}$ by Proposition 3.4.31. Since $\mathsf{r}_{\mathcal{K}}$ and $\mathsf{r}_{\mathcal{L}}$ are nontrivial by Lemma 3.4.32, we can find for each action $a \in \mathsf{Act}$ a stochastic relation $m_a : M_a \rightsquigarrow M_a$ and strong morphisms $k_a \xrightarrow{f_a} m_a \xleftarrow{g_a} \ell_a$.

2. Construct the amalgamated equivalence relation $\mathsf{r}_{\mathcal{K}} \diamond \mathsf{r}_{\mathcal{L}}$ according to Section 2.2 (see p. 74); then we infer from Proposition 1.6.25 that the state space M_a can be chosen as

$$M_a := \{\langle s,t\rangle \in S \times T \mid [s]_{r_\mathcal{K} \diamond r_\mathcal{L}} = [t]_{r_\mathcal{K} \diamond r_\mathcal{L}}\}$$

with $f_a : M_a \to S$ and $g_a : M_a \to T$ as the respective projections. Thus all these components may be chosen independently of the action $a \in \mathsf{Act}$. Consequently we have constructed a stochastic Kripke model $\mathcal{M} = (M, (m_a)_{a\in\mathsf{Act}})$ and strong morphisms $\mathcal{K} \xrightarrow{f} \mathcal{M} \xleftarrow{g} \mathcal{L}$. □

Summing up, we obtain as a consequence the following general result. It is visualized already in Figure 3.1 on page 133.

Theorem 3.4.34. *Let $\mathcal{K}$ and $\mathcal{L}$ be nondegenerate stochastic Kripke models over analytic spaces. Consider these statements:*

a. $\mathcal{K}$ and $\mathcal{L}$ are logically equivalent.
b. $\mathcal{K}$ and $\mathcal{L}$ are distributionally equivalent.
c. $\mathcal{K}$ and $\mathcal{L}$ are bisimilar.
d. There exist $\mathfrak{L}$-ergodic morphisms $\Psi : \mathcal{K} \curvearrowright^{\mathsf{e}} \mathcal{L}$ and $\Xi : \mathcal{L} \curvearrowright^{\mathsf{e}} \mathcal{K}$.
e. $\mathcal{K}$ and $\mathcal{L}$ are behaviorally equivalent.
f. $\mathcal{K}$ and $\mathcal{L}$ are weakly behaviorally equivalent.

Then

i. $a \Leftrightarrow b \Leftrightarrow c \Leftrightarrow d \Leftrightarrow e$.
ii. $b \Rightarrow f$.

Proof 1. $a \Leftrightarrow c \Leftrightarrow e$ is established in Theorem 2.3.11. The implication $e \Rightarrow f$ is trivial.

2. Proposition 3.4.17 establishes the equivalence $a \Leftrightarrow b$, Proposition 3.4.33 shows $b \Leftrightarrow c$, and the equivalence $b \Leftrightarrow d$ is proven in Proposition 3.4.21. Finally, the implication $b \Rightarrow f$ is shown in Proposition 3.4.24. □

Looking back at the development, it becomes clear that smooth equivalence relations, their lifting to the space of all subprobabilities, and the associated structure theorems play a leading rôle. Through using these devices it becomes possible to capitalize in this scenario — which is given by randomized morphisms — on the development for strong morphisms.

3.5 Bibliographic Notes

The Giry Monad

The construction of the Giry monad in Section 3.2 originates from M. Giry's work [38] on a categorical foundation of the measure-theoretic parts of probability theory. Most results on this monad are established in Giry's paper. The monad received fresh attention first through Panangaden's early work in this area [64]. It is investigated in a slightly more general setting in [20, Chapter 2.3], where also some parallels to the Manes monad are drawn. The latter monad has the powerset functor as its functorial part, and it relates to set-theoretic relations as the Giry monad does to stochastic relations. A complete characterization of the Eilenberg-Moore algebras for the monad over Polish spaces is found in [18, 26].

Randomized Morphisms

These morphisms are defined analogously to randomized policies in stochastic dynamic programming. A policy assigns each state an action to take; a randomized policy maps each state to a probability distribution over actions [47]. Similarly here. From an algebraic point of view, these morphisms are closely associated with the Kleisli construction for the Giry monad. This construction is discussed at length in Mac Lane's classic [57, Chapter VI] and in Borceux's survey [12, Section 4.2]; the framework there, however, is that of adjunctions and algebras for monads.

Hennessy-Milner Logic

This logic has been derived from a version that was originally introduced by Hennessy and Milner [44] for the study of bisimulations. It was modified by Larsen and Skou [58] in their seminal paper on testing, where a very close connection between bisimilarity and logical equivalence (accepting the same sets of formulas) for models was formulated in the context of discrete labeled Markov processes. The generalization to models over general analytic spaces with an eye towards bisimilarity is then fairly straightforward, see [22] for a discussion. Some side tracks are found in discussions on continuous time stochastic logics; see the Bibliographic Notes at the end of Chapter 2. The results presented in Section 3.4 are new.

The bulk of the work undertaken here assumes the underlying spaces to be analytic or even Polish. This is so because these spaces have on the one hand some pleasant measure-theoretic properties, and on the other hand it is difficult to conceive of an application that is not based on an analytic space. One notable exception should be mentioned: Moss and Viglizzo [63] derive terminal sequences and final coalgebras from functors over general measurable spaces, see also the discussions in [39, 86].

Smooth equivalence relations were again the leading actors in this play; most basic material on these relations may be found in Srivastava's book [79] on Borel sets or in Kechris' monograph [49] on analytic sets. These texts provide the reader also with the basic tools for investigating Polish and analytic spaces. The discussion on ergodic measures is taken from Arveson's monograph [2, Chapter 3].

Chapter 4
Coalgebraic Logic

4.1 Introduction

Investigating equivalent behavior of stochastic Kripke models for modal logics or their close cousins such as the tree logics used for model checking follows usually this pattern: the state space is partitioned into states that satisfy exactly the same formulas. The existence of a bisimulation, behavioral equivalence, or identification of a minimal set of formulas to test against follows then from an investigation of this equivalence relation, exploiting characteristic properties that are handed down from the logic. A stochastic Kripke model is usually based on a coalgebra, and the analysis of the behavior may usually be reduced to an investigation of congruence properties for this coalgebra. When having a closer look at the logic, one sees that the composition rules for the formulas may usually be partitioned into algebraic rules (such as conjunction or negation) and coalgebraic rules (such as those given through the modal operators). If the logic operates on two levels as, e.g., μ**CSL** and **CTL*** do, distinguishing state formulas from path formulas, then there is usually a bridge between them.

Coalgebraic logic [61, 15, 75, 66] investigates behavioral properties of models in terms of coalgebras and predicate liftings. Assume that (S, γ) is a coalgebra for a functor in the category of sets with maps as morphisms. The general idea is that a modal formula $\langle\lambda\rangle\phi$ is valid for a state $s \in S$ iff the set $[\![\phi]\!]$ of states for which formula ϕ is valid is transformed through predicate lifting λ into the set $\lambda_S([\![\phi]\!])$ which contains $\gamma(s)$ as a member, so that $[\![\langle\lambda\rangle\phi]\!] = \{s \in S \mid \gamma(s) \in \lambda_S([\![\phi]\!])\}$, or, equivalently, $[\![\langle\lambda\rangle\phi]\!] = \gamma^{-1} \circ \lambda_S([\![\phi]\!])$. Here predicate liftings play the rôle of modal operators, a predicate lifting being a natural transformation for the contravariant power set functor and the functor governing the coalgebra. It can be shown that the usual semantic operations can be formulated in terms of suitably chosen predicate liftings. The coalgebraic approach permits a clearer view of the semantic mechanisms underlying the logic: it becomes clear which properties are attributed to the

E.-E. Doberkat, *Stochastic Coalgebraic Logic*, Monographs in Theoretical Computer Science. An EATCS Series, DOI 10.1007/978-3-642-02995-0_4,

coalgebra, and which ones are due to the modal structure, which in turn is modelled through predicate liftings.

The present discussion translates these ideas into the realm of stochastic systems for an investigation of bisimilarity and behavioral equivalence of stochastic $\mathfrak{T}$-coalgebras. We propose using a logic for this that is based essentially on predicate liftings; the logic works like a tree logic for model checking on two levels. Because each stochastic $\mathfrak{T}$-coalgebra is a stochastic relation, we use tools from stochastic relations, in particular congruences. This requires that the collection of predicate liftings enjoys a certain selectivity similar to the separation properties proposed by Pattinson [66]. Since stochastic relations are bisimilar provided there are congruences defined on them that are simulation equivalent, simulation equivalence gets to be the center of attention. It is shown that logical equivalence and bisimilarity are equivalent for stochastic $\mathfrak{T}$-coalgebras provided the functor is compatible with the congruences, and it is shown that logical equivalence and observational equivalence are the same provided the functor distributes over the congruences. Logical equivalence means as usual that the states' theories are identical, so that the logic cannot separate the corresponding states.

We address coalgebras for the functor $\mathfrak{S} \circ \mathfrak{T}$, transporting results from stochastic relations to the case at hand. This was possible because coalgebras of this type are special cases of these relations. We call this kind of coalgebra *stochastic left coalgebra*, indicating the position of the functor $\mathfrak{T}$ relative to the subprobability functor $\mathfrak{S}$. These coalgebras are closely related to the *generative systems* discussed by A. Sokolova [78, Def. 2.2.1]. But consider a Markov transition system over a state space S with a set Act of actions. These systems are modelled as coalgebras through the functor $\mathfrak{S}^{\mathsf{Act}} : S \mapsto \left(\mathfrak{S}\left(S\right)\right)^{\mathsf{Act}}$, giving rise to a *stochastic right coalgebra*, i.e., to a coalgebra for a functor $\mathfrak{T} \circ \mathfrak{S}$ (related to *reactive systems* [78, Def. 2.2.1]). An easy reduction to stochastic relations is not available here, so we need to find another approach. We first characterize congruences in terms of morphisms (a first step has been done already in Proposition 1.7.21) and discuss under which conditions the equivalence generated by the logic is a congruence. This is characterized through a separation property for the set of predicate liftings. The crucial step in the investigation of the relationship between models induced by the logic lies usually in establishing the existence of a span from a cospan, which happens in this context usually by showing that a semi-pullback exists in the category one works in. This need arises here as well, and we show that a simple selection property for maps — nota bene, not for morphisms — is sufficient for our purposes. This selection property is then established through a selection theorem for set-valued measurable sets, opening an avenue to investigate the relationships between models. The equivalence proven addresses the global situation: we have a look at the bisimilarity of models, and not of individual states. A sophisticated construction due to Ch. Schubert shows, however, that the global characterization can be put to good use when investigating bisimilarity of two logically equivalent states. This result is derived as well.

We discuss different types of logics here: for left coalgebras, it seems sensible to discuss a two-level approach. These algebras are used for an interpretation of logics which have state formulas as well as formulas that would be called path formulas for tree logics; in the absence of paths we call them $\mathfrak{T}$-formulas. Modal operators are replaced here with operators which are derived from predicate liftings; bridges are built between the two types of formulas through natural transformations which we call bridge operators. Conjunction is an important operator for reasons which become clear through the π-λ-Theorem; all other Boolean operators can be represented through bridge operators. For right coalgebras, however, we have neither need nor occasion for introducing a two-level approach. Here we work with bridge operators, but also with infinitesimal operators which permit representing fixed points.

We will have a look at the definition and first properties of predicate liftings in Section 4.2. Section 4.3 defines and investigates left coalgebras; it is shown that modal logics can be treated in this framework as well; Section 4.4 discusses right coalgebras. An application of the general result is the characterization of bisimilar states in Section 4.4.5.

4.2 Predicate Liftings

Let us fix some notations first and introduce names for categories which have remained anonymous until now (poor things). Denote by **ANL** the category of analytic spaces with surjective Borel maps as morphisms. Thus $f : A \to B$ is a morphism in **ANL** iff f is onto, and in addition, $f^{-1}[Q] \in \mathcal{B}(A)$ holds whenever $Q \in \mathcal{B}(B)$ is a Borel set. Let the category **BOR** have as objects the Borel sets $\mathcal{B}(A)$ of analytic spaces A; a morphism between objects $\mathcal{B}(A)$ and $\mathcal{B}(B)$ in **BOR** is a map $\mathcal{B}(A) \to \mathcal{B}(B)$; thus in particular an inverse $f^{-1} : \mathcal{B}(A) \to \mathcal{B}(B)$ for a morphism $f : B \to A$ in **ANL** is a morphism in **BOR**. Denote by $\mathfrak{B}$ the contravariant functor **ANL** $\to$ **BOR** that sends each analytic space to its Borel sets, and each morphism $f : A \to B$ into $f^{-1} : \mathcal{B}(B) \to \mathcal{B}(A)$. We will write f^{-1} or $\mathfrak{B}(f)$, depending on the situation. Let $\mathfrak{T}$: **ANL** $\to$ **ANL** be a functor; then define the functor $\mathfrak{T}^{\mathrm{R}}$ as $\mathfrak{T} \circ \mathfrak{S}$, ${}^{\mathrm{L}}\mathfrak{T}$ is defined as $\mathfrak{S} \circ \mathfrak{T}$.

When interpreting logics with the stochastic coalgebras that will be introduced in later sections, we perceive predicate liftings as rather abstract devices for modelling properties of formulas. Take for example a Markov transition system $(S, (k_a)_{a \in \mathsf{Act}})$ over an analytic space S interpreting the very simple Hennessy-Milner logic $\mathfrak{L}$ in Section 2.6; $k_a : S \rightsquigarrow S$ is a stochastic relation for each action $a \in \mathsf{Act}$. The intuition behind an interpretation of the modal formula $\langle a \rangle_q \phi$ is that this formula holds in a state $s \in S$ if it is possible to enter a state in which ϕ holds with at least probability q upon action a. Formally,

$$
\begin{aligned}
[\![\langle a\rangle_q\phi]\!] &= \{s \in S \mid k_a(s)([\![\phi]\!]) \geq q\} \\
&= \mathfrak{B}\,(k_a)\,(\{\mu \in \mathfrak{S}\,(S) \mid \mu([\![\phi]\!]) \geq q\}) \\
&= \mathfrak{B}\,(k_a)\,(\varpi_{q,S}([\![\phi]\!])) \\
&= \big(k_a^{-1} \circ \varpi_{q,S}\big)([\![\phi]\!]) \\
&= \big(\mathfrak{B}\,(k_a) \circ \varpi_{q,S}\big)([\![\phi]\!]),
\end{aligned}
$$

where $\varpi_{q,S}(A) := \{\mu \in \mathfrak{S}\,(S) \mid \mu(A) \geq q\}$ whenever $A \in \mathcal{B}(S)$ is a Borel set, so that $\varpi_{q,S}(A) \in \mathcal{B}(\mathfrak{S}\,(S))$ by the definition of the weak*-σ-algebra. In the last line, k_a^{-1} is perceived as a map $\mathcal{B}(\mathfrak{S}\,(S)) \to \mathcal{B}(S)$ from the Borel sets of $\mathfrak{S}\,(S)$ to the Borel sets of S; hence we are dealing here really with $\mathfrak{B}\,(k_a)$.

Lemma 4.2.1. $\varpi_q : \mathfrak{B} \stackrel{\bullet}{\to} \mathfrak{B} \circ \mathfrak{S}$ *is a natural transformation.*

Proof Let $f : T \to S$ be Borel measurable; then

$$
\begin{aligned}
\mu \in \varpi_{q,T}\big(\mathfrak{B}\,(f)\,(C)\big) &\Leftrightarrow \mu\big(\mathfrak{B}\,(f)\,(C)\big) \geq q \\
&\Leftrightarrow \mathfrak{S}\,(f)\,(\mu)(C) \geq q \\
&\Leftrightarrow \mu \in \mathfrak{B}\,(\mathfrak{S}\,(f))\,(\varpi_{q,S}(C))
\end{aligned}
$$

for each Borel set $C \in \mathcal{B}(S)$; consequently, $\varpi_q \circ \mathfrak{B}\,(f) = \big(\mathfrak{B} \circ \mathfrak{S}\big)(f) \circ \varpi_q$, from which the assertion follows (remember that $\mathfrak{B}$ is a contravariant functor). □

Interpreting k_a as the dynamics of an $\mathfrak{S}$-coalgebra, we represent $[\![\langle a\rangle_q\phi]\!]$ as the image of $[\![\phi]\!]$ under the composition of the image of this dynamics under functor $\mathfrak{B}$ and a natural transformation $\mathfrak{B} \stackrel{\bullet}{\to} \mathfrak{B} \circ \mathfrak{S}$.

This consideration leads to the definition of a predicate lifting.

Definition 4.2.2. *A natural transformation* $\mathfrak{B} \stackrel{\bullet}{\to} \mathfrak{B} \circ \mathfrak{F}$ *is called a* predicate lifting *for functor* $\mathfrak{F}$: **ANL** $\to$ **ANL**; *an* $\mathfrak{F}$-predicate lifting *is a natural transformation* $\mathfrak{B} \circ \mathfrak{F} \stackrel{\bullet}{\to} \mathfrak{B} \circ {}^L\mathfrak{F}$.

Given a predicate lifting λ for functor $\mathfrak{F}$ and an analytic space A, the component λ_A maps each Borel set $Q \in \mathcal{B}(A)$ into a Borel set $\lambda_A(Q) \in \big(\mathfrak{B} \circ \mathfrak{S} \circ \mathfrak{F}\big)(A)$. Similarly, an $\mathfrak{F}$-predicate lifting ϑ maps a Borel set $R \in \big(\mathfrak{B} \circ \mathfrak{F}\big)(A) = \mathcal{B}(\mathfrak{F}\,(A))$ into a Borel set $\vartheta_A(R) \in \big(\mathfrak{B} \circ {}^{\mathrm{L}}\mathfrak{F}\big)(A)$ whenever A is an analytic space. These liftings will be needed when modelling the analogue of state formulas and path formulas from μ**CSL** in Section 4.3.2.

Note that $\gamma(a) \in \mathfrak{S}\,(\mathfrak{F}\,(A))$ for a stochastic coalgebra (A, γ), and that $\mathfrak{B}\,(\gamma)\,(R) = \gamma^{-1}\,[R] \in \mathcal{B}(A)$ whenever $R \in \big(\mathfrak{B} \circ {}^{\mathrm{L}}\mathfrak{F}\big)(A) = \mathcal{B}(\mathfrak{S}\,(\mathfrak{F}\,(A)))$ is a Borel subset of all subprobabilities on $\mathfrak{F}\,(A)$. This conforms to the discussion above. The interplay between Borel sets visible here renders predicate liftings interesting for interpreting logics through a stochastic coalgebra.

Example 4.2.3. Define functor $\mathfrak{Z}$ through $\mathfrak{Z}\,(A) := (A \times \mathbb{R}_+)^\infty$ with $f : A \to B$ being mapped into $\mathfrak{Z}\,(f)\,(\langle a_n, t_n\rangle_{n\in\mathbb{N}}) := \langle f(a_n), t_n\rangle_{n\in\mathbb{N}}$. This functor is used for interpreting the continuous stochastic time logics μ**CSL** in Section 2.4; in the present discussion it will be called upon from time to time for illustration.

As a first example we discuss the transformation which underlies path quantification $\mathcal{P}$ in μ**CSL**. Put for the real number q with $0 \leq q \leq 1$ and for the Borel set $Q \in \mathcal{B}(\mathfrak{Z}(A))$

$$\vartheta_{\ltimes q,A}(Q) := \{\mu \in \mathfrak{S}(\mathfrak{Z}(A)) \mid \mu(Q) \ltimes q\};$$

then $\vartheta_{\ltimes q}$ is an $\mathfrak{Z}$-predicate lifting when relation $\ltimes$ assumes as usual the values $\{<, \leq, \geq, >\}$. Arguing as in Lemma 4.2.1, one first notes that $\vartheta_{\ltimes q,A}$ maps $\mathcal{B}(\mathfrak{Z}(A))$ to $\mathcal{B}({}^{\mathrm{L}}\mathfrak{Z}(A))$. If $f : B \to A$ is a surjective Borel map, the diagram

$$\begin{array}{ccc}
(\mathfrak{B} \circ \mathfrak{Z})(A) & \xrightarrow{(\mathfrak{B}\circ\mathfrak{Z})(f)} & (\mathfrak{B} \circ \mathfrak{Z})(B) \\
{\scriptstyle \vartheta_{\ltimes q,A}}\downarrow & & \downarrow{\scriptstyle \vartheta_{\ltimes q,B}} \\
(\mathfrak{B} \circ {}^{\mathrm{L}}\mathfrak{Z})(B) & \xrightarrow[(\mathfrak{B}\circ^{\mathrm{L}}\mathfrak{Z})(f)]{} & (\mathfrak{B} \circ {}^{\mathrm{L}}\mathfrak{Z})(B)
\end{array}$$

commutes by the definition of the way in which $\mathfrak{S}$ acts on Borel maps. Let $Q \in \mathcal{B}(\mathfrak{Z}(A))$ be a Borel set; then $(\mathfrak{S} \circ \mathfrak{Z})(f)(\mu) \in \vartheta_{q,A}(Q)$ iff $\mu \in \vartheta_{q,B}\big(\mathfrak{B}(\mathfrak{Z}(f))(Q)\big)$. This establishes naturalness of ϑ_q. $\dashv$

The next example is defined through functor $\mathfrak{Z}$ as well.

Example 4.2.4. Let $Q \in \mathcal{B}(A)$ be a Borel set; then

$$\mathcal{S}_{\ltimes p,A}(Q) := \{\mu \in \mathfrak{S}(\mathfrak{Z}(A)) \mid \lim_{t\to\infty} \mu(L_{t,A}(Q)) \text{ exists and is } \ltimes p\}.$$

Here $L_{t,A}(Q) := \{r \in \mathfrak{Z}(A) \mid r@t \in Q\}$ indicates all paths the state of which at time t is an element of C. The @-operator $\langle s_0, t_0, s_1, t_1 \ldots\rangle@t$ indicates by the state s_k the smallest index k for which the accumulated residence times for the path $\langle s_0, t_0, s_1, t_1 \ldots\rangle \in (A \times \mathbb{R}_+)^\infty$ exceeds t; see page 89. It is fairly straightforward to see that $r \mapsto r@t$ is defined on a measurable subset of $\mathfrak{Z}(A)$, and that it furnishes a measurable map to A on this subset.

Then $\mathcal{S}_{\ltimes p}$ defines a natural transformation $\mathcal{S}_{\ltimes p} : \mathfrak{B} \stackrel{\bullet}{\to} \mathfrak{B} \circ {}^{\mathrm{L}}\mathfrak{Z}$ for each real number $p \in [0, 1]$. In order to see this, let $f : B \to A$ be a Borel map with A, B analytic spaces. If $Q \in \mathcal{B}(A)$, we observe for the sequence $r \in \mathfrak{Z}(B)$ that $\mathfrak{Z}(f)(r)@t \in Q$ iff $f(r@t) \in Q$. This is so since $\mathfrak{Z}$ does not transform the timing information. Thus we infer for $\mu \in (\mathfrak{S} \circ \mathfrak{Z})(B)$ that

$$(\mathfrak{S} \circ \mathfrak{Z})(f)(\mu)(L_{t,A}(Q)) = \mu\big(\mathfrak{B}(\mathfrak{Z}(f))(L_{t,A}(Q))\big) = \mu\big(L_{t,B}(f^{-1}[Q])\big).$$

This implies $\mathfrak{B}((\mathfrak{S} \circ \mathfrak{Z})(f))(\mathcal{S}_{\ltimes p,A}(Q)) = \mathcal{S}_{\ltimes p,B}(\mathfrak{B}(f)(Q))$, so that this diagram commutes:

$$\begin{array}{ccc}
\mathfrak{B}(A) & \xrightarrow{\mathfrak{B}(f)} & \mathfrak{B}(B) \\
{\scriptstyle \mathcal{S}_{\ltimes p,A}}\downarrow & & \downarrow{\scriptstyle \mathcal{S}_{\ltimes p,B}} \\
(\mathfrak{B} \circ {}^{\mathrm{L}}\mathfrak{Z})(A) & \xrightarrow[(\mathfrak{B} \circ {}^{\mathrm{L}}\mathfrak{Z})(f)]{} & (\mathfrak{B} \circ {}^{\mathrm{L}}\mathfrak{Z})(B)
\end{array}$$

This establishes the naturalness of $\mathcal{S}_{\ltimes p}$. $\dashv$

The next example deals with the functor $-^{\mathsf{Act}}$ for some countable set Act of labels. It is of interest when modelling Markov transition systems.

Example 4.2.5. Define for A analytic the functor $\mathfrak{T}$ through $\mathfrak{T}(A) := A^{\mathsf{Act}}$; the measurable map $f : A \to B$ is mapped to $\mathfrak{T}(f) : (x_a)_{a \in \mathsf{Act}} \mapsto \big(f(x_a)\big)_{a \in \mathsf{Act}}$. Because Act is countable, $\mathfrak{T}$ is an endofunctor on **ANL**. Now define for $b \in \mathsf{Act}$ and the real number p the map $\lambda_{b,p,A} : \mathfrak{B}(A) \to \big(\mathfrak{B} \circ \mathfrak{T} \circ \mathfrak{S}\big)(A)$ through

$$\lambda_{b,p,A}(C) := \{(\mu_a)_{a \in \mathsf{Act}} \mid \mu_b(C) \geq p\}.$$

The definition of the weak*-σ-algebra makes sure that $\lambda_{b,p,A}(C) \in \mathcal{B}(\mathfrak{T}(A))$. Let $f : B \to A$ be a Borel map; then we obtain directly from the definition

$$\begin{aligned}
(\mu_a)_{a \in \mathsf{Act}} \in \mathfrak{B}\left((\mathfrak{T} \circ \mathfrak{S})(f)\right)(\lambda_{b,p,A}(C)) &\Leftrightarrow (\mathfrak{T} \circ \mathfrak{S})(f)\big((\mu_a)_{a \in \mathsf{Act}}\big) \in \lambda_{b,p,A}(C) \\
&\Leftrightarrow \big(\mathfrak{S}(f)(\mu_a)_{a \in \mathsf{Act}}\big) \in \lambda_{b,p,A}(C) \\
&\Leftrightarrow \mathfrak{S}(f)(\mu_b)(C) \geq p \\
&\Leftrightarrow (\mu_a)_{a \in \mathsf{Act}} \in \lambda_{b,p,B}\big(\mathfrak{B}(f)(C)\big).
\end{aligned}$$

But this means that this diagram commutes:

$$\begin{array}{ccc}
\mathfrak{B}(A) & \xrightarrow{\mathfrak{B}(f)} & \mathfrak{B}(B) \\
{\scriptstyle \lambda_{b,p,A}}\downarrow & & \downarrow{\scriptstyle \lambda_{b,p,B}} \\
(\mathfrak{B} \circ \mathfrak{T}^{\mathrm{R}})(A) & \xrightarrow[(\mathfrak{B} \circ \mathfrak{T}^{\mathrm{R}})(f)]{} & (\mathfrak{B} \circ \mathfrak{T}^{\mathrm{R}})(B)
\end{array}$$

Hence $\lambda_{b,p} : \mathfrak{B} \stackrel{\bullet}{\to} \mathfrak{B} \circ \mathfrak{T}^{\mathrm{R}}$ is a predicate lifting for functor $\mathfrak{T}^{\mathrm{R}}$. $\dashv$

Each logics has Boolean operations, which are represented also through natural transformations. In addition logics have sometimes bridge operators that transform one class of formulas into another. Let us have a look at μ**CSL**. The until-operator renders two state formulas into a path formula; similarly, a path quantifier takes a path formula and returns a state formula. These formulas are represented through natural transformations as well.

Definition 4.2.6. *A* bridge operator *of arity* $n \in \mathbb{N}$ *is a natural transformation* $\mathfrak{B}^n \stackrel{\bullet}{\to} \mathfrak{B}$*; a* $\mathfrak{T}$-bridge operator *of arity* $n \in \mathbb{N}$ *is a natural transformation* $\mathfrak{B}^n \stackrel{\bullet}{\to} \mathfrak{B} \circ \mathfrak{T}$.

Here $\mathfrak{B}^n : \mathbf{ANL} \to \mathbf{BOR}^n$ assigns to each analytic space A the set $\mathcal{B}(A)^n$, and if $f : B \to A$ is a Borel map, then $\mathfrak{B}^n(f)(C_1, \ldots, C_n) := \langle \mathfrak{B}(f)(C_1), \ldots, \mathfrak{B}(f)(C_n)\rangle$.

Example 4.2.7. Define for the analytic space A and the Borel sets $C_1, C_2 \in \mathcal{B}(A)$ the maps V_A and D_A through $V_A(C_1, C_2) := C_1 \cup C_2$ and $D_A(C_1, C_2) := C_1 \setminus C_2$. Then both D and V are bridge operators of arity 2. Complementation is a bridge operator of arity 1. This is straightforward, because the inverse image of a map is compatible with the Boolean set operations. ⊣

We illustrate $\mathfrak{T}$-bridge operators through functor $\mathfrak{Z}$. Fix an open or closed interval $\emptyset \neq I \subseteq \mathbb{R}_+$ for this.

Example 4.2.8. Define for the analytic space A the map $N_{I,A}$ through

$$N_{I,A}(C) := \{\langle t_0, a_0, t_1, a_1, \ldots\rangle \in \mathfrak{Z}(A) \mid t_0 \in I \text{ and } a_1 \in C\}.$$

This corresponds to the next operator in μ**CSL**. Because

$$N_{I,A}(C) = I \times A \times \mathbb{R}_+ \times C \times (\mathbb{R}_+ \times A)^\infty,$$

we see that $N_{I,A} : \mathcal{B}(A) \to \mathcal{B}(\mathfrak{Z}(A))$. Because $\mathfrak{Z}$ does not modify the timing information, it is easy to see that

$$N_{I,B}\big(\mathfrak{B}(f)(C)\big) = \mathfrak{B}(\mathfrak{Z}(f))\big(N_{I,A}(C)\big)$$

whenever $f : B \to A$ is measurable and $C \in \mathcal{B}(A)$. Consequently, this defines a natural transformation $N_I : \mathfrak{B} \stackrel{\bullet}{\to} \mathfrak{B} \circ \mathfrak{Z}$. ⊣

The next example deals with modelling the until operator of μ**CSL**.

Example 4.2.9. Define for the analytic space A the map $U_{I,A}$ through

$$U_{I,A}(C_1, C_2) := \{r \in \mathfrak{Z}(A) \mid \exists t \in I : r@t \in C_2 \wedge \forall t' \in [0, t[: r@t' \in C_1\}.$$

Recall that for any given $t \in \mathbb{R}_+$ there exists $\hat{t} \in \mathbb{Q}$ with $r@t = r@\hat{t}$; hence quantification over an uncountable range like $\mathbb{R}_+$ may be replaced here by one over the countable set of positive rationals (so things are not as bad as they look). Put

$$C_t := \{r \in \mathfrak{Z}(A) \mid r@t \in C\};$$

then $C_t \in \mathcal{B}(\mathfrak{Z}(A))$, provided $C \in \mathcal{B}(A)$. This is due to the measurability of $r \mapsto r@t$. The set $U_{I,A}(C_1, C_2)$ may be represented as

$$U_{I,A}(C_1, C_2) = \bigcup_{t \in \mathbb{Q} \cap I} \Big(C_{2,t} \cap \bigcap_{t' \in \mathbb{Q} \cap [0,t[} C_{1,t'}\Big).$$

Because only countable operations with Borel sets are involved, it follows that $U_{I,A} : \mathcal{B}(A)^2 \to \mathcal{B}(\mathfrak{Z}(A))$. We show that $U_I : \mathfrak{B}^2 \stackrel{\bullet}{\to} \mathfrak{B} \circ \mathfrak{Z}$ is a $\mathfrak{Z}$-bridge operator of arity 2. Let $f : B \to A$ be a Borel map with A, B analytic spaces. We know for $r \in \mathfrak{Z}(B)$ that $\mathfrak{Z}(f)(r)@t \in Q$ iff $f(r@t) \in Q$ (see Example 4.2.4). This implies for given $C_1, C_2 \in \mathcal{B}(A)$

$$U_{I,B}(\mathfrak{B}(f)(C_1), \mathfrak{B}(f)(C_2)) = \mathfrak{B}(\mathfrak{Z}(f))(U_{I,A}(C_1, C_2)).$$

But this means that the diagram

$$\begin{array}{ccc}
\mathfrak{B}^2(A) & \xrightarrow{\mathfrak{B}(f)} & \mathfrak{B}^2(B) \\
{\scriptstyle U_{I,A}}\big\downarrow & & \big\downarrow{\scriptstyle U_{I,B}} \\
(\mathfrak{B} \circ \mathfrak{Z})(A) & \xrightarrow[\mathfrak{B}\circ\mathfrak{Z}(f)]{} & (\mathfrak{B} \circ \mathfrak{Z})(B)
\end{array}$$

commutes; hence $U_I : \mathfrak{B}^2 \stackrel{\bullet}{\rightarrow} \mathfrak{B} \circ \mathfrak{Z}$ is a natural transformation. ⊣

The bridge operators are in bijection with the Boolean combinations of their arguments, as we will show now. This permits also counting them.

Proposition 4.2.10. *The bridge operators of arity $n \in \mathbb{N}$ are exactly the Boolean combinations of n symbols. In particular, there are exactly 2^{2^n} of them.*

Proof (Sketch) Let **Anl** be the category of analytic spaces with Borel maps as morphisms, so we lift for the moment to requirement of morphisms between analytic spaces to be onto.

We show first that $\mathfrak{B}$ is naturally isomorphic to $\mathbf{Anl}(-, \mathbf{2})$ with $\mathbf{2}$ as the discrete analytic space over the object set $\{0, 1\}$. In fact, define $\eta_A(C) := \chi_C$, the latter being the indicator function of set C. This defines a bijection $\eta_A : \mathfrak{B}(A) \rightarrow \mathbf{Anl}(A, \mathbf{2})$ which is easily seen to be natural. Consequently (this is the sketchy part), $\mathfrak{B}^n$ is naturally isomorphic to $\mathbf{Anl}(-, \mathbf{2})^n$ which in turn is naturally isomorphic to $\mathbf{Anl}(-, \mathbf{2}^n)$. On the other hand, the Yoneda Lemma 1.2.1 shows that $\mathbf{Anl}(-, \mathbf{2}^n) \rightarrow \mathbf{Anl}(-, \mathbf{2})$ is in bijection with $\mathbf{Anl}(\mathbf{2}^n, \mathbf{2})$. Because the analytic spaces concerned in the latter homset are discrete, there exists a bijection between it and the power set of a set with 2^n elements. □

4.3 Left Coalgebras

Consider μ**CSL** again. This logic is interpreted through models the underlying coalgebras of which are determined through the functor

$$A \mapsto \mathfrak{S}((\mathbb{R}_+ \times A)^\infty) = (\mathfrak{S} \circ \mathfrak{Z})(A).$$

Hence each state a is assigned a subprobability $\gamma(a)$ over the infinite paths of residence times and states, so that $\gamma(a)$ yields the distribution of all mixed trajectories which start in state a. A. Sokolova writes "... in a generative system probabilities are distributed over all outgoing transitions from a state" [78, p. 30], so that in her classification this functor characterizes a generative system (in contrast to reactive systems, which will be discussed in the next section). This characterization is valid for each coalgebra the functor of which has the shape $A \mapsto (\mathfrak{S} \circ \mathfrak{T})(A)$ for a suitable functor $\mathfrak{T}$ on the category of analytic spaces. The outgoing transitions are in this case determined through functor $\mathfrak{T}$; their distribution is then represented algebraically through the subprobability functor. The corresponding coalgebras are called *left coalgebras*, to indicate lexically the position of the subprobability functor.

This section will discuss aspects of coalgebraic logic for left coalgebras. We have seen that they arise, e.g., through logics which lift μ**CSL**, and we will see in Section 4.3.4 that modal logics may be represented through these coalgebras as well. In Section 4.3.1, these coalgebras are defined formally. A helpful reduction shows that for investigating the properties we are interested in it suffices to consider left coalgebras over Polish spaces. Section 4.3.2 will introduce the logics. We model the logic along μ**CSL**, which gives a distinction between state formulas and path formulas. Accordingly, we investigate formulas the validity of which is determined through the state space A of the interpreting coalgebra, and those formulas which hold for elements of $\mathfrak{T}(A)$; this space is akin to the space of all trajectories. Observing the interplay of these two kinds of formulas, we see that we need some kind of a bridge between them, producing $\mathfrak{T}$-formulas from state formulas. Bisimilarity and behavioral equivalence are defined, and the structure of the factor spaces is investigated when the logic is used to partition the states. We have a look at conditions which make the functor $\mathfrak{T}$ compatible with these relations and derive from this the circumstances under which logical equivalence, bisimilarity, and behavioral equivalence are the same. This is done in Section 4.3.3; it provides a general set of criteria, and we show that modal logics as introduced in Section 2.3 can be considered a special case. This requires encoding a modal logic into a coalgebraic logic for a suitable functor, and here some ideas related to generating functions from combinatorics are applied.

4.3.1 Stochastic Left Coalgebras

We fix for the rest of this section an endofunctor $\mathfrak{T}$ on **ANL**, and we assume that $\mathfrak{T}(A)$ is a Standard Borel space whenever A is.

Definition 4.3.1. *Let A and B be analytic spaces.*

a. A stochastic left $\mathfrak{T}$-coalgebra $\mathsf{F} = (A, \gamma)$ for $\mathfrak{T}$ *is a measurable map* $\gamma : A \to {}^{L}\mathfrak{T}(A)$.

b. A coalgebra morphism $f : \mathsf{F} \to \mathsf{G}$ *for the stochastic left $\mathfrak{T}$-coalgebras* $\mathsf{F} = (A, \gamma)$ *and* $\mathsf{G} = (B, \delta)$ *is a morphism* $f : A \to B$ *in* **ANL** *such that* $\delta \circ f = {}^{L}\mathfrak{T}(f) \circ \gamma$.

Thus a stochastic left $\mathfrak{T}$-coalgebra $\mathsf{F} = (A, \gamma)$ for $\mathfrak{T}$ is based on a stochastic relation $(A, \mathfrak{T}(A), \gamma)$; in particular, $\gamma : A \to \mathfrak{S}(\mathfrak{T}(A))$ is a Borel map from A to the subprobabilities on $\mathfrak{T}(A)$. Hence (A, γ) can also be perceived as a ${}^{L}\mathfrak{T}$-coalgebra in the sense of Section 1.2. The lexical position of the subprobability functor $\mathfrak{S}$ is fixed to the left of the functor $\mathfrak{T}$, explaining the attribute *left.* Because we will deal only with one functor $\mathfrak{T}$ we will usually only talk about *stochastic left coalgebras* or just *left coalgebras* in this section, omitting $\mathfrak{T}$ from the notation altogether.

A morphism $f : \mathsf{F} \to \mathsf{G}$ with $\mathsf{F} = (A, \gamma)$ and $\mathsf{G} = (B, \delta)$ renders the diagram

$$\begin{array}{ccc} A & \xrightarrow{\quad f \quad} & B \\ {\scriptstyle \gamma}\downarrow & & \downarrow{\scriptstyle \delta} \\ {}^{L}\mathfrak{T}(A) & \xrightarrow[{}^{L}\mathfrak{T}(f)]{} & {}^{L}\mathfrak{T}(B) \end{array}$$

commutative. Stochastic left coalgebras are based on stochastic relations, morphisms of which are comprised of pairs of surjective Borel maps. Because $\mathfrak{T}$ is an endofunctor on **ANL**, and because the morphisms in this category are based on surjective maps, we know that $\mathfrak{T}(f)$ is onto whenever f is by Proposition 1.6.19. Hence a morphism (f, g) for the stochastic relations $(A, \mathfrak{T}(A), \gamma)$ and $(B, \mathfrak{T}(B), \delta)$ in the sense of Definition 1.6.24 is a coalgebra morphism for the left coalgebras (A, γ) and (B, δ), provided $g = \mathfrak{T}(f)$. Conversely, a coalgebra morphism $f : (A, \gamma) \to (B, \delta)$ is always a morphism $(f, \mathfrak{T}(f)) : (A, \mathfrak{T}(A), \gamma) \to (B, \mathfrak{T}(B), \delta)$ for the stochastic relations.

Bisimilarity and behavioral equivalence are defined coalgebraically through the existence of a span respectively a cospan of morphisms; see Section 1.2.1. Thus, if coalgebras F and G are bisimilar, we can find for each $a \in A$ an element $c \in C$ in the mediating coalgebra C and an element b in G with $a = f(c), b = g(c)$, and vice versa. If they are behaviorally equivalent then we can find for $a \in A$ an element $b \in B$ with $f(a) = g(b)$, and vice versa. This is treated similarly to the cases dealt with so far.

Relating bisimulations for stochastic relations and coalgebras is fairly straightforward: a bisimulation

$$(A, \mathfrak{T}(A), \gamma) \xleftarrow{(f, f')} (C, C', \varepsilon) \xrightarrow{(g, g')} (B, \mathfrak{T}(B), \delta)$$

for stochastic relations can be made into a bisimulation for the left coalgebras provided $C' = \mathfrak{T}(C)$, and provided both $f' = \mathfrak{T}(f)$ and $g' = \mathfrak{T}(g)$ hold. Of course, each bisimulation between these coalgebras can be interpreted as a span of morphisms for the associated stochastic relations. Note, however, that this is not necessarily a bisimulation of the stochastic relations, since the requirement of having nontrivial common events needs to be satisfied as well; see Definition 2.2.1. Similarly, cospans for exploring behavioral equivalence are related.

The fact that each analytic space is the image of a Polish space under a Borel map is extended to stochastic relations: given an analytic stochastic relation L, we construct in Proposition 1.6.26 a stochastic relation K over Polish spaces and a morphism $\mathsf{f} : \mathsf{K} \to \mathsf{L}$. This extends to stochastic coalgebras, as we will see now.

Proposition 4.3.2. *If* F *is a stochastic coalgebra over an analytic space, then there exists a stochastic coalgebra* F_0 *over a Polish space and a morphism* $f : \mathsf{F}_0 \to \mathsf{F}$.

Proof Let $\mathsf{F} = (A, \gamma)$. We find a Polish space A_0 and a surjective Borel map $f : A_0 \to A$ for A, because A is an analytic space. Because $\mathfrak{T}$ is an endofunctor on **ANL**, $\mathfrak{T}(f) : \mathfrak{T}(A_0) \to \mathfrak{T}(A)$ is a surjective Borel map, and $\mathfrak{T}(A_0)$ is Standard Borel, and hence may be assumed to be Polish. From Proposition 1.6.26 we obtain a stochastic relation $\gamma_0 : A_0 \rightsquigarrow \mathfrak{T}(A)$ such that $(f, \mathfrak{T}(f)) : (A_0, \mathfrak{T}(A_0), \gamma_0) \to (A, \mathfrak{T}(A), \gamma)$ is a morphism between these stochastic relations. By the remarks just made, $f : \mathsf{F}_0 \to \mathsf{F}$ is a coalgebra morphism. □

This observation will be helpful later because it sometimes permits reducing a discussion from coalgebras on analytic spaces to those on Polish spaces without losing expressiveness; see Proposition 4.3.17.

4.3.2 The Logic

The logic $\mathfrak{L}^\flat = \mathfrak{L}^\flat(\Lambda, \Theta, \Gamma)$ we are about to investigate distinguishes state formulas that hold on the state space A from formulas that hold on $\mathfrak{T}(A)$. We call the latter $\mathfrak{T}$-*formulas*, in contrast to state formulas. This two-level approach has been studied in the context of μ**CSL** with fairly specific operators (steady state, next, until, path quantification; see Section 2.4), and we will investigate a generalization now that is based solely on predicate liftings.

For the remainder of this section we fix at most countable sets Λ and Θ of predicate liftings respectively $\mathfrak{T}$-predicate liftings.

This is the definition of the formulas in logic $\mathfrak{L}^\flat(\Lambda, \Theta, \Gamma)$.

1. *State formulas* are given according to the following grammar:

$$\phi ::= \top \mid \phi_1 \wedge \phi_2 \mid \langle\lambda\rangle\phi \mid \langle\vartheta\rangle\psi$$

   Here $\lambda \in \Lambda$ is a predicate lifting and $\vartheta \in \Theta$ is a $\mathfrak{T}$-predicate lifting.

2. $\mathfrak{T}$-*formulas* are given according to the following grammar:

$$\psi ::= \widetilde{\top} \mid \psi_1 \wedge \psi_2 \mid b(\phi_1, \ldots, \phi_{\mathfrak{ar}(b)})$$

We take b from a set Γ of $\mathfrak{T}$-*bridge operators* that maintain a bridge from state formulas to $\mathfrak{T}$-formulas; $\mathfrak{ar}(b)$ is the arity of $b \in \Gamma$; $\phi_1, \ldots, \phi_{\mathfrak{ar}(b)}$ are state formulas.

Both state and $\mathfrak{T}$-formulas are closed under conjunction, but neither is closed under negation.

The logic is quite terse and may be used as a kernel logic, to be extended by additional expressions for formulas. For example, atomic propositions could be added; so could negation and disjunction. We will see, however, that the logic contains the essentials for a discussion of bisimilarity and for behavioral equivalence, and it turns out that conjunction is an essential technical requirement. We will return to extending this logic when discussing possible extensions in Section 4.3.5.

For interpreting the logic, let $\mathsf{F} = (A, \gamma)$ be a stochastic coalgebra. Assume furthermore that we have for each bridge operator $b \in \Gamma$ a natural transformation $N_b : \mathfrak{B}^{\mathfrak{ar}(b)} \xrightarrow{\bullet} \mathfrak{B} \circ \mathfrak{T}$, so that N_b is an $\mathfrak{T}$-bridge operator of arity $\mathfrak{ar}(b) = n$. For example, let $\mathfrak{ar}(b) = 1$; then N_b can be given for each analytic space A through a surjective Borel map $n_{b,A} : \mathfrak{T}(A) \to A$ upon setting $N_{b,A} = \mathfrak{B}(n_{b,A}) : C \mapsto \mathfrak{B}(n_{b,A})(C)$ whenever $C \in \mathcal{B}(A)$ is a Borel set in A. The requirement that N_b is a natural transformation is evidently more general. Bridge operator b will be interpreted according to N_b, whereas the interpretation of a predicate lifting depends always on the coalgebraic structure. This is why bridge operators are maintained in parallel to predicate liftings.

Example 4.3.3. Let $\mathfrak{Z} = (- \times \mathbb{R}_+)^\infty$; we will define a negation-free variant of $\mu\mathbf{CSL}(\emptyset, \emptyset, \emptyset)$ (see page 84). Define the respective sets of predicate liftings, $\mathfrak{Z}$-predicate liftings, and $\mathfrak{Z}$-bridge operators through

$$\begin{aligned}
\Lambda &:= \{\mathcal{S}_{\ltimes p} \mid p \in \mathbb{Q} \cap [0,1], \ltimes \in \{<, \leq, \geq, >\}\}, \\
\Theta &:= \{\theta_{\ltimes q} \mid p \in \mathbb{Q} \cap [0,1], \ltimes \in \{<, \leq, \geq, >\}\}, \\
\Gamma &:= \{N_I \mid I \subseteq \mathbb{R}_+ \text{ rational interval}\} \cup \{U_I \mid I \subseteq \mathbb{R}_+ \text{ rational interval}\}.
\end{aligned}$$

Here $\mathcal{S}_{\ltimes p}$ defines the steady state operator (Example 4.2.4), $\theta_{\ltimes q}$ defines path quantification (Example 4.2.3), N_I interprets the next operator (Example 4.2.8), and finally U_I is in charge of representing the until operator (Example 4.2.9). $\dashv$

Given the stochastic coalgebra $\mathsf{F} = (A, \gamma)$, define recursively for state formulas ϕ and for $\mathfrak{T}$-formulas ψ the sets $[\![\phi]\!]_{\mathsf{F}}$ respectively $[\![\psi]\!]_{\mathsf{F}}$ with $\lambda \in \Lambda, \vartheta \in \Theta, b \in \Gamma$:

$$
\begin{aligned}
[\![\top]\!]_{\mathsf{F}} &= A\\
[\![\phi_1 \wedge \phi_2]\!]_{\mathsf{F}} &= [\![\phi_1]\!]_{\mathsf{F}} \cap [\![\phi_2]\!]_{\mathsf{F}}\\
[\![\langle\lambda\rangle\phi]\!]_{\mathsf{F}} &= \big(\mathfrak{B}(\gamma) \circ \lambda_A\big)([\![\phi]\!]_{\mathsf{F}})\\
[\![\langle\vartheta\rangle\psi]\!]_{\mathsf{F}} &= \big(\mathfrak{B}(\gamma) \circ \vartheta_A\big)([\![\psi]\!]_{\mathsf{F}})\\
[\![\widetilde{\top}]\!]_{\mathsf{F}} &= \mathfrak{T}(A)\\
[\![\psi_1 \wedge \psi_2]\!]_{\mathsf{F}} &= [\![\psi_1]\!]_{\mathsf{F}} \cap [\![\psi_2]\!]_{\mathsf{F}}\\
[\![b(\phi_1, \ldots, b_{\mathfrak{ar}(b)})]\!]_{\mathsf{F}} &= N_{b,A}([\![\phi_1]\!]_{\mathsf{F}}, \ldots, [\![\phi_{\mathfrak{ar}(b)}]\!]_{\mathsf{F}})
\end{aligned}
$$

We define the satisfaction relation $\models$ between states or elements of $\mathfrak{T}(A)$ and formulas through $\mathsf{F}, a \models \phi$ iff $a \in [\![\phi]\!]_{\mathsf{F}}$ whenever $a \in A$ or $\mathsf{F}, t \models \psi$ iff $t \in [\![\psi]\!]_{\mathsf{F}}$ for $t \in \mathfrak{T}(A)$. As above, the *theory* is defined through the formulas that are satisfied; formally

$$Th_{\mathsf{F}}(a) := \{\phi \mid \phi \text{ is a state formula}, \mathsf{F}, a \models \phi\}$$

and

$$Th_{\mathsf{F}}(t) := \{\psi \mid \psi \text{ is a } \mathfrak{T}\text{-formula}, \mathsf{F}, t \models \psi\}.$$

We overload the symbols $[\![\cdot]\!]_{\mathsf{F}}$, $\models$ and $Th_{\mathsf{F}}(\cdot)$, trusting that the context makes it clear whether state formulas or $\mathfrak{T}$-formulas are addressed.

An easy consequence of the construction yields that we are working with Borel sets.

Lemma 4.3.4. *Whenever ϕ is a state formula, $[\![\phi]\!]_{\mathsf{F}}$ is a Borel set in A, and $[\![\psi]\!]_{\mathsf{F}}$ is a Borel set in $\mathfrak{T}(A)$ for each $\mathfrak{T}$-formula ψ.*

Proof The proof proceeds by induction and observes that $\lambda_A(C)$ is a Borel set in $\mathfrak{S}(\mathfrak{T}(A))$ for each Borel set $C \in \mathcal{B}(A)$, and each predicate lifting λ. Consequently,

$$\big(\mathfrak{B}(\gamma) \circ \lambda_A\big)(C) = \{a \in A \mid \gamma(a) \in \lambda_A(C)\}$$

is a Borel set in A on account of γ being a stochastic relation, hence a Borel measurable map $A \to \mathfrak{S}(\mathfrak{T}(A))$. The argumentation for a $\mathfrak{T}$-predicate lifting is analogous. □

Morphisms between coalgebras preserve the meaning of formulas.

Proposition 4.3.5. *Let $\mathsf{G} = (B, \delta)$ be another stochastic left coalgebra, and $f : \mathsf{F} \to \mathsf{G}$ be a morphism. Then*

a. $\mathfrak{B}(f)([\![\phi]\!]_{\mathsf{G}}) = [\![\phi]\!]_{\mathsf{F}}$ *for each state formula* ϕ,
b. $\mathfrak{B}(\mathfrak{T}(f))([\![\psi]\!]_{\mathsf{G}}) = [\![\psi]\!]_{\mathsf{F}}$ *for each $\mathfrak{T}$-formula* ψ.

Proof 0. The proof proceeds by induction on the formulas' structure. Assume that the assertion is established for the state formula ϕ and for the $\mathfrak{T}$-formula ψ.

1. Let λ be a predicate lifting; then

$$\begin{aligned}
\mathfrak{B}(f)([\![\langle\lambda\rangle\phi]\!]_{\mathsf{G}}) &= \big(\mathfrak{B}(f)\circ\mathfrak{B}(\delta)\circ\lambda_B\big)([\![\phi]\!]_{\mathsf{G}})\\
&= \big(\mathfrak{B}(\delta\circ f)\circ\lambda_B\big)([\![\phi]\!]_{\mathsf{G}})\\
&\overset{(\dagger)}{=} \big(\mathfrak{B}(\mathfrak{S}(\mathfrak{T}(f))\circ\gamma)\circ\lambda_B\big)([\![\phi]\!]_{\mathsf{G}})\\
&= \big(\mathfrak{B}(\gamma)\circ\mathfrak{B}(\mathfrak{S}(\mathfrak{T}(f)))\circ\lambda_B\big)([\![\phi]\!]_{\mathsf{G}})\\
&\overset{(\ddagger)}{=} \big(\mathfrak{B}(\gamma)\circ\lambda_A\circ\mathfrak{B}(f)\big)([\![\phi]\!]_{\mathsf{G}})\\
&= \big(\mathfrak{B}(\gamma)\circ\lambda_A\big)(\mathfrak{B}(f)([\![\phi]\!]_{\mathsf{G}}))\\
&= \big(\mathfrak{B}(\gamma)\circ\lambda_A\big)([\![\phi]\!]_{\mathsf{F}}).
\end{aligned}$$

Equation (†) holds because f is a coalgebra morphism, and Equation (‡) derives from this diagram, which is commutative because λ is natural:

$$\begin{array}{ccc}
\mathcal{B}(B) & \xrightarrow{\mathfrak{B}(f)} & \mathcal{B}(A)\\
\Big\downarrow{\scriptstyle \lambda_B} & & \Big\downarrow{\scriptstyle \lambda_A}\\
\mathcal{B}(\mathfrak{S}(\mathfrak{T}(B))) & \xrightarrow[\mathfrak{B}(\mathfrak{S}(\mathfrak{T}(f)))]{} & \mathcal{B}(\mathfrak{S}(\mathfrak{T}(A)))
\end{array}$$

In a similar way we see that for a $\mathfrak{T}$-predicate lifting ϑ

$$\begin{aligned}
\mathfrak{B}(f)([\![\langle\vartheta\rangle\psi]\!]_{\mathsf{G}}) &= \big(\mathfrak{B}(\gamma)\circ\mathfrak{B}(\mathfrak{S}(\mathfrak{T}(f)))\circ\vartheta_B\big)([\![\psi]\!]_{\mathsf{G}})\\
&= \big(\mathfrak{B}(\gamma)\circ\vartheta_A\big)(\mathfrak{B}(\mathfrak{S}(\mathfrak{T}(f)))([\![\psi]\!]_{\mathsf{G}}))\\
&= \big(\mathfrak{B}(\gamma)\circ\vartheta_A\big)([\![\psi]\!]_{\mathsf{F}})\\
&= [\![\langle\vartheta\rangle\psi]\!]_{\mathsf{F}}
\end{aligned}$$

holds.

2. Take for simplicity $\mathfrak{ar}(b) = 1$ for the bridge operator $b \in \Gamma$. Because N_b is natural, we obtain

$$[\![b(\phi)]\!]_{\mathsf{F}} = N_{b,A}([\![\phi]\!]_{\mathsf{F}}) = N_{b,A}(\mathfrak{B}(f)([\![\phi]\!]_{\mathsf{G}})) = \\ \mathfrak{B}(\mathfrak{T}(f))(N_{b,B}([\![\phi]\!]_{\mathsf{G}})) = \mathfrak{B}(\mathfrak{T}(f))([\![b(\phi)]\!]_{\mathsf{G}}).$$

□

We assume for the rest of the section that the interpretation through a left coalgebra F is *nontrivial* in the sense that we can always find a $\mathfrak{T}$-formula ψ such that $\emptyset \neq [\![\psi]\!]_{\mathsf{F}} \neq \mathfrak{T}(A)$. This assumption is necessary in order to prevent pathological cases to creep in; see the discussion in Section 2.2.

4.3.3 Congruences

Congruences are, as we have seen, at the core of the investigation of bisimilarity and behavioral equivalence. Hence we want to find out under what conditions the pair of equivalences defined through the logic forms a congruence for the stochastic relations associated with the respective left coalgebras, and, in a second step, to investigate under which conditions logically equivalent coalgebras induce congruences that simulate each other. Simulation is a rather strong relation between congruences, entailing the probabilistic behavior of one congruence being sufficient for a characterization of the behavior of the other. Simulation equivalence, i.e., mutual simulation of the congruences, is a sufficient condition for the bisimilarity of stochastic relations by Proposition 2.2.4. The mediating relation in a bisimulation needs not be associated with a left coalgebra; this is so since the problem is solved on the level of stochastic relations. It turns out, however, that the mediator can be tuned to conform to the restrictions of a coalgebra under suitable conditions for the functor. This then yields a coalgebra as a mediator, so that logical equivalent coalgebras are bisimilar.

Then the focus of the discussion shifts to behavioral equivalence. It is quite obvious that coalgebras that are behaviorally equivalent have the same theories, and we show that the converse also holds, provided the functor is well behaved. We follow basically the same pattern in the construction: The problem is solved first for the stochastic relations associated with the coalgebras, and then this solution is massaged into one for coalgebras.

At the very center of the discussion lies the fact that the smooth equivalence relations induced by the logic form a congruence. This does not hold in general; it requires the concept of separation for the predicate liftings, which is discussed first.

Intuitively, separation means for a set of liftings that it provides sufficiently many values for different probabilities to be perceived as being different indeed.

Definition 4.3.6. *The set Θ of $\mathfrak{T}$-predicate liftings is said to* separate the logic $\mathfrak{L}^{\flat}$ *iff for every left coalgebra* $\mathsf{F} = (A, \gamma)$*, for each $\mathfrak{T}$-formula ψ, and for arbitrary $\mu, \mu' \in {}^{L}\mathfrak{T}(A)$*

$$\left[\forall \vartheta \in \Theta : \mu \in \vartheta_A([\![\psi]\!]_{\mathsf{F}}) \Leftrightarrow \mu' \in \vartheta_A([\![\psi]\!]_{\mathsf{F}})\right] \Rightarrow \mu([\![\psi]\!]_{\mathsf{F}}) = \mu'([\![\psi]\!]_{\mathsf{F}}).$$

Consequently, if Θ separates $\mathfrak{L}^{\flat}$, then two measures μ, μ' on ${}^{L}\mathfrak{T}(A)$ yield different values $\mu([\![\psi]\!]_{\mathsf{F}})$ and $\mu'([\![\psi]\!]_{\mathsf{F}})$ on the extension $[\![\psi]\!]_{\mathsf{F}}$ of some $\mathfrak{T}$-formula, provided we can find a predicate lifting $\vartheta \in \Theta$ so that $\vartheta_A([\![\psi]\!]_{\mathsf{F}})$ contains exactly one of μ and μ'. Hence if the set $\{\vartheta_A([\![\psi]\!]_{\mathsf{F}} \mid \vartheta \in \Theta\}$ cannot distinguish between μ and μ', then these measures must be identical on the Borel set generated by $[\![\psi]\!]_{\mathsf{F}}$.

For each $\vartheta \in \Theta$, the set $\vartheta_A([\![\psi]\!]_{\mathsf{F}})$ is a Borel subset of $\mathfrak{S}(\mathfrak{T}(S))$ in the weak*-σ-algebra. This does not bind the values $\{\mu([\![\psi]\!]_{\mathsf{F}}) \mid \mu \in \vartheta_A([\![\psi]\!]_{\mathsf{F}})\} \subseteq$

$\mathbb{R}_+$ in any particular way. But through separation some discerning statements can be made.

We will encounter examples for this concept when discussing modal logics in Section 4.3.4 and continuous time stochastic logic as we go. For illustrating the concept, we discuss briefly strong separation, which makes use of witness sets.

Definition 4.3.7. *Call a set $\mathcal{S}$ of subsets of a set X* separating in X *iff given two different elements $q, q' \in X$ there exists $S \in \mathcal{S}$ such that either $q \in S$ or $q' \in S$.*

Consequently, $\mathcal{S}$ is separating iff $q = q'$, provided $q \in S \Leftrightarrow q' \in S$ holds for all $S \in \mathcal{S}$, or equivalently, iff the equivalence relation induced by $\mathcal{S}$ on X is the identity. Consider as an example the collection $\mathcal{S}_1 := \{]p, q[| \ p, q \in \mathbb{Q}, p < q\}$ of open intervals with rational endpoints, or the collection of all closed interval $\mathcal{S}_2 := \{[p, +\infty[| \ p \in \mathbb{Q}\}$ that are unbounded to the right with a rational left endpoint. Both are separating in $\mathbb{R}$, because the rationals $\mathbb{Q}$ are dense in $\mathbb{R}$. More generally, a subbase for a topology on a set X is separating iff the topology is Hausdorff.

Definition 4.3.8. *The set Θ of $\mathfrak{T}$-predicate liftings is said to* strongly separate the logic *iff the following holds:*

a. given a $\mathfrak{T}$-predicate lifting $\vartheta \in \Theta$ and a $\mathfrak{T}$-formula ψ there exists a witness set $S_{\vartheta,\psi} \in \mathcal{B}([0,1])$ such that

$$\vartheta_A([\![\psi]\!]_{\mathsf{F}}) = \{\mu \in \mathfrak{S}(\mathfrak{T}(A)) \mid \mu([\![\psi]\!]_{\mathsf{F}}) \in S_{\vartheta,\psi}\},$$

whenever A is an analytic space,

b. for each $\mathfrak{T}$-formula ψ, the collection $\{S_{\vartheta,\psi} \mid \vartheta \in \Theta\}$ of witness sets is separating in $[0,1]$.

Consequently, the witness sets are independent of a specific analytic space; this will permit comparing values for probabilities across coalgebras. If Θ strongly separates the logic, then it separates the logic; this follows directly from the definition of a separating set.

Example 4.3.9. Consider path quantification and the operator underlying it in Example 4.2.3

$$\Theta := \{\vartheta_{\ltimes q} \mid q \in \mathbb{Q} \cap [0,1]\}.$$

Define for $\theta := \vartheta_{\geq q} \in \Theta$ the set $S_\theta := [q, 1]$ uniformly for each $\mathfrak{T}$-formula ψ. Evidently $\mu([\![\psi]\!]_{\mathsf{F}}) \in S_\theta$ iff $\mu \in \theta_A([\![\psi]\!]_{\mathsf{F}})$ for each left coalgebra F over the state space A. These witness sets are separating in $[0,1]$ since $\mathbb{Q} \cap [0,1]$ is dense in $[0,1]$. Hence Θ strongly separates the logic. $\dashv$

Let again $\mathsf{F} = (A, \gamma)$ be a stochastic left coalgebra. Define on A respectively $\mathfrak{T}(A)$ the equivalence relations

$$a_1 \mathrel{\mathsf{r_F}} a_2 \text{ iff } \big[\mathsf{F}, a_1 \models \phi \Leftrightarrow \mathsf{F}, a_2 \models \phi\big] \text{ holds for all state formulas } \phi$$

and

$$t_1 \mathrel{\widetilde{\mathsf{r_F}}} t_2 \text{ iff } \big[\mathsf{F}, t_1 \models \psi \Leftrightarrow \mathsf{F}, t_2 \models \psi\big] \text{ holds for all } \mathfrak{T}\text{-formulas } \psi.$$

Thus $a_1 \mathrel{\mathsf{r_F}} a_2$ iff the states a_1 and a_2 satisfy exactly the same state formulas, so the logic cannot separate these states. Because we have a countable number of state formulas, this equivalence relation is smooth. Technically, $\mathsf{r_F}$ is determined by the set $\{[\![\phi]\!]_{\mathsf{F}}, \mid \phi \text{ is a state formula}\}$ in the sense of Proposition 1.7.12; similarly for $\widetilde{\mathsf{r_F}}$, which is thus smooth as well.

We talk a bit loosely about a congruence for a stochastic coalgebra F in the sequel when we actually deal with a congruence for the associated stochastic relation $(A, \mathfrak{T}(A), \gamma)$.

Proposition 4.3.10. *If the set Θ of $\mathfrak{T}$-predicate liftings separates the logic, then $\mathsf{c_F} := (\mathsf{r_F}, \widetilde{\mathsf{r_F}})$ is a congruence for the left coalgebra $\mathsf{F} = (A, \gamma)$.*

Proof 0. We have to show that $\gamma(a_1)(D) = \gamma(a_2)(D)$ holds whenever $a_1 \mathrel{\mathsf{r_F}} a_2$ and $D \in \mathbf{\Sigma}\big(\mathcal{B}(\mathfrak{T}(A)), \widetilde{\mathsf{r_F}}\big)$. Fix $a_1, a_2 \in A$ with $a_1 \mathrel{\mathsf{r_F}} a_2$ and put

$$\mathcal{D} := \{D \in \mathbf{\Sigma}\big(\mathcal{B}(\mathfrak{T}(A)), \widetilde{\mathsf{r_F}}\big) \mid \gamma(a_1)(D) = \gamma(a_2)(D)\}.$$

1. From the elementary properties of a measure we see that $\mathcal{D}$ is closed under countable disjoint unions. From the definition of the equivalence relation $\mathsf{r_F}$ it is inferred that in particular

$$\mathsf{F}, a_1 \models \langle\vartheta\rangle\psi \Leftrightarrow \mathsf{F}, a_2 \models \langle\vartheta\rangle\psi$$

for all predicate liftings $\vartheta \in \Theta$ and for an arbitrary $\mathfrak{T}$-formula ψ. Hence we infer for an arbitrary $\mathfrak{T}$-formula ψ

$$\forall\vartheta \in \Theta : \gamma(a_1) \in \vartheta_A([\![\psi]\!]_{\mathsf{F}}) \Leftrightarrow \gamma(a_2) \in \vartheta_A([\![\psi]\!]_{\mathsf{F}}).$$

Because Θ separates the logic, we infer that $\gamma(a_1)([\![\psi]\!]_{\mathsf{F}}) = \gamma(a_2)([\![\psi]\!]_{\mathsf{F}})$. This implies that

$$\mathcal{E} := \{[\![\psi]\!]_{\mathsf{F}} \mid \psi \text{ is a } \mathfrak{T}\text{-formula}\}$$

is a subset of $\mathcal{D}$.

2. Since $\mathfrak{T}$-formulas are closed under conjunction, $\mathcal{E}$ is closed under intersections. Since $[\![\widetilde{\top}]\!]_{\mathsf{F}} = \mathfrak{T}(A)$, we infer that $\mathfrak{T}(A) \in \mathcal{D}$; thus additivity of a measure yields that $\mathcal{D}$ is closed under complementation, too. From the π-λ-Theorem 1.3.1 we infer now $\sigma(\mathcal{E}) \subseteq \mathcal{D}$. But since $\mathbf{\Sigma}\big(\mathcal{B}(\mathfrak{T}(A)), \widetilde{\mathsf{r_F}}\big) = \sigma(\mathcal{E})$, this implies the assertion. □

Looking back at the proof, a subtle point which again emphasizes the rôle of conjunction in $\mathfrak{L}^\flat$ should not go unnoticed. The family $\mathcal{D}$ of sets is closed under complementation, countable disjoint unions, and under intersection. These properties are needed for applying the π-λ-Theorem. Closure under finite intersections may be derived since the logic is closed under conjunction, closedness under complementation and countable disjoint unions, however, comes from additivity of the measures and not from properties of the logic. The asymmetry was noticed above in more specialized logics and surfaces in this more general context again. It is somewhat remarkable.

Definition 4.3.11. *Call the stochastic coalgebras* $\mathsf{F} = (A, \gamma)$ *and* $\mathsf{G} = (B, \delta)$ *logically equivalent iff both* $\{Th_{\mathsf{F}}(a) \mid a \in A\} = \{Th_{\mathsf{G}}(b) \mid b \in B\}$ *and* $\{Th_{\mathsf{F}}(s) \mid s \in \mathfrak{T}(A)\} = \{Th_{\mathsf{G}}(t) \mid t \in \mathfrak{T}(B)\}$ *hold.*

Thus F and G are logically equivalent iff the following hold:

1. Given state $a \in A$ there exists a state $b \in B$ such that
$$\mathsf{F}, a \models \phi \Leftrightarrow \mathsf{G}, b \models \phi$$
is true for all state formulas ϕ; hence $Th_{\mathsf{F}}(a) = Th_{\mathsf{G}}(b)$ holds, and vice versa.
2. Given $s \in \mathfrak{T}(A)$ there exists $t \in \mathfrak{T}(B)$ such that
$$\mathsf{F}, s \models \psi \Leftrightarrow \mathsf{G}, t \models \psi$$
is true for all $\mathfrak{T}$-formulas ψ; hence $Th_{\mathsf{F}}(s) = Th_{\mathsf{G}}(t)$ holds, and vice versa.

Equality of theories entails the logic not being able to distinguish these points.

A necessary condition for the bisimilarity of stochastic relations is given by the existence of simulation equivalent congruences on them. This relationship between the congruences generated by the logic on the state spaces of the coalgebras will be discussed now.

Proposition 4.3.12. *Let* $\mathsf{F} = (A, \gamma)$ *and* $\mathsf{G} = (B, \delta)$ *be logically equivalent. Then* $\mathsf{r_F}$ *and* $\mathsf{r_G}$ *spawn each other, and so do* $\widetilde{\mathsf{r_F}}$ *and* $\widetilde{\mathsf{r_G}}$.

Proof 0. Because the discussion is symmetric for $\mathsf{r_F}$ and for $\mathsf{r_G}$, the proof deals with the first case only. Put

$$\mathcal{G}_0 := \{[\![\phi]\!]_{\mathsf{F}} \mid \phi \text{ is a state formula}\},$$

and define $\Upsilon : A/\mathsf{r_F} \to B/\mathsf{r_G}$ through $\Upsilon([a]_{\mathsf{r_F}}) := [b]_{\mathsf{r_G}}$, provided $Th_{\mathsf{F}}(a) = Th_{\mathsf{G}}(b)$. We claim then that $\mathsf{r_F}$ spawns $\mathsf{r_G}$ via $(\Upsilon, \mathcal{G}_0)$.

1. Because $[a_1]_{\mathsf{r_F}} = [a_2]_{\mathsf{r_F}}$ implies $Th_{\mathsf{F}}(a_1) = Th_{\mathsf{F}}(a_2)$, it is clear that Υ is well defined. Since state formulas are closed under conjunction, $\mathcal{G}_0$ is closed under intersections, and from Proposition 1.7.12 it is inferred that $\sigma(\mathcal{G}_0) = \mathbf{\Sigma}(\mathcal{B}(A), \mathsf{r_F})$.

2. Let ϕ be a state formula; then

$$\begin{aligned}\Upsilon_{[\phi]_{\mathsf{F}}} &= \bigcup\{\Upsilon([a]_{\mathsf{r_F}}) \mid \mathsf{F}, a \models \phi\}\\ &= \bigcup\{[b]_{\mathsf{r_G}} \mid \mathsf{F}, a \models \phi, Th_{\mathsf{F}}(a) = Th_{\mathsf{G}}(b)\}\\ &= \bigcup\{[b]_{\mathsf{r_G}} \mid \mathsf{G}, b \models \phi\}\\ &= [\![\phi]\!]_{\mathsf{G}}.\end{aligned}$$

Since the set $\{[\![\phi]\!]_{\mathsf{G}} \mid \phi \text{ is a state formula}\}$ generates $\mathbf{\Sigma}(\mathcal{B}(B), \mathsf{r_G})$ the assertion follows.

3. Turning to $\mathfrak{T}(A)$ and $\mathfrak{T}(B)$, put

$$\mathcal{H}_0 := \{[\![\psi]\!]_{\mathsf{F}} \mid \psi \text{ is a } \mathfrak{T}\text{-formula}\},$$

and define $\Xi : \mathfrak{T}(A)/\widetilde{\mathsf{r_F}} \to \mathfrak{T}(B)/\widetilde{\mathsf{r_G}}$ through $\Xi([s]_{\widetilde{\mathsf{r_F}}}) := [t]_{\widetilde{\mathsf{r_G}}}$, provided $Th_{\mathsf{F}}(s) = Th_{\mathsf{G}}(t)$. Then $\widetilde{\mathsf{r_F}}$ spawns $\widetilde{\mathsf{r_G}}$ via $(\Xi, \mathcal{H}_0)$. This is established through the same argumentation. □

Interestingly, logical equivalence is reflected in the Borel structure of the factor spaces; see Lemma 2.6.10:

Corollary 4.3.13. *Assume that the stochastic coalgebras (A, γ) and (B, δ) are logically equivalent. Then both $A/\mathsf{r_F}$ and $B/\mathsf{r_G}$ are Borel isomorphic to $(A + B)/\mathsf{r_F} \diamond \mathsf{r_G}$ as analytic spaces. Similarly, $\mathfrak{T}(A)/\widetilde{\mathsf{r_F}}$ and $\mathfrak{T}(B)/\widetilde{\mathsf{r_G}}$ are both Borel isomorphic to $\big(\mathfrak{T}(A) + \mathfrak{T}(B)\big)/\widetilde{\mathsf{r_F}} \diamond \widetilde{\mathsf{r_G}}$.* □

Now that we have shown that under the conditions of separation the logic defines a congruence on the algebra, we relate the congruences to each other. This expresses that knowing one congruence suffices essentially for knowing the other one.

Proposition 4.3.14. *Assume that the stochastic congruences $\mathsf{F} = (A, \gamma)$ and $\mathsf{G} = (B, \delta)$ are logically equivalent, and that the set Θ of $\mathfrak{T}$-predicate liftings separates the logic. Then $\mathsf{c_F}$ and $\mathsf{c_G}$ are simulation equivalent congruences.*

Proof We show that $\mathsf{c_F}$ simulates $\mathsf{c_G}$; by symmetry, the assertion will subsequently follow. For convenience the notations from the proof for Proposition 4.3.12 are used. Let $a \in A$ and $b \in \Upsilon([a]_{\mathsf{r_F}})$; thus $Th_{\mathsf{F}}(a) = Th_{\mathsf{G}}(b)$. Because $\Xi_{[\psi]_{\mathsf{F}}} = [\![\psi]\!]_{\mathsf{G}}$, it is sufficient to show that $\gamma(a)([\![\psi]\!]_{\mathsf{F}}) = \delta(b)([\![\psi]\!]_{\mathsf{G}})$ for each $\mathfrak{T}$-formula ψ.

Fix ψ. Since $\mathsf{F}, a \models \phi \Leftrightarrow \mathsf{G}, b \models \phi$ for each state formula ϕ, considering the state formula $\langle\vartheta\rangle\psi$ we know in particular that

$$\gamma(a) \in \vartheta_A([\![\psi]\!]_{\mathsf{F}}) \Leftrightarrow \delta(b) \in \vartheta_B([\![\psi]\!]_{\mathsf{G}})$$

for each $\mathfrak{T}$-predicate lifting $\vartheta \in \Theta$. Because Θ separates the logic, we may infer as in the proof of Proposition 4.3.10 that $\gamma(a)([\![\psi]\!]_{\mathsf{F}}) = \delta(b)([\![\psi]\!]_{\mathsf{G}})$. □

We formulate a condition which relates the functor to the congruences.

Definition 4.3.15. *Let* $\mathsf{F} = (A, \gamma)$ *and* $\mathsf{G} = (B, \delta)$ *be stochastic coalgebras so that the associated congruences* c_F *and* c_G *are simulation equivalent. We say that* functor $\mathfrak{T}$ distributes *over* c_F *and* c_G *iff there exists a morphism* ζ *in* **ANL** *making this diagram commutative (the* π*s are the projections):*

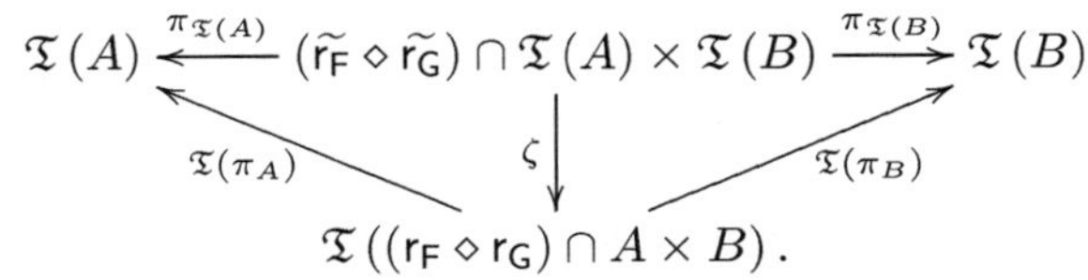

Thus, if $\mathfrak{T}$ distributes over c_F and c_G, then there is a morphism

$$\{\langle s, t\rangle \in \mathfrak{T}(A) \times \mathfrak{T}(B) \mid [s]_{\widetilde{\mathsf{r}_\mathsf{F}} \diamond \widetilde{\mathsf{r}_\mathsf{G}}} = [t]_{\widetilde{\mathsf{r}_\mathsf{F}} \diamond \widetilde{\mathsf{r}_\mathsf{G}}}\} \to \mathfrak{T}\left(\{\langle a, b\rangle \in A \times B \mid [a]_{\mathsf{r}_\mathsf{F} \diamond \mathsf{r}_\mathsf{G}} = [b]_{\mathsf{r}_\mathsf{F} \diamond \mathsf{r}_\mathsf{G}}\}\right)$$

relating $\widetilde{\mathsf{r}_\mathsf{F}} \diamond \widetilde{\mathsf{r}_\mathsf{G}}$-equivalent members of $\mathfrak{T}(A)$ and $\mathfrak{T}(B)$ to the $\mathfrak{T}$-images of the $\mathsf{r}_\mathsf{F} \diamond \mathsf{r}_\mathsf{G}$-equivalent elements in A and B. In this way, the simulation equivalent congruences c_F and c_G are related to the functor $\mathfrak{T}$. Structurally, the condition serves as a replacement for the condition on the functor for which the coalgebras are formulated to preserve weak pullbacks, which is customary in coalgebraic reasoning [72].

As a technical point we note that a smooth equivalence relation over a Standard Borel space or an analytic space X constitutes a Borel set respectively an analytic subset of $X \times X$. With this in mind, we see that $(\widetilde{\mathsf{r}_\mathsf{F}} \diamond \widetilde{\mathsf{r}_\mathsf{G}}) \cap \mathfrak{T}(A) \times \mathfrak{T}(B)$ is Standard Borel if both A and B are Standard Borel spaces, and it is analytic provided A and B are.

Example 4.3.16. Consider functor $\mathfrak{Z} := (- \times \mathbb{R}_+)^\infty$ from Example 4.2.3, and take as a logic the negation-free variant of μ**CSL** without variables from Example 4.3.3. Assume that the dynamics of both coalgebras F and G is represented through a projective limit; see Definition 1.6.32 on page 41. This means for $\mathsf{F} = (A, \gamma)$ with A Polish and γ probabilistic that there exists for each $n \in \mathbb{N}$ a stochastic relation $\gamma_n : A \rightsquigarrow (A \times \mathbb{R}_+)^n$ such that

$$\gamma(a)\big(D \times (A \times \mathbb{R}_+)^\infty\big) = \gamma_n(a)(D)$$

holds whenever $D \in \mathcal{B}((A \times \mathbb{R}_+)^n)$. Similarly for G. We infer from Proposition 2.4.9 that $\big(\langle t_n, a_n\rangle\big)_{n \in \mathbb{N}} \ \widetilde{\mathsf{r}_\mathsf{F}} \ \big(\langle t'_n, a'_n\rangle\big)_{n \in \mathbb{N}}$ holds iff $t_n = t'_n$ and $a_n \ \mathsf{r}_\mathsf{F} \ a'_n$ for all $n \in \mathbb{N}$. Consequently, $\big(\langle t_n, z_n\rangle\big)_{n \in \mathbb{N}} \ \widetilde{\mathsf{r}_\mathsf{F}} \diamond \widetilde{\mathsf{r}_\mathsf{G}} \ \big(\langle t'_n, z'_n\rangle\big)_{n \in \mathbb{N}}$ is true iff $\forall n \in \mathbb{N} : t_n = t'_n$ and $z_n \ \mathsf{r}_\mathsf{F} \diamond \mathsf{r}_\mathsf{G} \ z'_n$. This implies in turn that

$$\zeta : \big\langle \big(\langle t_n, z_n\rangle\big)_{n \in \mathbb{N}}, \big(\langle t'_n, z'_n\rangle\big)_{n \in \mathbb{N}}\big\rangle \mapsto \big(\langle t_n, z_n, z'_n\rangle\big)_{n \in \mathbb{N}}$$

will satisfy the requirements of Definition 4.3.15. Hence $\mathfrak{Z}$ distributes over the equivalent congruences c_F and c_G. $\dashv$

Proposition 4.3.17. *Assume that the set Θ of $\mathfrak{T}$-predicate liftings separates the logic, and let* F *and* G *be stochastic left coalgebras with associated congruences* c_F *and* c_G. *Consider the following statements:*

a. The coalgebras F *and* G *are bisimilar.*
b. F *and* G *are logically equivalent.*

Then $a \Rightarrow b$ holds always, and if the functor $\mathfrak{T}$ distributes over c_F *and* c_G, *also $b \Rightarrow a$ holds.*

Proof 1. $a \Rightarrow b$ follows immediately from Proposition 4.3.5.

2. Turning to $b \Rightarrow a$, let $\mathsf{F} = (A, \gamma)$ and $\mathsf{G} = (B, \delta)$; we may and do assume that both A and B are Polish spaces (Propositions 4.3.2 and 4.3.5). We know from Proposition 4.3.14 that the congruences c_F and c_G are simulation equivalent; hence we may infer from Proposition 2.2.4 that there is a mediating stochastic relation (E, F, K). A closer analysis of the components' construction E and F in the proof of that proposition reveals that in this instance

$$E = \{\langle a, b\rangle \in A \times B \mid [a]_{\mathsf{r}_\mathsf{F} \diamond \mathsf{r}_\mathsf{G}} = [b]_{\mathsf{r}_\mathsf{F} \diamond \mathsf{r}_\mathsf{G}}\}$$

holds; similarly,

$$F = \{\langle s, t\rangle \in \mathfrak{T}(A) \times \mathfrak{T}(B) \mid [s]_{\widetilde{\mathsf{r}_\mathsf{F}} \diamond \widetilde{\mathsf{r}_\mathsf{G}}} = [t]_{\widetilde{\mathsf{r}_\mathsf{F}} \diamond \widetilde{\mathsf{r}_\mathsf{G}}}\}$$

is inferred. The proof shows also that the corresponding morphisms are just the projections $\pi_A : E \to A$ and $\pi_B : E \to B$, and that both E and F are Standard Borel spaces. Put $\varepsilon := \mathfrak{S}(\zeta) \circ K$; then $\mathsf{E} := (E, \varepsilon)$ is a stochastic coalgebra with coalgebra morphisms $\mathsf{F} \xleftarrow{\pi_A} \mathsf{E} \xrightarrow{\pi_B} \mathsf{G}$. □

Turning to behavioral equivalence, we relate it in a similar way to logical equivalence. This observation follows directly from Proposition 4.3.5:

Lemma 4.3.18. *If stochastic left coalgebras are behaviorally equivalent, they are logically equivalent.* □

The converse is — as in the case of bisimilarity — a little more complicated to establish; it will involve the assumption on the functor $\mathfrak{T}$ that it is compatible with factors of sums, so that a sum absorbs the factor under $\mathfrak{T}$. To be specific:

Definition 4.3.19. *Let (α, β) be a congruence for the stochastic left $\mathfrak{T}$-coalgebra (A, γ). Then the functor $\mathfrak{T}$ is said to* absorb (α, β) *iff there exists an* **ANL**-*morphism $\omega_A : \mathfrak{T}(A)/\beta \to \mathfrak{T}(A/\alpha)$ that makes the diagram*

$$\begin{array}{ccc}
\mathfrak{T}(A) & \xrightarrow{\eta_\beta} & \mathfrak{T}(A)/\beta \\
 & \searrow^{\mathfrak{T}(\eta_\alpha)} & \downarrow{\scriptstyle \omega_A} \\
 & & \mathfrak{T}(A/\alpha)
\end{array}$$

commutative.

Thus an absorbing functor relates the quotient $\mathfrak{T}(A)/\beta$ to the $\mathfrak{T}$-image of the quotient $\mathfrak{T}(A/\alpha)$ by factoring the quotient maps.

Simulation equivalent congruences are either both absorbed by the functor, or neither is:

Lemma 4.3.20. *Let* $\mathsf{F} = (A, \gamma)$ *and* $\mathsf{G} = (B, \delta)$ *be stochastic left coalgebras with simulation equivalent congruences* $\mathsf{c_F}$ *respectively* $\mathsf{c_G}$. *Then* $\mathsf{c_F}$ *is absorbed by* $\mathfrak{T}$ *iff* $\mathsf{c_G}$ *is.*

Proof Corollary 4.3.13 gives Borel isomorphisms

$$A/\mathsf{r_F} \cong (A+B)/\mathsf{r_F} \diamond \mathsf{r_G} \cong B/\mathsf{r_G}$$

and

$$\mathfrak{T}(A)/\widetilde{\mathsf{r_F}} \cong \big(\mathfrak{T}(A) + \mathfrak{T}(B)\big)/\widetilde{\mathsf{r_F}} \diamond \widetilde{\mathsf{r_G}} \cong \mathfrak{T}(B)/\widetilde{\mathsf{r_G}}.$$

□

This, then, yields a more symmetric picture. It says essentially that $\widetilde{\mathsf{r_F}} \diamond \widetilde{\mathsf{r_G}}$ coincides with $\widetilde{\mathsf{r_F} \diamond \mathsf{r_G}}$.

Corollary 4.3.21. *If the functor* $\mathfrak{T}$ *absorbs* $\mathsf{c_F}$ *or* $\mathsf{c_G}$, *then there exists a morphism* ν *which renders this diagram commutative:*

$$\begin{array}{ccccc}
\mathfrak{T}(A) & \xrightarrow{\eta_{\widetilde{\mathsf{r_F}}} \circ i_{\mathfrak{T}(A)}} & \big(\mathfrak{T}(A) + \mathfrak{T}(B)\big)/\big(\widetilde{\mathsf{r_F}} \diamond \widetilde{\mathsf{r_G}}\big) & \xleftarrow{\eta_{\widetilde{\mathsf{r_G}}} \circ i_{\mathfrak{T}(B)}} & \mathfrak{T}(B) \\
& \searrow^{\mathfrak{T}(\eta_{\rho_\gamma} \circ i_A)} & \downarrow \nu & \swarrow^{\mathfrak{T}(\eta_{\mathsf{r_G}} \circ i_B)} & \\
& & \mathfrak{T}\big((A+B)/(\mathsf{r_F} \diamond \mathsf{r_G})\big) & &
\end{array}$$

Proof Immediate. □

Using absorption, we may now state the second characterization: behavioral and logical equivalence are the same.

Proposition 4.3.22. *Assume that the set* Θ *of* $\mathfrak{T}$*-predicate liftings separates the logic, and let* F *and* G *be stochastic left coalgebras with associated congruences* $\mathsf{c_F}$ *and* $\mathsf{c_G}$. *Consider the following statements:*

a. The coalgebras F *and* G *are behaviorally equivalent.*
b. F *and* G *are logically equivalent.*

Then $a \Rightarrow b$ *holds always, and if the functor* $\mathfrak{T}$ *absorbs* $\mathsf{c_F}$ *and* $\mathsf{c_G}$, *also* $b \Rightarrow a$ *holds.*

Proof 1. We need only to take care of $b \Rightarrow a$ due to Lemma 4.3.18.

2. Let $\mathsf{F} = (A, \gamma)$, $\mathsf{G} = (B, \delta)$. From Proposition 4.3.14 it is inferred that the congruences $\mathsf{c_F}$ and $\mathsf{c_G}$ are simulation equivalent. From the proof of Proposition 2.2.4 we obtain from the stochastic relations $\mathsf{K} := (A, \mathfrak{T}(A), \gamma)$ and $\mathsf{L} := (B, \mathfrak{T}(B), \delta)$ a stochastic relation

$$\mathsf{M} := \big((A+B)/(\mathsf{r_F} \diamond \mathsf{r_G}), (\mathfrak{T}(A) + \mathfrak{T}(B))/(\widetilde{\mathsf{r_F}} \diamond \widetilde{\mathsf{r_G}}), K\big)$$

with morphisms

$$\mathsf{K} \xrightarrow{(\eta_{\mathsf{r_F}} \circ i_A, \eta_{\widetilde{\mathsf{r_F}}} \circ i_{\mathfrak{T}(A)})} \mathsf{M} \xleftarrow{(\eta_{\mathsf{r_G}} \circ i_B, \eta_{\widetilde{\mathsf{r_G}}} \circ i_{\mathfrak{T}(B)})} \mathsf{L}.$$

Consequently, $\mathsf{H} := ((A+B)/(\mathsf{r_F} \diamond \mathsf{r_G}), \mathfrak{S}\,(\nu) \circ K)$ is a stochastic left coalgebra with coalgebra morphisms

$$\mathsf{F} \xrightarrow{\eta_{\mathsf{r_F}} \circ i_A} \mathsf{H} \xleftarrow{\eta_{\mathsf{r_G}} \circ i_B} \mathsf{G}$$

□

4.3.4 Application to Modal Logic

The results are now applied to modal logic. The development so far had the logic operate on two stages through state formulas and $\mathfrak{T}$-formulas; thus we will need to mimic this approach. So we will introduce a suitable functor $\mathfrak{T}$, but we will concentrate on state formulas, using the additionally introduced class of formulas as auxiliary devices for studying the behavior of the algebras and the models involved. The bridge operators that come with the $\mathfrak{T}$-formulas will ease the transition between classical modal logic and its coalgebraic cousin.

We define stochastic Kripke models, relate these models to stochastic $\mathfrak{T}$-coalgebras, and show that the properties which we study for Kripke models are mirrored in the properties of the coalgebras. The functor is shown to observe the pleasant properties that permit formulating the relevant equivalences.

We define as in Section 2.3 a similarity type τ as a collection of modal operators; each modal operator $\triangle \in \tau$ has an arity $\mathfrak{ar}(\triangle) > 0$. For simplicity, we will not deal with modal operators of arity 0 and not with propositional letters. Both may be added without much ado. The similarity type τ is assumed to be at most countable; it will be fixed in this section.

The *extended modal language* $\mathcal{L}_\tau$ is defined through the syntax

$$\phi ::= \top \mid \phi_1 \wedge \phi_2 \mid \triangle_q(\phi_1, \ldots \phi_{\mathfrak{ar}(\triangle)}),$$

where $\triangle \in \tau$ is a modal operator, and $q \in \mathbb{Q} \cap [0,1]$ is a rational number.

We recall the definition of a stochastic Kripke model from Section 2.3; in Section 2.3.2, logical equivalence, bisimilarity, and behavioral equivalence are defined for these models.

An interpretation of modal logics in left coalgebras goes like this. Define the functor $\mathfrak{T}$ upon defining for an analytic space A

$$\mathfrak{T}(A) := \bigcup_{\triangle \in \tau} \{\triangle\} \times A^{\mathfrak{ar}(\triangle)}.$$

Put for $f : A \to B$

$$\mathfrak{T}(f) : \langle \triangle, \langle s_1, \ldots, s_{\mathfrak{ar}(\triangle)} \rangle \rangle \mapsto \langle \triangle, \langle f(s_1), \ldots, f(s_{\mathfrak{ar}(\triangle)}) \rangle \rangle .$$

Because **ANL** is closed under finite products and countable unions, $\mathfrak{T}(A)$ is an analytic space provided A is, and a countable union carries as a σ-algebra the sum of the individual σ-algebras. Since $\mathfrak{T}(f)$ is a surjective Borel map, if f is, $\mathfrak{T} : \mathbf{ANL} \to \mathbf{ANL}$ constitutes a functor. If A is a Borel space, $\mathfrak{T}(A)$ is one; consequently the functor meets the general requirements.

We associate with each $\mathfrak{T}$-coalgebra a stochastic Kripke model, and vice versa, in such a way that the corresponding morphisms are identical. To be specific, let $\mathsf{F} = (A, \gamma)$ be a stochastic $\mathfrak{T}$-coalgebra; then define for the modal operator $\triangle \in \tau$, $s \in A$ and the Borel set $C \subseteq A^{\mathfrak{ar}(\triangle)}$

$$K^{\gamma}_{\triangle}(s)(C) := \gamma(s)(\{\triangle\} \times C).$$

Then

$$\mathcal{K}^{\mathsf{F}} := \left(A, (K^{\gamma}_{\triangle})_{\triangle \in \tau}\right)$$

is evidently a stochastic Kripke model. Conversely, we want to encode the countable family of stochastic relations into one single entity with the requirements that the encoding permits us to set up a stochastic left $\mathfrak{T}$-coalgebra, and that we can recover the individual members of the family from the encoded relation; moreover, that morphisms remain undisturbed.

Having a look at generating functions from combinatorics is helpful. There a sequence $(x_n)_{n \in \mathbb{N}}$ of real or complex numbers is encoded into the generating function $\mathcal{G}(z) := \sum_{n \in \mathbb{N}} x_n \cdot z^n$; for example, the sequence $(1)_{n \in \mathbb{N}}$ is encoded into $1/(1 - z)$, or the sequence $(2^n)_{n \in \mathbb{N}}$ into $1/(1 - 2 \cdot z)$. From $\mathcal{G}$ one can recover the sequence upon differentiation (see [83] for an illuminating yet entertaining exposition, and [52, 41] for combinatorial aspects).

We assume that $\tau = \{\triangle_n \mid n \in \mathbb{N}\}$ with a fixed enumeration of the operators; we additionally fix a sequence $(w_n)_{n \in \mathbb{N}}$ of strictly positive real numbers with $\sum_{n \in \mathbb{N}} w_n \leq 1$. Let $\mathcal{K} = \left(A, (K_{\triangle_n})_{n \in \mathbb{N}})\right)$ be a stochastic Kripke model, and put for the nth modal operator $\triangle_n$, $s \in A$ and the Borel set $C \in \mathcal{B}(A^{\mathfrak{ar}(\triangle_n)})$

$$\gamma^{\mathcal{K}}(s)(\{\triangle_n\} \times C) := w_n \cdot K_{\triangle_n}(s)(C).$$

Define for $C \in \mathcal{B}(\mathfrak{T}(A))$

$$\gamma^{\mathcal{K}}(s)(C) := \sum_{n \in \mathbb{N}} \gamma^{\mathcal{K}}(s)\big(C \cap (\{\triangle_n\} \times A^{\mathfrak{ar}(\triangle_n)})\big).$$

Then it is trivial that $\gamma^{\mathcal{K}} : A \rightsquigarrow \mathfrak{T}(A)$. Put $\mathsf{F}^{\mathcal{K}} := (A, \gamma^{\mathcal{K}})$.

From $\gamma^{\mathcal{K}}$ we are able to recover the Kripke model, as we will see in the proof of Lemma 4.3.23. This construction keeps morphisms invariant, so that one can switch between model morphisms and $\mathfrak{T}$-morphisms without losing information.

Lemma 4.3.23. *Let $\mathsf{F} = (A, \gamma)$ and $\mathsf{G} = (B, \delta)$ be left $\mathfrak{T}$-coalgebras with A and B analytic spaces. Assume that $f : A \to B$ is a surjective Borel map. Then*

a. If $f : \mathsf{F} \to \mathsf{G}$ is a $\mathfrak{T}$-coalgebra morphism, then $f : \mathcal{K}^{\mathsf{F}} \to \mathcal{K}^{\mathsf{G}}$ is a model morphism.
b. If $f : \mathcal{K} \to \mathcal{L}$ is a Kripke model morphism, then $f : \mathsf{F}^{\mathcal{K}} \to \mathsf{F}^{\mathcal{L}}$ is a $\mathfrak{T}$-morphism.

Proof 1. For establishing part a, we have to show that $K^{\delta}_{\triangle} \circ f = \mathfrak{S}\left(f^{\mathfrak{ar}(\triangle)}\right) \circ K^{\gamma}_{\triangle}$ holds for each modal operator $\triangle \in \tau$, provided $f : \mathsf{F} \to \mathsf{G}$ is a $\mathfrak{T}$-coalgebra morphism. Since $\triangle \in \tau$, there exists $n \in \mathbb{N}$ with $\triangle = \triangle_n$. Now let $G \subseteq B^{\mathfrak{ar}(\triangle)}$ be a Borel set; then

$$\begin{aligned}
\left(K^{\delta}_{\triangle} \circ f\right)(s)(G) &= K^{\delta}_{\triangle}(f(s))(G) \\
&= \delta(f(s))\big(\{\triangle\} \times G\big) \\
&= \gamma(s)\big(\{\triangle\} \times \mathfrak{B}\left(\left(f^{\mathfrak{ar}(\triangle)}\right)\right)(G)\big) \\
&= K^{\gamma}_{\triangle}(s)\big(\mathfrak{B}(f^{\mathfrak{ar}(\triangle)})(G)\big) \\
&= \big(\mathfrak{S}(f^{\mathfrak{ar}(\triangle)}) \circ K^{\gamma}_{\triangle}\big)(s)(G).
\end{aligned}$$

2. Part b is established similarly. □

Now take a modal operator $\triangle \in \tau$, and a threshold $q \in \mathbb{Q} \cap [0, 1]$. Put for the analytic space A and the Borel set $D \in \mathcal{B}(\mathfrak{T}(A))$

$$\vartheta_{\triangle,q,S}(D) := \{\mu \in \mathfrak{S}(\mathfrak{T}(A)) \mid \mu \blacktriangleright \{\triangle\} \times S^{\mathfrak{ar}(\triangle)}, \mu(F) \geq q\}.$$

We say that a measure $\mu \in \mathfrak{S}(\mathfrak{T}(A))$ *lives on a Borel set* B ($\mu \blacktriangleright B$) iff $\mu(G) = 0$, provided the Borel set G is disjoint to B, or, equivalently, $\mu(B) = \mu(\mathfrak{T}(A))$, so that all the μ-mass concentrates on B. In a Polish space we could formulate this conveniently through the support C_{μ} of a measure (Lemma 3.4.25), but this construction is not available in a general analytic space.

Lemma 4.3.24. *$\vartheta_{\triangle,q}$ constitutes a natural transformation $\mathfrak{B} \circ \mathfrak{T} \overset{\bullet}{\to} \mathfrak{B} \circ \mathfrak{S} \circ \mathfrak{T}$.*

Proof 1. The definition of the weak*-σ-algebra yields that $\{\mu \in \mathfrak{S}(\mathfrak{T}(A)) \mid \mu$ lives on $G\}$ is a Borel set in $\mathfrak{S}(\mathfrak{T}(A))$ whenever G is a Borel set in $\mathfrak{T}(A)$. Thus we may conclude that $\vartheta_{\triangle,q,A}$ maps $\mathcal{B}(\mathfrak{T}(A))$ to $\mathcal{B}(\mathfrak{S}(\mathfrak{T}(A)))$ whenever A is an analytic space.

2. Now let $f : A \to B$ be an **ANL**-morphism. Because a subprobability μ on $\mathfrak{T}(A)$ lives on A^k iff its image $\mathfrak{S}(\mathfrak{T}(f))(\mu)$ lives on B^k, we obtain for a Borel set $F \in \mathcal{B}(\mathfrak{T}(B))$

$$
\begin{aligned}
&\vartheta_{\triangle,q,S}(\mathfrak{B}(\mathfrak{T}(f))(D)) = \\
&\qquad \{\mu \in \mathfrak{S}(\mathfrak{T}(A)) \mid \mu \blacktriangleright \{\triangle\} \times A^{\mathfrak{ar}(\triangle)}, \mu(\mathfrak{B}(\mathfrak{T}(f))(D)) \geq q\} = \\
&\qquad \{\mu \in \mathfrak{S}(\mathfrak{T}(A)) \mid \mu \blacktriangleright \{\triangle\} \times A^{\mathfrak{ar}(\triangle)}, \mathfrak{S}(\mathfrak{T}(f))(\mu)(D) \geq q\} = \\
&\quad \mathfrak{B}(\mathfrak{S}(\mathfrak{T}(f)))(\{\nu \in \mathfrak{S}\left(\mathfrak{T}(B)) \mid \nu \blacktriangleright \{\triangle\} \times B^{\mathfrak{ar}(\triangle)}, \nu(D) \geq q\}\right) = \\
&\qquad\qquad\qquad\qquad\qquad\qquad \mathfrak{S}(\mathfrak{T}(f))(\vartheta_{\triangle,q,B}(F)).
\end{aligned}
$$

Thus $\vartheta_{\triangle,q} : \mathfrak{B} \circ \mathfrak{T} \xrightarrow{\bullet} \mathfrak{B} \circ \mathfrak{S} \circ \mathfrak{T}$ is natural. □

Let $\Theta := \{\vartheta_{\triangle,q} \mid \triangle \in \tau, q \in \mathbb{Q} \cap [0,1]\}$ be the corresponding set of predicate liftings; Θ is countable. Furthermore, let b_k be a bridge operator of arity k for each arity k of a modal operator in τ with Γ as the set of all bridge operators. We define the logic $\mathcal{L}_\tau^c := \mathfrak{L}^b(\emptyset, \Theta, \Gamma)$ through state formulas and $\mathfrak{T}$-formulas as a special case of the logic $\mathfrak{L}^b$ defined in Section 4.3.2 on page 167. To be specific:

- *State formulas* ϕ are given according to this grammar

$$\phi ::= \top \mid \phi_1 \wedge \phi_2 \mid \langle\vartheta\rangle\psi$$

  for $\vartheta \in \Theta$ and the $\mathfrak{T}$-formula ψ.
- *$\mathfrak{T}$-formulas* ψ are given according to this grammar

$$\psi ::= \widetilde{\top} \mid \psi_1 \wedge \psi_2 \mid b_\triangle(\phi_1, \ldots, \phi_{\mathfrak{ar}(\triangle)}),$$

  for the state formulas $\phi_1, \ldots, \phi_{\mathfrak{ar}(\triangle)}$, $b_\triangle$ being a bridge operator with arity $\mathfrak{ar}(\triangle)$ for each modal operator $\triangle \in \tau$.

Associate with each bridge operator $b_\triangle \in \Gamma$ the natural transformation $N_\triangle : \mathfrak{B}^{\mathfrak{ar}(\triangle)} \xrightarrow{\bullet} \mathfrak{B} \circ \mathfrak{T}$ through

$$N_{\triangle,A}(C_1, \ldots, C_{\mathfrak{ar}(\triangle)}) := \{\triangle\} \times C_1 \times \cdots \times C_{\mathfrak{ar}(\triangle)}.$$

Now let $\mathsf{F} = (A, \gamma)$ be a stochastic $\mathfrak{T}$-algebra; then we have according to Section 4.3.2 for $s \in A$ and $t \in \mathfrak{T}(A)$

$$\mathsf{F}, s \models \langle\vartheta\rangle\psi \Leftrightarrow \gamma(s) \in \vartheta_S([\![\psi]\!]_{\mathsf{F}})$$

and

$$
\begin{aligned}
\mathsf{F}, t \models b_\triangle(\phi_1, \ldots, \phi_{\mathfrak{ar}(\triangle)}) &\Leftrightarrow t \in N_\triangle([\![\phi_1]\!]_{\mathsf{F}}, \ldots, [\![\phi_{\mathfrak{ar}(\triangle)}]\!]_{\mathsf{F}}) \\
&\Leftrightarrow t \in \{\triangle\} \times [\![\phi_1]\!]_{\mathsf{F}} \times \cdots \times [\![\phi_{\mathfrak{ar}(\triangle)}]\!]_{\mathsf{F}}.
\end{aligned}
$$

Lemma 4.3.25. *Let* $\mathsf{F} = (A, \gamma)$ *and* $\mathsf{G} = (T, \delta)$ *be* $\mathfrak{T}$*-coalgebras. Then the following statements are equivalent for* $x, y \in \mathfrak{T}(A)$ *with* $x = \langle \triangle, \langle s_1, \ldots, s_n \rangle \rangle$ *and* $y = \langle \triangle', \langle t_1, \ldots, t_m \rangle \rangle$.

a. $Th_{\mathsf{F}}(x) = Th_{\mathsf{G}}(y)$.
b. $\triangle = \triangle'$, $n = \mathfrak{ar}(\triangle) = \mathfrak{ar}(\triangle') = m$, *and* $Th_{\mathsf{F}}(s_i) = Th_{\mathsf{G}}(t_i)$ *for each index* i, $1 \leq i \leq \mathfrak{ar}(\triangle)$.

Proof 1. One notes for the proof of $a \Rightarrow b$ that both $\triangle = \triangle'$ and $n = \mathfrak{ar}(\triangle) = \mathfrak{ar}(\triangle') = m$ follow immediately from the construction. Let $1 \leq i \leq \mathfrak{ar}(\triangle)$, and for simplicity $i = 1$. Assume $\phi_1 \in Th_{\mathsf{F}}(s_1)$; thus $x \models_\gamma b_\triangle(\phi_1, \top, \ldots, \top)$, by assumption $y \models_\delta b_\triangle(\phi_1, \top, \ldots, \top)$; consequently $\phi_1 \in Th_{\mathsf{G}}(t_1)$.

2. For establishing $b \Rightarrow a$, suppose that there exists $\psi \in Th_{\mathsf{F}}(x)$ with $\psi \notin Th_{\mathsf{G}}(y)$. Select a formula ψ of minimal length; then we can find state formulas $\phi_1, \ldots, \phi_n$ such that $\psi = b_\triangle(\phi_1, \ldots, \phi_n)$. We conclude from $\psi \in Th_{\mathsf{F}}(x)$ that $\phi_i \in Th_{\mathsf{F}}(s_i)$ for $1 \leq i \leq n$, which in turn implies $\psi \in Th_{\mathsf{G}}(y)$. This is a contradiction. Hence $Th_{\mathsf{F}}(x) \subseteq Th_{\mathsf{G}}(y)$; the other inclusion is established in the same way. □

Let $\mathsf{F} = (A, \gamma)$ be a $\mathfrak{T}$-coalgebra with r_{F} and $\widetilde{\mathsf{r}_{\mathsf{F}}}$ as the smooth equivalence relations induced on A respectively $\mathfrak{T}(A)$ by the logic. The characterization of the theories yields as an immediate consequence a Borel isomorphism between the factor spaces.

Proposition 4.3.26. *Let* $\mathsf{F} = (A, \gamma)$ *be a stochastic* $\mathfrak{T}$*-coalgebra with analytic state space* A. *Then* $\mathfrak{T}(A)/\widetilde{\mathsf{r}_{\mathsf{F}}}$ *and* $\mathfrak{T}(A/\mathsf{r}_{\mathsf{F}})$ *are Borel isomorphic.*

Proof 1. Define

$$\varpi : \mathfrak{T}(A)/\widetilde{\mathsf{r}_{\mathsf{F}}} \ni [\langle \triangle, \langle s_1, \ldots, s_k \rangle \rangle]_{\widetilde{\mathsf{r}_{\mathsf{F}}}} \mapsto \langle \triangle, \langle [s_1]_{\mathsf{r}_{\mathsf{F}}}, \ldots, [s_k]_{\mathsf{r}_{\mathsf{F}}} \rangle \rangle \in \mathfrak{T}(A/\mathsf{r}_{\mathsf{F}});$$

then Lemma 4.3.25 implies that ϖ is both well defined and one-to-one. It is also easy to see that ϖ is onto, so it remains to show that the map is a Borel isomorphism.

2. Define

$$\mathcal{D} := \{G \in \mathcal{B}(\mathfrak{T}(A/\mathsf{r}_{\mathsf{F}})) \mid \mathfrak{B}(\varpi)(G) \in \mathcal{B}(\mathfrak{T}(A)/\widetilde{\mathsf{r}_{\mathsf{F}}})\};$$

then $\mathcal{D}$ is a σ-algebra. We show that $G = \{\triangle\} \times G_1 \times \cdots \times G_n \in \mathcal{D}$, provided $G_i \in \mathcal{B}(A/\mathsf{r}_{\mathsf{F}})$ for each i. Since these sets G generate the σ-algebra $\mathcal{B}(\mathfrak{T}(A/\mathsf{r}_{\mathsf{F}}))$, this will establish Borel measurability of ϖ.

3. In fact, if $G_i \in \mathcal{B}(A/\mathsf{r}_{\mathsf{F}})$, then there exists $H_i \in \mathbf{\Sigma}(\mathcal{B}(A), \mathsf{r}_{\mathsf{F}})$ such that $G_i = \eta_{\mathsf{r}_{\mathsf{F}}}[H_i]$. It is fairly easy to see that $\{\triangle\} \times H_1 \times \cdots \times H_n \in \mathbf{\Sigma}(\mathcal{B}(\mathfrak{T}(A)), \widetilde{\mathsf{r}_{\mathsf{F}}})$, and that

$$\mathfrak{B}(\varpi)(G) = \eta_{\widetilde{\mathsf{r}_{\mathsf{F}}}}[\{\triangle\} \times H_1 \times \cdots \times H_n],$$

so that $\mathfrak{B}(\varpi)(G) \in \mathcal{B}(\mathfrak{T}(A)/\widetilde{\mathsf{r}_{\mathsf{F}}})$ by Proposition 1.7.12.

4. A similar argument using the characterization of Borel sets in factor spaces shows that $\varpi\left[G\right] \in \mathcal{B}(\mathfrak{T}\left(A/\mathsf{r}_\mathsf{F}\right))$, provided $G \in \mathcal{B}(\mathfrak{T}\left(A\right)/\widetilde{\mathsf{r}_\mathsf{F}})$. □

As a consequence we obtain that the functor $\mathfrak{T}$ absorbs the congruence which is generated through the logic on the coalgebra.

Lemma 4.3.27. *Define* $\mathsf{c}_\mathsf{F} := (\mathsf{r}_\mathsf{F}, \widetilde{\mathsf{r}_\mathsf{F}})$ *for the stochastic left coalgebra* $\mathsf{F} = (A, \gamma)$ *with analytic* A. *Then*

a. c_F *is a congruence for* F.
b. $\mathfrak{T}$ *absorbs* c_F.
c. If $\mathsf{G} = (B, \delta)$ *is a stochastic coalgebra with analytic* T *such that* c_F *and* $\mathsf{c}_\mathsf{G} := (\mathsf{r}_\mathsf{G}, \widetilde{\mathsf{r}_\mathsf{G}})$ *are simulation equivalent, then* $\mathfrak{T}$ *distributes over* c_F *and* c_G

Proof 1. For establishing part a we need to show that Θ separates the logic; then the assertion will follow from Proposition 4.3.10. In fact, assume that we know for a $\mathfrak{T}$-formula ψ $\mu([\![\psi]\!]_\mathsf{F}) \geq q \Leftrightarrow \mu'([\![\psi]\!]_\mathsf{F}) \geq q$ for every rational q with $0 \leq q \leq 1$. Then $\mu([\![\psi]\!]_\mathsf{F}) = \mu'([\![\psi]\!]_\mathsf{F})$ follows immediately due to the rationals being dense.

2. Property b is an easy consequence of Proposition 4.3.26.

3. Because $[s]_{\mathsf{r}_\mathsf{F} \diamond \mathsf{r}_\mathsf{G}} = [t]_{\mathsf{r}_\mathsf{F} \diamond \mathsf{r}_\mathsf{G}}$ for $s \in A, t \in B$ is equivalent to $Th_\mathsf{F}(s) = Th_\mathsf{G}(t)$, and, similarly, $[x]_{\widetilde{\mathsf{r}_\mathsf{F}} \diamond \widetilde{\mathsf{r}_\mathsf{G}}} = [y]_{\widetilde{\mathsf{r}_\mathsf{F}} \diamond \widetilde{\mathsf{r}_\mathsf{G}}}$ for $x \in \mathfrak{T}\left(A\right), y \in \mathfrak{T}\left(B\right)$ iff $Th_\mathsf{F}(x) = Th_\mathsf{G}(y)$. Thus we infer from Lemma 4.3.25 that

$$\zeta : \langle\langle \triangle, \langle s_1, \ldots, s_k\rangle\rangle, \langle \triangle, \langle t_1, \ldots, t_k\rangle\rangle\rangle \mapsto \langle \triangle, \langle\langle s_1, t_1\rangle, \ldots, \langle s_k, t_k\rangle\rangle\rangle$$

constitutes a bijection

$$\zeta : \{\langle x, y\rangle \in \mathfrak{T}\left(A\right) \times \mathfrak{T}\left(B\right) \mid [x]_{\widetilde{\mathsf{r}_\mathsf{F}} \diamond \widetilde{\mathsf{r}_\mathsf{G}}} = [y]_{\widetilde{\mathsf{r}_\mathsf{F}} \diamond \widetilde{\mathsf{r}_\mathsf{G}}}\} \to \mathfrak{T}\left(\{\langle s, t\rangle \in S \times T \mid [s]_{\mathsf{r}_\mathsf{F} \diamond \mathsf{r}_\mathsf{G}} = [t]_{\mathsf{r}_\mathsf{F} \diamond \mathsf{r}_\mathsf{G}}\}\right).$$

Using the argument from the proof to Proposition 4.3.26 it is not difficult (but a wee bit tedious) to establish Borel measurability of ζ, and both $\mathfrak{T}\left(\pi_A\right) \circ \zeta = \pi_{\mathfrak{T}(A)}$ and $\mathfrak{T}\left(\pi_B\right) \circ \zeta = \pi_{\mathfrak{T}(B)}$ are apparent. □

Wrapping things up, we obtain the following.

Theorem 4.3.28. *Assume that* $\mathcal{K}$ *and* $\mathcal{K}'$ *are Kripke models over analytic spaces; then the following conditions are equivalent:*

a. $\mathcal{K}$ *and* $\mathcal{K}'$ *are bisimilar.*
b. $\mathcal{K}$ *and* $\mathcal{K}'$ *are behaviorally equivalent.*
c. $\mathcal{K}$ *and* $\mathcal{K}'$ *are logically equivalent.*

Proof Construct for the Kripke models the corresponding $\mathfrak{T}$-coalgebras. The associated smooth equivalence relations are congruences, since the set of $\mathfrak{T}$-liftings is separating. Then Lemma 4.3.27 shows that the functor behaves benevolently with respect to these congruences. An application of Proposition 4.3.17 and Proposition 4.3.22 yields the desired equivalence for the coalgebras, and Lemma 4.3.23 shows that it holds also for the Kripke models we started with. □

Theorem 4.3.28 contains the fairly general characterization of bisimilarity in [22, 20] and the equivalence of bisimilarity and logical equivalence from Theorem 2.3.11 for stochastic Kripke models as special cases. Using this coalgebraic approach gives a result with a much broader range of applicability.

4.3.5 A Remark on Extending the Logic

The logic contains only the bare minimum to make it interesting: formulas may be composed through conjunction; there are bridge operators generating $\mathfrak{T}$-formulas from state formulas. In addition, there are modal operators for each predicate lifting $\lambda : \mathfrak{B} \xrightarrow{\bullet} \mathfrak{B} \circ \mathfrak{S} \circ \mathfrak{T}$ and each $\mathfrak{T}$-predicate lifting, $\vartheta : \mathfrak{B} \circ \mathfrak{T} \xrightarrow{\bullet} \mathfrak{B} \circ \mathfrak{S} \circ \mathfrak{T}$ so that both $\langle \lambda \rangle \phi$ and $\langle \vartheta \rangle \psi$ are state formulas whenever ϕ is a state formula and ψ is a $\mathfrak{T}$-formula. Technically, these liftings serve the interpretation of the logic. We illustrate these concepts with examples from continuous time stochastic logics **CSL** and μ**CSL**.

It is rather straightforward to add to the logic atomic sentences for states and for the action space. Let Σ and Π be constant functors representing states and action spaces, respectively. Incorporating atomic sentences then requires adding transformations $Æ_{\Sigma} : \Sigma \xrightarrow{\bullet} \mathfrak{B}$ respectively $Œ_{\Pi} : \Pi \xrightarrow{\bullet} \mathfrak{B} \circ \mathfrak{T}$ to the interpretation. This construction takes care of assigning each state $\sigma \in \Sigma$ a Borel subset $Æ_{\Sigma,A}(\sigma) \in \mathcal{B}(A)$ of the state space A respectively a Borel subset $Œ_{\Pi,A}(\pi) \in \mathcal{B}(\mathfrak{T}(A))$ for each $\pi \in \Pi$. It appears to be more adequate than defining the semantics of atomic sentences also through a predicate lifting, because it does not involve the coalgebra (which is truly not required here).

The idea using bridge operators for transforming state formulas into $\mathfrak{T}$-formulas may obviously be extended to reversing the bridge, and to performing algebraic operations on the respective class of formulas. For example, disjunction can be added to state formulas together with the interpretation through the natural transformation $V : \mathfrak{B} \times \mathfrak{B} \xrightarrow{\bullet} \mathfrak{B}$ with the obvious definition $V_A(C_1, C_2) := C_1 \cup C_2$, the important point being that this transformation is natural. Similarly, negation can be added; see Example 4.2.7. The same argumentation applies to natural transformations $(\mathfrak{B} \circ \mathfrak{T})^n \xrightarrow{\bullet} \mathfrak{B} \circ \mathfrak{T}$ for defining the semantics of n-ary Boolean or algebraic operations of $\mathfrak{T}$-formulas, going a bit beyond Proposition 4.2.10.

We did fix here as well as in earlier chapters only the conjunction of formulas as an algebraic operation. Conjunction translates in the semantics into intersection, and we required closedness under intersections for making sure that the Borel sets $\{[\![\psi]\!] \mid \psi \text{ is a } \mathfrak{T}\text{-formula}\}$ generate the crucial σ-algebras of invariant Borel sets. They in turn are constitutive for the equivalences defined through formulas to be congruences. Consequently, when modifying the logic, it is difficult just to replace conjunction by disjunction: the whole machinery would then need to be overhauled.

4.4 Right Coalgebras

The discussion in Section 4.3 referred to logics which are interpreted through coalgebras the functor of which has lexically the form $\mathfrak{S} \circ \mathfrak{T}$ with a suitable functor $\mathfrak{T}$. Hence the dynamics of the coalgebra is modelled through a probability distribution: whenever we are in state $a \in A$, the new state is described through a subprobability $\gamma(a)$ over the transformed state space $\mathfrak{T}(A)$. As we have seen, the interpretation of a coalgebraic logic makes noticeable use of probabilistic arguments, since functor $\mathfrak{T}$ is dominated by the subprobability functor: what comes out of $\mathfrak{T}$ will be filtered through $\mathfrak{S}$. Architecturally speaking, the subprobabilty functor provides the user interface; the internal structure is given through $\mathfrak{T}$.

The present section will reverse the rôles of the functors by basing the interpretation of the logics on coalgebras for the functor $\mathfrak{T} \circ \mathfrak{S}$: the probabilistic outcome will be transformed through functor $\mathfrak{T}$. Look at Markov transition systems for an example: the functor in charge is $A \mapsto (\mathfrak{S}(A))^{\mathsf{Act}}$. Being in state $a \in A$, the transition systems reacts by selecting for each action $j \in A$ a subprobability distribution $\gamma_j(a)$; all these distributions are then composed to produce the overall reaction $\left(\gamma_j(a)\right)_{j \in \mathsf{Act}}$. Quoting A. Sokolova again: "In a reactive system actions are treated as *input* actions being provided by the environment. When a reactive system receives input from the environment, then it acts probabilistically by choosing the next state according to a probability distribution assigned to its input." [78, p. 30]. Generalizing, we take the subprobability functor as the innermost functor, being responsible for determining the next state, and transform the new state according to another functor $\mathfrak{T}$, so that the coalgebra is now formulated for the functor $\mathfrak{T} \circ \mathfrak{S}$. Speaking in terms of the design patterns of Software Engineering [37], the facade of the behavior of the coalgebra is provided by functor $\mathfrak{T}$ while the local behavior is determined through $\mathfrak{S}$.

We will consider logics the modalities of which are given through predicate liftings for functor $\mathfrak{T} \circ \mathfrak{S}$, so that a modal formula like $\langle\lambda\rangle\phi$ is interpreted over a state space A using Borel subsets of $\mathfrak{T}(\mathfrak{S}(A))$; the case $\mathfrak{T} := -^{\mathsf{Act}}$ requires for example Borel sets of $\mathfrak{S}(A)^{\mathsf{Act}}$. Apart from predicate liftings we will make sure that conjunction belongs to the basic operators, but we will also be able to cater to other Boolean operators by providing a set of bridge operators again. In addition to that we want to show that also infinitesimal operators may be added without much ado — fixed-point operators are examples. It will turn out that the choice of the bridge operators and of the infinitesimal operators does not really have an impact on the way we tackle the problems. We have of course to make provisions for their proper interpretation by extending coalgebras to models. Before that, we introduce and discuss in Section 4.4 right coalgebras, and we characterize equivalence relations as congruences through by looking at their factor spaces; moreover we show that right coalgebras are closed under countable products and coproducts; this will be helpful later. Models are introduced in Section 4.4.2, where we also define our logics and the corresponding interpretations. We require to have a sufficient number of predicate liftings at our disposal, so that we will be able to separate the logic. This has as a first consequence that the canonical equivalence relation defined by the logic on the state space of a coalgebra is in fact a congruence, which in turn gives some pleasant structural properties of the factor models. Behavioral and logical equivalence are studied in Section 4.4.3. This requires the construction of a span from a cospan, and we assume that the functor has a property which we call completing, which means that we have enough maps at our disposal. The step from maps to measurable maps then requires some selection techniques (just as in Section 2.1). We also exercise — as in previous sections — a kind of reduction strategy by showing the desired results first for the case of Polish state spaces, and then providing some tools for deriving the general case of analytic spaces from this.

Finally, we give a remarkable application of this technique due to Ch. Schubert to the corresponding local problem: instead of having a look at the conditions under which *models* are bisimilar, we seek conditions under which individual *states* are bisimilar. This problem is discussed in Section 4.4.5; it entails among other things a new criterion under which states in Markov transition systems are bisimilar.

4.4.1 Stochastic Right Coalgebras

Let $\mathfrak{F} : \mathbf{ANL} \to \mathbf{ANL}$ be an endofunctor. The functor $\mathfrak{T}^{\mathrm{R}} = \mathfrak{T} \circ \mathfrak{S}$ will serve as a basis for the coalgebras from which models will be defined. Before we define models, however, we have a brief look at congruences and at the factor structure they entail.

Definition 4.4.1. *Let A, B be analytic spaces.*

a. A stochastic right $\mathfrak{T}$-coalgebra F *is a pair (A, γ) where $\gamma : A \to \mathfrak{T}^R(A)$ is measurable.*
b. A morfism $f : \mathsf{F} \to \mathsf{G}$ *with a stochastic right $\mathfrak{T}$-coalgebra $\mathsf{G} = (B, \delta)$ is a measurable map $f : A \to B$ such that $\delta \circ f = \mathfrak{T}^R(f) \circ \gamma$.*

Thus a stochastic right $\mathfrak{T}$-coalgebra $\mathsf{F} = (A, \gamma)$ is usually not a stochastic relation of some sort. Given $a \in A$, $\gamma(a)$ is a member of $\mathfrak{T}(\mathfrak{S}(A))$, letting the functor $\mathfrak{T}$ transform (or filter) the probabilistic information. This means that probabilistic aspects carry a weight which is indirect, the dominating partner being functor $\mathfrak{T}$. If $f : \mathsf{F} \to \mathsf{G}$ is a morfism (note that we do not necessarily require morfisms to be onto; so we adhere to the lexical convention suggested in Section 2.6); then the diagram

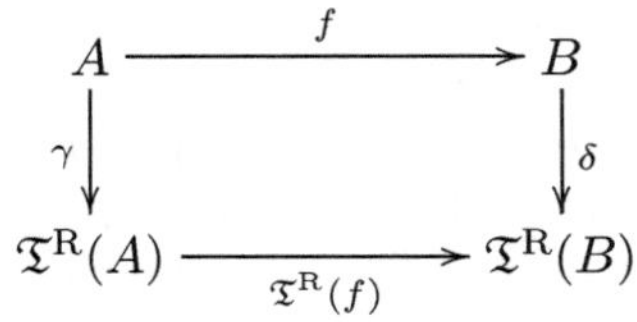

commutes. We will talk about *stochastic right coalgebras* or simply about *right coalgebras* in this section, omitting the functor $\mathfrak{T}$ from the notation whenever it appears convenient.

Let us have a look at some examples.

Example 4.4.2. The identity functor $\mathfrak{T}$ on **ANL** leads to stochastic relations. Thus a $\mathfrak{T}^{\mathrm{R}}$-coalgebra γ is in this case a Borel measurable map $\gamma : A \to \mathfrak{S}(A)$, interpreting $\gamma(a)(B)$ as usual as the probability that the next state will be a member of Borel set $B \in \mathcal{B}(A)$ after some action modelled by γ in state $a \in A$. $\dashv$

Example 4.4.3. If Act is a countable set of actions, $\mathfrak{T} = -^{\mathsf{Act}}$ is the functor underlying labeled Markov transition systems; see Example 4.2.5. In fact, such a transition system $\big(A, (k_\ell)_{\ell \in \mathsf{Act}}\big)$ has the analytic space A as its state space, and k_ℓ is for each action $\ell \in \mathsf{Act}$ a stochastic relation on A. The dynamics of the coalgebra is given by $a \mapsto \big(k_\ell(a)\big)_{\ell \in \mathsf{Act}}$. $\dashv$

Surjective morfisms have the pleasant property that they are final in the following sense.

Lemma 4.4.4. *Let* $\mathsf{F} = (A, \gamma), \mathsf{G} = (B, \delta)$*, and* $\mathsf{H} = (C, \epsilon)$ *be right coalgebras; assume that the morfism* $f : \mathsf{F} \to \mathsf{G}$ *is onto, and that we know for a Borel map* $g : B \to C$ *that* $g \circ f$ *is a morfism. Then* g *is a morfism.*

Proof Look at this diagram;

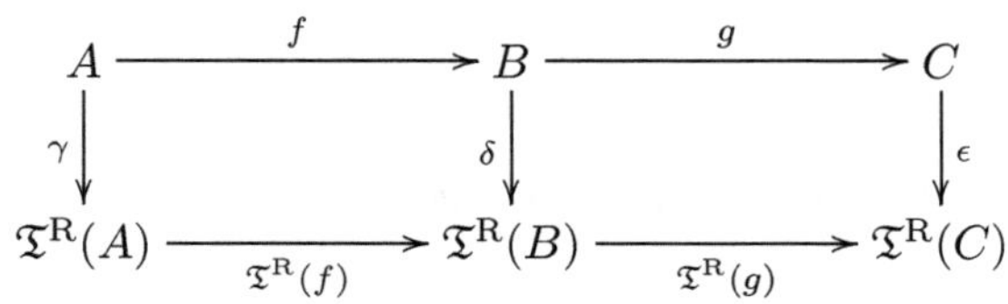

We know that

$$
\begin{aligned}
(\epsilon \circ g) \circ f &= \mathfrak{T}^{\mathrm{R}}(g) \circ \mathfrak{T}^{\mathrm{R}}(f) \circ \gamma \\
&= (\delta \circ \mathfrak{T}^{\mathrm{R}}(g)) \circ f.
\end{aligned}
$$

Since f is onto, it may be cancelled from the right; thus we infer $\epsilon \circ g = \delta \circ \mathfrak{T}^{\mathrm{R}}(g)$, as desired. $\square$

We will investigate congruences and at most countable coproducts of these coalgebras. Starting with congruences, we note that we cannot directly carry over the notion of a congruence via invariant sets, as we did for stochastic relations and for left coalgebras. We resort, however, to properties of the factor space (see Proposition 1.7.21).

Lemma 4.4.5. *Let ρ be a smooth equivalence relation on the analytic space A of the right coalgebra $\mathsf{F} = (A, \gamma)$. Then the following statements are equivalent.*

a. $\rho \subseteq \mathsf{ker}\left(\mathfrak{T}^R(\eta_\rho)\right)$.
b. There exists a measurable map $\gamma_\rho : A/\rho \to \mathfrak{T}^R\big(A/\rho\big)$ *with* $\gamma_\rho \circ \eta_\rho = \mathfrak{T}^R \circ \gamma$.

Proof The implication $b \Rightarrow a$ is trivial; so we take care of $a \Rightarrow b$. Define $\gamma_\rho([a]_\rho) := \mathfrak{T}^{\mathrm{R}}(\eta_\rho)(\gamma(a))$, for $a \in A$; then $\gamma_\rho : A/\rho \to \mathfrak{T}^{\mathrm{R}}\big(A/\rho\big)$ is well defined, since by assumption $a \ \rho \ a'$ implies $\mathfrak{T}^{\mathrm{R}}(\eta_\rho)(\gamma(a)) = \mathfrak{T}^{\mathrm{R}}(\eta_\rho)(\gamma(a'))$. Because A/ρ carries the final σ-algebra with respect to η_ρ, and since $\gamma_\rho \circ \eta_\rho$ equals the measurable map $\mathfrak{T}^{\mathrm{R}}(\eta_\rho) \circ \gamma$, γ_ρ is measurable by Lemma 1.3.3. □

This leads to the definition of a congruence for a right coalgebra. Note that we stick to equivalence relations that are smooth. This is so since otherwise the factor space will not be analytic, and so the functor cannot be applied to it.

Definition 4.4.6. *Let $\mathsf{F} = (A, \gamma)$ be a stochastic right $\mathfrak{T}$-coalgebra. The smooth equivalence relation ρ on A is a* congruence *for F iff there exists a measurable map $\gamma_\rho : A/\rho \to \mathfrak{T}^R\big(A/\rho\big)$ with $\gamma_\rho \circ \eta_\rho = \mathfrak{T}^R(\eta_\rho) \circ \gamma$. The* factor coalgebra *$(A/\rho, \gamma_\rho)$ is denoted by F/ρ.*

Consequently, if ρ is a congruence for F, then $\eta_\rho : \mathsf{F} \to \mathsf{F}/\rho$ is a surjective morfism, and the diagram

$$\begin{array}{ccc} A & \xrightarrow{\quad \eta_\rho \quad} & A/\rho \\ {\scriptstyle\gamma}\big\downarrow & & \big\downarrow{\scriptstyle\gamma_\rho} \\ \mathfrak{T}^{\mathrm{R}}(A) & \xrightarrow[\mathfrak{T}^{\mathrm{R}}(\eta_\rho)]{} & \mathfrak{T}^{\mathrm{R}}\big(A/\rho\big) \end{array}$$

commutes. Let us have a look at some examples.

Example 4.4.7. Continuing Example 4.4.2, we infer from Lemma 4.4.5 that ρ is a congruence for the right coalgebra induced by the identity functor iff $\rho \subseteq \mathsf{ker}\left(\mathfrak{S}\left(\eta_\rho\right) \circ \gamma\right)$. Thus $a \ \rho \ a'$ implies

$$\gamma(a)(\eta_\rho^{-1}\left[D\right]) = \big(\mathfrak{S}\left(\eta_\rho\right) \circ \gamma\big)(a)(D) = \big(\mathfrak{S}\left(\eta_\rho\right) \circ \gamma\big)(a')(D) = \gamma(a')(\eta_\rho^{-1}\left[D\right])$$

for all Borel sets $D \in \mathcal{B}(A/\rho)$. Since the inverse images of the Borel sets in A/ρ are exactly the ρ-invariant Borel sets on A by Proposition 1.7.12, we infer that $\gamma(a)(G) = \gamma(a')(G)$ for all $G \in \mathbf{\Sigma}\big(\mathcal{B}(A), \rho\big)$. Thus a congruence for the right coalgebra is in this case a congruence for the corresponding stochastic relation, and vice versa. Proposition 1.7.21 states this for the more general case of a stochastic relation between two analytic spaces. ⊣

Example 4.4.8. The functor $\mathfrak{T} = -^{\mathsf{Act}}$ has been discussed in Example 4.4.3. We see from Lemma 4.4.5 that $a \ \rho \ a'$ implies that

$$\big(\mathfrak{S}\,(\eta_\rho) \circ \gamma_\ell\big)(a) = (\mathfrak{S}\,(\eta_\rho) \circ \gamma_\ell)(a')$$

holds for each $\ell \in \mathsf{Act}$, provided $\gamma = (\gamma_\ell)_{\ell \in \mathsf{Act}}$, where $\gamma_\ell : A \rightsquigarrow A$ is for $\ell \in \mathsf{Act}$ the stochastic relation from which γ is composed. Thus we conclude from Example 4.4.7 that ρ is a congruence iff it is a congruence for each stochastic relation γ_ℓ. ⊣

Turning to coproducts, we show that the category of right coalgebras is closed under countable coproducts. The restriction of the index set being countable is given by the observation that analytic spaces are closed under countable, but not under arbitrary, coproducts. We will not need this result in full generality but rather for finite sums only; alas, it does not cost much to establish a slightly more general result.

Lemma 4.4.9. *The category of stochastic right $\mathfrak{T}$-coalgebras is closed under countable coproducts.*

Proof 1. Let $\big(\mathsf{F}_k = (A_k, \gamma_k)\big)_{k \in K}$ be a family of right coalgebras, and assume that K is countable. Then $A := \coprod_{k \in K} A_k$ is an analytic space provided it is equipped with the measurable structure which is final with respect to the injections $i_k : A_k \to A$. Put $\gamma(i_k(a_k)) := \big(\mathfrak{T}^{\mathrm{R}}(i_k) \circ \gamma_k\big)(a_k)$ whenever $a_k \in A_k$; then $\mathsf{F} := (A, \gamma)$ is a stochastic right $\mathfrak{T}$-coalgebra, and the injections $i_k : \mathsf{F}_k \to \mathsf{F}$ are morfisms.

2. Let $\mathsf{G} = (B, \delta)$ be a right coalgebra with morphisms $j_k : \mathsf{F}_k \to \mathsf{G}$. Then $j(i_k(a_k)) := j_k(a_k)$ defines the unique morfism $j : \mathsf{F} \to \mathsf{G}$ with $j_k = j \circ i_k$ for all $k \in K$. □

4.4.2 Models, Lifting Valuations

Right coalgebras serve as a basis for models. They will be introduced now by adding a valuation to a coalgebra. Let V be a countable set of variables. A model based on a right coalgebra (A, γ) will assign each variable from V a Borel subset of A; formally:

Definition 4.4.10. *A* model $\mathcal{M} = (A, \gamma, V)$ *consists of a $\mathfrak{T}^R$-coalgebra (A, γ) together with a* valuation V*, i.e., a map $V : \mathsf{V} \to \mathcal{B}(A)$. A* morfism $f : \mathcal{M} \to$ *$\mathcal{N}$ with $\mathcal{N} = (B, \delta, W)$ is given by a coalgebra morfism $f : (A, \gamma) \to (B, \delta)$ which satisfies $\mathfrak{B}\,(f) \circ W = V$.*

Thus in addition to being a coalgebra morfism, a model morfism $f : \mathcal{M} \to \mathcal{N}$ satisfies $f^{-1}[W(x)] = V(x)$ for each variable $x \in \mathsf{V}$; thus $a \in V(x)$ iff $f(a) \in W(x)$. This implies in particular that each set $V(x)$ is f-invariant, which occurs as a reasonable property for a model morfism.

This observation suggests the definition of a congruence on a model: since the factor map should be a morfism, we might wish to observe that the valuation takes this into account.

Definition 4.4.11. *The smooth equivalence relation ρ is a* model congruence *for the model $\mathcal{M} = (A, \gamma, V)$ iff ρ is a congruence for the right coalgebra (A, γ), and if $V(x)$ is ρ-invariant for each variable $x \in \mathsf{V}$.*

Let $x \in \mathsf{V}$ be a variable. Because the set $V(x)$ is ρ-invariant, and because it is a Borel set, we conclude that $V(x) \in \boldsymbol{\Sigma}(\mathcal{B}(A), \rho)$. But this implies through Proposition 1.7.12 that $V_\rho(x) := \eta_\rho[V(x)]$ is a Borel set in the factor space A/ρ, and $\eta_\rho^{-1}[V_\rho(x)] = V(x)$. Because we know that $\eta_\rho : (A, \gamma) \to (A/\rho, \gamma_\rho)$ is a morfism for the underlying right coalgebras (Lemma 4.4.5), we conclude that $\eta_\rho : \mathcal{M} \to \mathcal{M}/\rho$ is a morfism, where $\mathcal{M}/\rho := (A/\rho, \gamma_\rho, V_\rho)$ is the factor model.

Given a model $\mathcal{M} = (A, \gamma, V)$, a variable $x \in \mathsf{V}$, and a Borel set $D \in \mathcal{B}(A)$, the model $\mathcal{M}[x\backslash D]$ is defined as the model (A, γ, V'), where $V'(x) := D$, and for all other variables x', $V'(x') = V(x')$; see Section 2.4.3. It is not difficult to see that model morfisms behave as one would expect when dealing with substitutions.

Lemma 4.4.12. *Let $f : \mathcal{M} \to \mathcal{N}$ be a model morfism, then*

$$f : \mathcal{M}[x\backslash f^{-1}[D]] \to \mathcal{N}[x\backslash D]$$

is a model morfism as well whenever $x \in \mathsf{V}$ is a variable, and D is a Borel set in $\mathcal{N}$. □

Like their coalgebraic cousins, surjective model morfisms have the pleasant property that they are final; Lemma 4.4.4 carries over.

Lemma 4.4.13. *Let $\mathcal{M} = (A, \gamma, V), \mathcal{N} = (B, \delta, W)$ and $\mathcal{Q} = (C, \epsilon, X)$ be models; assume that the model morfism $f : \mathcal{M} \to \mathcal{N}$ is onto, and we know for a Borel map that $g : B \to C$ that $g \circ f$ is a morfism. Then g is a model morfism.* □

Assume that $f : \mathsf{F} \to \mathsf{G}$ is a morfism for the right coalgebras $\mathsf{F} = (A, \gamma)$ and $\mathsf{G} = (B, \delta)$. Suppose that we have a valuation $V : \mathsf{V} \to \mathcal{B}(A)$ on F. We show that it is possible under a very simple condition to establish that there exists a valuation $W : \mathsf{V} \to \mathfrak{B}(B)$ such that

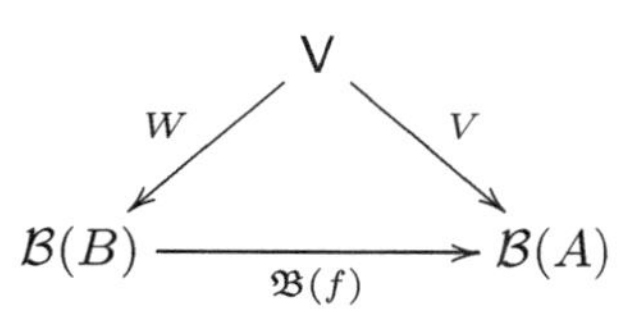

commutes. Clearly, uniqueness of such a valuation W is equivalent to $\mathfrak{B}(f)$ being injective, and hence to f being surjective. Moreover, if such a W exists, each $V(x)$ has to be f-invariant. In fact, the converse is true as well:

Proposition 4.4.14. *Suppose that $f : (A, \gamma) \to (B, \delta)$ is a surjective morfism and $V : \mathsf{V} \to \mathfrak{B}(A)$ is a valuation. If each $V(x)$ is f-invariant, then there exists a unique valuation W on B with $\mathfrak{B}(f) \circ W = V$.*

Proof Uniqueness follows from injectivity of $\mathfrak{B}(f)$. For any $x \in \mathsf{V}$, $V(x)$ is f-invariant; thus $f[V(x)] \subseteq B$ is a Borel set by Corollary 1.4.32. Thus, the map $W : \mathsf{V} \to \mathfrak{B}(B)$ defined as $W(x) := f[V(x)]$ is a valuation. Moreover, we have

$$\mathfrak{B}(f) \circ W(x) = f^{-1}[f[V(x)]] = V(x),$$

again by invariance of $V(x)$. □

Corollary 4.4.15. *Let $(f_i : X_i \to Y)_{i \in I}$ be a family of surjective measurable maps between analytic spaces, and $\big(V_i : \mathsf{V} \to \mathfrak{B}(X_i)\big)_{i \in I}$ be a family of valuations. Then these statements are equivalent:*

a. there exists a valuation $W : \mathsf{V} \to \mathfrak{B}(Y)$ such that $\mathfrak{B}(f_i) \circ W = V_i$ for all $i \in I$.
b. each $V_i(x)$ is f_i-invariant and we have $f_i[V_i(x)] = f_j[V_j(x)]$ for all $i, j \in I$, $x \in \mathsf{V}$. □

Turning our attention to coproducts, we extend Lemma 4.4.9 to models.

Lemma 4.4.16. *The category of stochastic models is closed under countable coproducts.*

Proof Let $(\mathcal{M}_k)_{k \in K}$ be a countable family of models with $\mathcal{M}_k = (A_k, \gamma_k, V_k)$, and define

$$(A, \gamma) := \coprod_{k \in K} (A_k, \gamma_k)$$

as the coproduct in the category of right coalgebras according to Lemma 4.4.9, with injections $i_{A_k} : (A_k, \gamma_k) \to (A, \gamma)$. Define for $x \in \mathsf{V}$ the valuation

$$V(x) := \bigcup_{k \in K} i_{A_k}[V_k(x)].$$

Then $\mathcal{M} := (A, \gamma, W)$ is a model with morfisms $i_{A_k} : \mathcal{M}_k \to \mathcal{M}$, and it is easy to see that $\mathcal{M}$ is the coproduct of $(\mathcal{M}_k)_{k \in K}$. □

We need a slight generalization of bridge operators which permit representing fixed points.

Definition 4.4.17. *An* infinitesimal operator *is a natural transformation $\iota : \mathfrak{B}^\infty \stackrel{\bullet}{\to} \mathfrak{B}$.*

Infinitesimal operators are used for modelling dynamical aspects. These are examples.

Example 4.4.18. Define

$$\iota^{(1)}(R_1, \ldots) := \bigcup_{n\in\mathbb{N}} \bigcap_{k\geq n} R_k = \limsup_{k\to\infty} R_k.$$

An element a is in $\iota^{(1)}(R_1, \ldots)$ iff it is eventually in all R_k; thus iff there exists $n \in \mathbb{N}$ such that $a \in R_k$ for all $k \geq n$. This permits modelling systems in which the convergence of behavior is at stake.

Similarly, put

$$\iota^{(2)}(R_1, \ldots) := \bigcap_{n\in\mathbb{N}} \bigcup_{k\geq n} R_k = \liminf_{k\to\infty} R_k.$$

Thus an element a is in $\iota^{(2)}(R_1, \ldots)$ iff it is infinitely often in one of the R_k: given an index n, there exists an index $k \geq n$ with $a \in R_k$. This permits modelling infinitely often recurring phenomena.

Finally, define

$$\iota^{(3)}(R_1, \ldots) := \bigcap_{n\in\mathbb{N}} R_n$$
$$\iota^{(4)}(R_1, \ldots) := \bigcup_{n\in\mathbb{N}} R_n.$$

These operators are oriented towards supplying operators for the smallest respectively the largest fixed point (akin to the μ- and the ν-operator, respectively).

Note that all these operators involve countable Boolean operations and nothing else. ⊣

This characterization of infinitesimal operators is proved exactly as is Proposition 4.2.10, using Yoneda's Lemma 1.2.1.

Proposition 4.4.19. *The infinitesimal operators are in one-to-one correspondence with the measurable maps* $\{0,1\}^\infty \to \{0,1\}$ *with* $\{0,1\}$ *as the discrete analytic space over the object set* $\{0,1\}$. □

We fix countable sets Λ of $\mathfrak{T}^{\mathrm{R}}$-predicate liftings, Br of bridge operators, and Inf of infinitesimal operators, as well as a countable set V of variables.

The Logics

These logics will be considered. The *(full) logic* $\mathfrak{L}^\sharp = \mathfrak{L}^\sharp(\Lambda, \mathsf{Br}, \mathsf{Inf}, \mathsf{V})$ is given by

$$\phi ::= \top \mid \phi_1 \wedge \phi_2 \mid \langle\lambda\rangle\phi \mid \beta(\phi_1, \ldots, \phi_{\partial(\beta)}) \mid x \mid \iota x.\phi$$

with $\lambda \in \Lambda$, $\beta \in \mathsf{Br}$, $x \in \mathsf{V}$, and $\iota \in \mathsf{Inf}$.

We also define the *basic* or *kernel logic* to be the special case $\mathfrak{L}_b^\sharp = \mathfrak{L}^\sharp(\Lambda, \emptyset, \emptyset, \emptyset)$. Thus, $\mathfrak{L}_b^\sharp$ is given by the grammar

$$\phi ::= \top \mid \phi_1 \wedge \phi_2 \mid \langle\lambda\rangle\phi$$

for $\lambda \in \Lambda$.

To interpret the extended logic, we require a right coalgebra and a valuation. Given model $\mathcal{M} = (A, \gamma, V)$, we define recursively an interpretation for every formula ϕ in $\mathfrak{L}^\sharp$ as follows:

$$\begin{aligned}
[\![\top]\!]_{\mathcal{M}} &= A \\
[\![\phi_1 \wedge \phi_2]\!]_{\mathcal{M}} &= [\![\phi_1]\!]_{\mathcal{M}} \cap [\![\phi_2]\!]_{\mathcal{M}} \\
[\![\langle\gamma\rangle\phi]\!]_{\mathcal{M}} &= \big(\mathfrak{B}(\gamma) \circ \lambda_A\big)([\![\phi]\!]_{\mathcal{M}}) \\
[\![x]\!]_{\mathcal{M}} &= V(x) \\
[\![\beta(\phi_1, \ldots, \phi_{\partial(\beta)})]\!]_{\mathcal{M}} &= \beta_A([\![\phi_1]\!]_{\mathcal{M}}, \ldots, [\![\phi_{\partial(\beta)}]\!]_{\mathcal{M}}) \\
[\![\iota x.\phi]\!]_{\mathcal{M}} &= \iota_A\big((R_n)_{n\in\mathbb{N}}\big)
\end{aligned}$$

where the sequence R_i is defined through $R_1 := [\![\phi]\!]_{\mathcal{M}[x\backslash\emptyset]}$, and $R_{n+1} := [\![\phi]\!]_{\mathcal{M}[x\backslash R_n]}$. Observe that by construction each $[\![\phi]\!]_{\mathcal{M}}$ is in $\mathcal{B}(A)$.

As usual, we write $\mathcal{M}, a \models \phi$ if $a \in [\![\phi]\!]_{\mathcal{M}}$ for a state $a \in A$. In this case, we say that ϕ is *valid* in a. The *theory* $Th_{\mathcal{M}}(a)$ of a state $a \in A$ consists of the formulas valid in it.

Observe that our definition of the full logic is a little redundant: there is no need to include finite conjunctions as separate entities; we could have instead demanded that Br always includes operators whose transformations are given through the usual Boolean operations. Nevertheless, we find it useful to include finite conjunctions explicitly since they are crucial for some of our results: they permit the application of the π-λ-Theorem; cf. Lemma 1.6.30.

Lemma 4.4.20. *If $f : \mathcal{M} \to \mathcal{N}$ is a model morfisms, then $[\![\phi]\!]_{\mathcal{M}} = \mathfrak{B}(f)([\![\phi]\!]_{\mathcal{N}})$ for all $\phi \in \mathfrak{L}^\sharp$.*

Proof The proof proceeds by induction of formula ϕ. It suffices to consider individual variables as well as modal, bridge, and infinitesimal operators.

For $x \in \mathsf{V}$, preservation of $[\![x]\!]_{\mathcal{M}}$ corresponds to the defining property of model-morphisms; for $\lambda \in \Lambda$, we have:

$$
\begin{aligned}
\mathfrak{B}(f)(\llbracket\langle\lambda\rangle\phi\rrbracket_{\mathcal{N}}) &= \big(\mathfrak{B}(f)\circ\mathfrak{B}(\delta)\circ\lambda_B\big)(\llbracket\phi\rrbracket_{\mathcal{N}})\\
&\stackrel{(*)}{=} \big(\mathfrak{B}(\delta\circ f)\circ\lambda_B\big)(\llbracket\phi\rrbracket_{\mathcal{N}})\\
&\stackrel{(\dagger)}{=} \big(\mathfrak{B}(\mathfrak{T}^{\mathrm{R}}(f)\circ\gamma)\circ\lambda_B\big)(\llbracket\phi\rrbracket_{\mathcal{N}})\\
&\stackrel{(*)}{=} \big(\mathfrak{B}(\gamma)\circ\mathfrak{B}(\mathfrak{T}^{\mathrm{R}}(f))\circ\lambda_B\big)(\llbracket\phi\rrbracket_{\mathcal{N}})\\
&\stackrel{(\ddagger)}{=} \big(\mathfrak{B}(\gamma)\circ(\lambda_A\circ\mathfrak{B}(f))\big)(\llbracket\phi\rrbracket_{\mathcal{N}})\\
&\stackrel{(\sharp)}{=} \mathfrak{B}(\gamma)\circ\lambda_A(\llbracket\phi\rrbracket_{\mathcal{M}})\\
&= \llbracket\langle\lambda\rangle\phi\rrbracket_{\mathcal{M}}.
\end{aligned}
$$

Equalities $(*)$ note that $\mathfrak{B}$ is contravariant, $(\dagger)$ holds because f is a morfism, $(\ddagger)$ holds because λ is natural, and finally, $(\sharp)$ uses the induction hypothesis. For an infinitesimal operator ι we write (compare Proposition 2.4.7)

$$
\begin{aligned}
R_1 &= \llbracket\phi\rrbracket_{\mathcal{M}[x\backslash\emptyset]}, & S_1 &= \llbracket\phi\rrbracket_{\mathcal{N}[x\backslash\emptyset]},\\
R_{n+1} &= \llbracket\phi\rrbracket_{\mathcal{M}[x\backslash R_n]}, & S_{n+1} &= \llbracket\phi\rrbracket_{\mathcal{N}[x\backslash S_n]}.
\end{aligned}
$$

Lemma 4.4.12 implies $R_n = f^{-1}[S_n]$ for all n. Thus we have

$$
\begin{aligned}
\llbracket\iota x.\phi\rrbracket_{\mathcal{M}} &= \iota_A\big((A_n)_{n\in\mathbb{N}}\big)\\
&= \iota_A\big((f^{-1}[S_n])_{n\in\mathbb{N}}\big)\\
&\stackrel{(\natural)}{=} f^{-1}\big[\iota_B\big((B_n)_{n\in\mathbb{N}}\big)\big]\\
&= \mathfrak{B}(f)(\llbracket\iota x.\phi\rrbracket_{\mathcal{N}}).
\end{aligned}
$$

In $(\natural)$ the naturality of ι enters. The proof for the bridge operators is done in exactly the same manner. □

We define now the relationships between models that are of interest in the present discussion. These definitions are oriented towards the general coalgebraic ones (see Section 1.2.1), we stipulate, however, that the corresponding morfisms be based on surjective maps. To be specific:

Definition 4.4.21. *We call models $\mathcal{M} = (A, \gamma, V)$ and $\mathcal{N} = (B, \delta, W)$*

1. behaviorally equivalent *iff there exists a cospan* $\mathcal{M} \xrightarrow{f} \mathcal{Q} \xleftarrow{g} \mathcal{N}$ *of* surjective *model morfisms for a suitable model $\mathcal{Q}$;*
2. bisimilar *iff there exists a model $\mathcal{Q}$ and a span* $\mathcal{M} \xleftarrow{f} \mathcal{Q} \xrightarrow{g} \mathcal{N}$ *of* surjective *model morfisms.*
3. logically equivalent *iff* $\{Th_{\mathcal{M}}(a) \mid a \in A\} = \{Th_{\mathcal{N}}(b) \mid b \in B\}$.

Thus we can find for two logical equivalent models for each state in one model a state in the other model so that exactly the same formulas are satisfied; this conforms to the logical equivalence for the logics considered so far. Behavioral equivalence and bisimilarity are defined just as in a coalgebraic setting. Note that bisimilar models for the identity functor $\mathfrak{I}$ are not necessarily bisimilar stochastic relations, since the latter require a nontrivial σ-algebra of common events; see Definition 2.2.1. Lemma 4.4.20 entails that both behavioral equivalence and bisimilarity of models implies their logical equivalence. We will show now that both implications can be reversed under suitable conditions.

4.4.3 Logical vs. Behavioral Equivalence

It will be demonstrated first that logical equivalence implies behavioral equivalence. This is done by investigating factoring through the equivalence relation induced by the logic. This relation is smooth, because the set of all formulas is countable. But this is not quite enough. For the factor space to preserve the properties of the logic we require the composite functor $\mathfrak{T}^{\mathrm{R}}$ to interface properly with the logic which can be expressed through a separation property.

We define for a model $\mathcal{M} = (A, \gamma, V)$ an equivalence relation $\mathsf{r}_{\mathcal{K}}$ on A by

$$a\ \mathsf{r}_{\mathcal{M}}\ a' \text{ iff } Th_{\mathcal{M}}(a) = Th_{\mathcal{M}}(a').$$

$\mathsf{r}_{\mathcal{M}}$ is countably generated since it is determined through the countable set $\{[\![\phi]\!]_{\mathcal{M}} \mid \phi \text{is a formula in } \mathfrak{L}^{\sharp}\}$; thus it is a smooth equivalence relation, and hence the quotient $A/\mathsf{r}_{\mathcal{M}}$ is an analytic space.

We will now investigate whether $\mathsf{r}_{\mathcal{M}}$ is a congruence of $\mathcal{M}$. To ensure this we need enough formulas to exist. This will be achieved by requiring the set Λ of predicate liftings to be separating; this will be defined now.

For $\lambda \in \Lambda$, we define an equivalence relation $\equiv^{\lambda}_{\mathcal{M}}$ on $\mathfrak{T}^{\mathrm{R}}(A)$ upon setting

$$t \equiv^{\lambda}_{\mathcal{M}} t' \text{ iff } \left[t \in \lambda_A([\![\phi]\!]_{\mathcal{M}}) \Leftrightarrow t' \in \lambda_A([\![\phi]\!]_{\mathcal{M}})\right] \text{ holds for each formula } \phi \text{ in } \mathfrak{L}^{\sharp},$$

and we write

$$\equiv^{\Lambda}_{\mathcal{M}} := \bigcap_{\lambda \in \Lambda} \equiv^{\lambda}_{\mathcal{M}} .$$

Definition 4.4.22. *The set Λ of predicate liftings* separates the logic *iff $\equiv^{\Lambda}_{\mathcal{M}} \subseteq \mathsf{ker}\left(\mathfrak{T}^{R}(\eta_{\mathsf{r}_{\mathcal{M}}})\right)$ holds for all models $\mathcal{M}$.*

Note that this definition of separation is conceptually different from the one proposed for left coalgebras in Definition 4.3.6 on page 171. This so because in the latter case we were able to argue directly with probabilities which we are not in the present case. There is one exception, though: if the functor under consideration is the identity, then both concepts of separation coincide, as we will see in the next example.

Example 4.4.23. Assume that $\mathfrak{T}$ is the identity functor on **ANL**, and put

$$\Lambda := \{\varpi_q \mid q \in \mathbb{Q} \cap [0,1]\}$$

with $\varpi_{q,A}(D) := \{\mu \in \mathfrak{S}(A) \mid \mu(A) \geq q\}$ as defined in Section 4.2. Let $\mathcal{M} = (A, \gamma, V)$ be a right coalgebra for the identity; thus $\gamma : A \rightsquigarrow A$ is a stochastic relation. We claim that Λ is separating. In fact, we know that $\langle \mu, \mu' \rangle \in \mathsf{ker}(\mathfrak{S}(\eta_{\mathcal{M}}))$ iff $\mu([\![\phi]\!]_{\mathcal{M}}) = \mu'([\![\phi]\!]_{\mathcal{M}})$ for each formula ϕ by Proposition 1.7.12 and Lemma 1.6.30. On the other hand, $\mu \equiv^{\Lambda}_{\mathcal{M}} \mu'$ is equivalent to $\mu \in \varpi_{q,A}([\![\phi]\!]_{\mathcal{M}})$ iff $\mu' \in \varpi_{q,A}([\![\phi]\!]_{\mathcal{M}})$ for all rational q, and all formulas ϕ. Consequently, $\mu([\![\phi]\!]_{\mathcal{M}}) = \mu'([\![\phi]\!]_{\mathcal{M}})$ for all ϕ. But this means $\langle \gamma(a), \gamma(a') \rangle \in \mathsf{ker}(\mathfrak{S}(\eta_{\mathcal{M}}))$. $\dashv$

Example 4.4.24. Let $\mathfrak{T} := \coprod_{n \in \mathbb{N}} -^n$. Then $\mathfrak{T} : \mathbf{ANL} \to \mathbf{ANL}$ is an endofunctor. Let $\mathcal{M} := (A, \gamma, V)$ be a right $\mathfrak{T}$-coalgebra. Put

$$\xi_{q,A}(D) := \{\langle \mu_1, \dots, \mu_k \rangle \in \mathfrak{T}^{\mathrm{R}}(A) \mid k \in \mathbb{N}, \max_{1 \leq j \leq k} \mu_j(D) \geq q\}$$

for the Borel set $D \in \mathcal{B}(A)$. Then $\xi_q : \mathfrak{B} \stackrel{\bullet}{\to} \mathfrak{B} \circ \mathfrak{T}^{\mathrm{R}}$ is a natural transformation (this is shown nearly verbatim as in the corresponding proof in Example 4.2.5). We claim that $\Xi := \{\xi_q \mid q \in [0,1]\}$ is not separating. First one notes that $\mathfrak{T}^{\mathrm{R}}(\eta_{\mathcal{M}})(\mu_1, \dots, \mu_k) = \mathfrak{T}^{\mathrm{R}}(\eta_{\mathcal{M}})(\mu'_1, \dots, \mu'_\ell)$ iff $k = \ell$ and $\mathfrak{S}(\eta_{\mathcal{M}})(\mu_i) = \mathfrak{S}(\eta_{\mathcal{M}})(\mu'_i)$ for $i \leq i \leq k$. This in turn is equivalent to $k = \ell$ and $\mu_i([\![\phi]\!]_{\mathcal{M}}) = \mu'_i([\![\phi]\!]_{\mathcal{M}})$ for all i and for all formulas ϕ. On the other hand, $\langle \mu_1, \dots, \mu_k \rangle \equiv^{\Xi}_{\mathcal{M}} \langle \mu'_1, \dots, \mu'_\ell \rangle$ iff $\max_{1 \leq i \leq k} \mu_i([\![\phi]\!]_{\mathcal{M}}) = \max_{1 \leq i \leq \ell} \mu'_i([\![\phi]\!]_{\mathcal{M}})$ for all ϕ. $\dashv$

Remark 4.4.25. Pattinson [66, Definition 5.2] defines a set Λ of predicate liftings for an endofunctor $\mathfrak{T}$ on the category of sets to be separating iff for all sets X the map $s \longmapsto \{\lambda_X(A) \mid \lambda \in \Lambda, A \subseteq X, s \in \lambda_X(A)\}$ is injective. Since the latter map is injective iff $s \mapsto \{(\lambda, A) \mid \lambda \in \Lambda, A \subseteq X, s \in \lambda_X(A)\}$ is, this definition coincides with the one given by Schröder [74, Definition 7]. Define in each case the equivalence relation ρ^λ_X on a set X through

$$x \; \rho^\lambda_X \; x' \iff \forall A \subseteq X : \big[x \in \lambda_X(A) \Leftrightarrow x' \in \lambda_X(A)\big],$$

and put

$$\rho^\Lambda_X := \bigcap_{\lambda \in \Lambda} \rho^\lambda_X.$$

Thus Λ is separating iff

$$\rho^\Lambda_X = \{\langle t, t' \rangle \mid t, t' \in \mathfrak{T}^{\mathrm{R}}(X)\}$$

for each set X. Consequently, the concept of separation proposed in [66, 74] is stricter than the one discussed here.

Nevertheless, separation serves as a means for establishing that a coalgebra on the factor space can be constructed. Analyzing, e.g., the proof of the fairly central property [74, Theorem 41], one sees that separation is used for establishing that the logical indistinguishability relation induces a coalgebra structure, i.e., is a congruence. This parallels our argumentation in Proposition 4.4.26 below, and it is of course the reason why separation has been introduced here as well. □

We now show that whenever Λ separates the logic, relation $\mathsf{r}_{\mathcal{M}}$ is in fact a congruence.

Proposition 4.4.26. *If Λ separates the logic, then $\mathsf{r}_{\mathcal{M}}$ is a congruence for any model $\mathcal{M}$.*

Proof First observe that $\mathsf{r}_{\mathcal{M}}$ is smooth; thus $A/\mathsf{r}_{\mathcal{M}}$ is an analytic space. Since each $V(x)$ is $\mathsf{r}_{\mathcal{M}}$-invariant, it remains to show that $\langle\gamma(a),\gamma(a')\rangle \in \mathsf{ker}\left(\mathfrak{T}^{\mathrm{R}}(\eta_{\mathsf{r}_{\mathcal{K}}})\right)$ whenever $\langle a, a'\rangle \in \mathsf{r}_{\mathcal{M}}$. Λ is separating; thus it suffices to show that $\gamma(a) \equiv^{\lambda}_{\mathcal{M}} \gamma(a')$ holds for all $\lambda \in \Lambda$. For such a λ and any $\phi \in \mathfrak{L}^{\sharp}$, we have

$$\gamma(a) \in \lambda_A([\![\phi]\!]_{\mathcal{M}}) \Leftrightarrow a \in [\![\langle\lambda\rangle\phi]\!]_{\mathcal{M}} \Leftrightarrow a' \in [\![\langle\lambda\rangle\phi]\!]_{\mathcal{M}} \Leftrightarrow \gamma(a') \in \lambda_A([\![\phi]\!]_{\mathcal{M}}).$$

So we see that in fact $\gamma(a) \equiv^{\lambda}_{\mathcal{M}} \gamma(a')$ holds. □

From now on, assume Λ to separate the logic.

Theorem 4.4.27. *If $\mathcal{M}$ and $\mathcal{N}$ are logically equivalent models, then $\mathcal{M}$ and $\mathcal{N}$ are behaviorally equivalent.*

Proof 1. Let $\mathcal{M} = (A, \gamma, V)$ and $\mathcal{N} = (B, \delta, W)$, and construct the sum $\mathcal{M}+\mathcal{N}$ according to Lemma 4.4.9. Denote the congruence induced on $\mathcal{M}+\mathcal{N}$ by $\approx$. We obtain from Proposition 4.4.26 a model $\mathcal{M}+\mathcal{N}/\approx$ together with model morfisms $i_{\mathcal{M}}, i_{\mathcal{N}}$ and $\eta_{\approx}$:

$$\begin{array}{ccccc} \mathcal{M} & \xrightarrow{\;i_{\mathcal{M}}\;} & \mathcal{M}+\mathcal{N} & \xleftarrow{\;i_{\mathcal{N}}\;} & \mathcal{N} \\ & & \big\downarrow{\scriptstyle \eta_{\approx}} & & \\ & & \mathcal{M}+\mathcal{N}/\approx & & \end{array}$$

2. The morfisms $\eta_{\approx} \circ i_{\mathcal{M}}$ is onto. In fact, take an equivalence class $t \in \mathcal{M}+\mathcal{N}/\approx$. If $t = [i_{\mathcal{M}}(c)]_{\approx}$ for some $c \in A$, then we are done; if, however, we have happened to have picked some $i_{\mathcal{N}}(c)$ with $c \in B$, then there exists an $a \in A$ such that $Th_{\mathcal{M}}(a) = Th_{\mathcal{N}}(b)$, so that $[i_{\mathcal{M}}(a)]_{\approx} = [i_{\mathcal{N}}(c)]_{\approx} = t$. Since both $i_{\mathcal{M}}$ and $i_{\mathcal{N}}$ are model morphisms, this implies that $Th_{\mathcal{M}+\mathcal{N}}(i_{\mathcal{M}}(a)) = Th_{\mathcal{M}+\mathcal{N}}(i_{\mathcal{N}}(c))$. Thus $[i_{\mathcal{M}}(a)]_{\approx} = [i_{\mathcal{N}}(c)]_{\approx} = t$. □

Suppose $\mathcal{M}$ and $\mathcal{N}$ are logically equivalent. Then the state spaces of their factor models are Borel isomorphic, so that logical equivalence has a structural implication for the respective state spaces. It turns out that even the factor models are isomorphic. As an alternative approach, this observation can be used for constructing the desired cospan; see for example Corollary 2.2.5, Corollary 2.4.11 and Lemma 2.6.10.

Proposition 4.4.28. *If $\mathcal{M}$ and $\mathcal{N}$ are logically equivalent models, then the models $\mathcal{M}/\mathsf{r}_{\mathcal{M}}$ and $\mathcal{N}/\mathsf{r}_{\mathcal{N}}$ are isomorphic.*

Proof 1. We use the notation from the proof for Theorem 4.4.27. Consider this diagram, where $\kappa : [a]_{\mathsf{r}_{\mathcal{M}}} \mapsto [a]_{\approx}$:

$$\begin{array}{ccc} A & \xrightarrow{\eta_{\mathsf{r}_{\mathcal{M}}}} & A/\mathsf{r}_{\mathcal{M}} \\ {\scriptstyle i_{\mathcal{M}}}\downarrow & & \downarrow{\scriptstyle \kappa} \\ A+B & \xrightarrow[\eta_{\approx}]{} & A+B/\approx \end{array}$$

Because $a \; \mathsf{r}_{\mathcal{M}} \; a'$ implies $i_{\mathcal{M}}(a) \approx i_{\mathcal{M}}(a')$, κ is well defined; similarly, κ is seen to be injective. Given $t \in A+B/\approx$, there exists $a \in A$ such that $t = [a]_{\approx} = \kappa([a]_{\mathsf{r}_{\mathcal{M}}})$, so that κ is onto.

2. By finality of $A/\mathsf{r}_{\mathcal{M}}$ with respect to $\eta_{\mathsf{r}_{\mathcal{M}}}$, the map κ is Borel measurable by Lemma 1.3.3, so that κ is a Borel isomorphism because of Proposition 1.4.33. Lemma 4.4.13 implies that κ is a model isomorphism $\mathcal{M}/\mathsf{r}_{\mathcal{M}} \rightarrow \mathcal{M}+\mathcal{N}/\approx$. In a similar way we obtain a model isomorphism $\mathcal{M}/\mathsf{r}_{\mathcal{M}} \rightarrow \mathcal{M}+\mathcal{N}/\approx$. Composing these isomorphisms, the assertion follows. □

4.4.4 Behavioral Equivalence vs. Bisimilarity

We will now show that logical equivalence of models implies bisimilarity under additional conditions. We will first reduce the problem by showing that we can find for every model $\mathcal{M}$ a model $\mathcal{C}$ based on a Standard Borel space and a surjectice morfism $\mathcal{C} \rightarrow \mathcal{M}$. For the latter class of models the existence of a mediating model is shown by a selection argument. This is done first for the case of coalgebras, which is in a second step lifted to the category of models. For this selection argument to work, we have to require $\mathfrak{F}$ to be completing, a property that we will introduce in Definition 4.4.31 below.

Definition 4.4.29. *A functor $\mathfrak{T}$: **ANL** $\rightarrow$ **ANL** is called* **SB**-admissible *iff $\mathfrak{T}(X)$ is Standard Borel whenever X is Standard Borel and if $\mathfrak{T}(f)$ is surjective, whenever f is surjective.*

The functor $\mathfrak{S}$ is **SB**-admissible and $\mathfrak{T}^{\mathrm{R}}$ is **SB**-admissible whenever $\mathfrak{T}$ is **SB**-admissible. Since Standard Borel spaces are closed under countable products in **ANL**, also $-^{\mathsf{Act}}$ is **SB**-admissible for any countable set Act.

Proposition 4.4.30. *Assume $\mathfrak{T}$ to be* **SB***-admissible. For every model $\mathcal{M}$ there exists a model $\mathcal{C} = (P, \delta, W)$ where P is a Standard Borel space and $e : \mathcal{C} \to \mathcal{M}$ is a surjective model morfism.*

Proof Write $\mathcal{M} = (A, \gamma, V)$ with A analytic. Thus there exists a Polish space P and a surjective Borel map $e : P \to A$ which defines the Borel structure of A. By assumption $\mathfrak{T}^{\mathrm{R}}(e) : \mathfrak{T}(P) \to \mathfrak{T}(A)$ is a Standard Borel cover. Using Proposition 1.5.4, we find a measurable function $\delta : P \to \mathfrak{T}^{\mathrm{R}}(P)$ such that this diagram

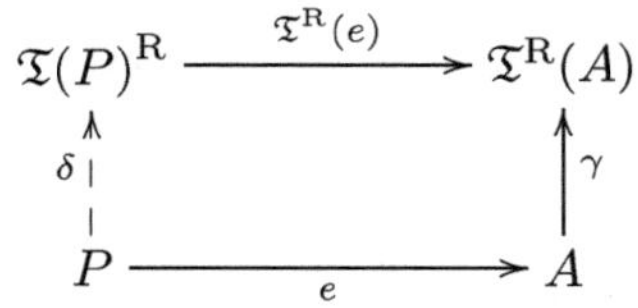

commutes; that is, $e : (P, \delta) \to (A, \gamma)$ is a coalgebra morfism. Finally, $W := \mathfrak{B}(s) \circ V$ turns $e : (P, \delta, W) \to \mathcal{M}$ into a model morfism. □

The Case of Coalgebras Based on Standard Borel Spaces

In this section, all coalgebras and models will be based on Standard Borel spaces.

Definition 4.4.31. *An endofunctor $\mathfrak{F}$:* **SB** $\to$ **SB** *on the category of Standard Borel spaces is called* completing *iff given a cospan*

$$(A, \gamma) \xrightarrow{f} (C, \epsilon) \xleftarrow{g} (B, \delta)$$

of surjective coalgebra morfisms there exists a map $s : S \to \mathfrak{T}^R(S)$ *that makes the diagram*

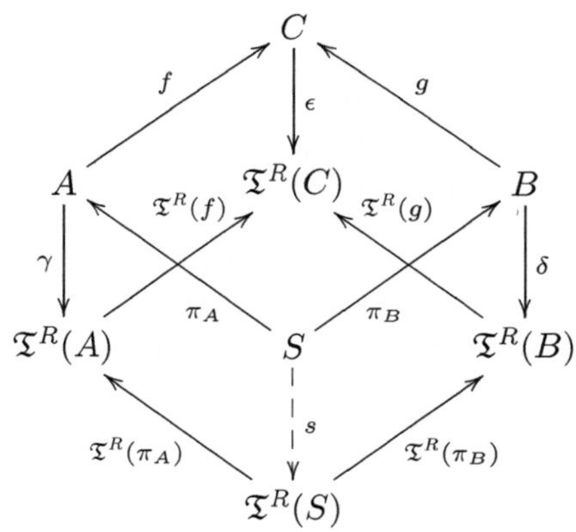

commute in category **Set***, where $S := \{\langle a, b\rangle \mid f(a) = g(b)\}$.*

The name *completing* for the functor — chosen for lack of a better one — is intended to indicate that the functor permits us to complete the cube above by providing the crucial edge. We postulate that the map s is merely a plain map without requiring any additional properties. So we do not assume that s has additional properties; in particular we do not assume that s is measurable, and *a fortiori*, we do not assume that s is compatible with the structure of the coalgebra, i.e., is a morfism.

Example 4.4.32. A constant functor is completing. In fact, assume that $\mathfrak{T}(A) = Z$ for all Standard Borel spaces (with $\mathfrak{T}(f) = id_Z$), and let the cospan

$$(A,\gamma) \xrightarrow{f} (C,\epsilon) \xleftarrow{g} (B,\delta)$$

be given. Hence we are requested to find a measurable map $s : S \to Z$ with

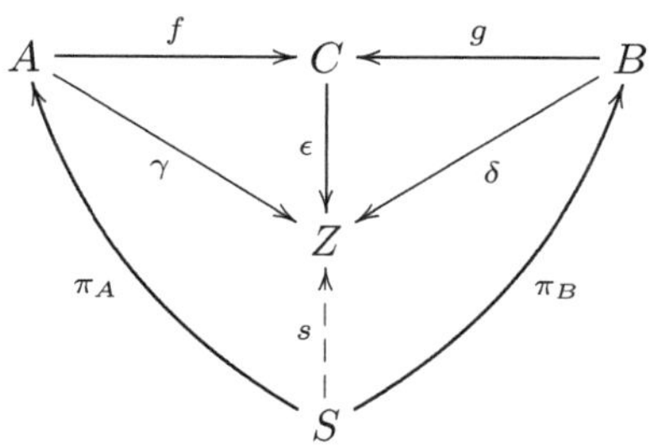

But we know

$$\gamma \circ \pi_A = \epsilon \circ f \circ \pi_A = \epsilon \circ g \circ \pi_B = \delta \circ \pi_B,$$

so that $s := \gamma \circ \pi_A$ will do. $\dashv$

The class of functors under consideration enjoys simple closure properties. We note first that the identity is completing; this is but a reformulation of a property that was used extensively, e.g., for making sure that semi pullbacks for stochastic relations exist; see Proposition 1.6.25.

Proposition 4.4.33. *The identity is completing.*

Proof Define $S := \{\langle a,b\rangle \mid f(a) = g(b)\}$, and take any $\langle a,b\rangle \in S$; hence $f(a) = g(b)$. Since $f \circ \pi_A$ is surjective, $\mathfrak{S}(f \circ \pi_A)$ is surjective as well by Proposition 1.6.19. Thus there exists $\mu \in \mathfrak{S}(S)$ such that $\mathfrak{S}(f \circ \pi_A)(\mu) = \epsilon(f(a))$. From $f(a) = g(b)$ we obtain

$$\big(\mathfrak{S}(f) \circ \gamma\big)(a) = (\epsilon \circ f)(a) = \mu = (\epsilon \circ g)(b) = \big(\mathfrak{S}(g) \circ \delta\big)(b);$$

thus we conclude for all $E \in f^{-1}[\mathcal{B}(B)]$ and for all $E' \in g^{-1}[\mathcal{B}(B)]$

$$\mathfrak{S}(\pi_A)(\mu)(E) = \gamma(a)(E) \text{ and } \mathfrak{S}(\pi_B)(\mu)(E') = \delta(b)(E').$$

By Proposition 1.6.3 there exists some $\tilde{\mu} \in \mathfrak{S}(S)$ with

$$\mathfrak{S}(\pi_A)(\tilde{\mu}) = \gamma(a) \text{ and } \mathfrak{S}(\pi_B)(\tilde{\mu}) = \delta(b).$$

But this means that

$$\Gamma(a,b) := \{\tilde{\mu} \in \mathfrak{S}(S) \mid \mathfrak{S}(\pi_A)(\tilde{\mu}) = \gamma(a) \text{ and } \mathfrak{S}(\pi_B)(\tilde{\mu}) = \delta(b)\} \neq \emptyset$$

for any $\langle a, b\rangle \in S$. Pick an element $s(a,b) \in \Gamma(a,b)$ for each $\langle a, b\rangle \in S$, which is possible by the axiom of choice. This yields a map $s : S \to \mathfrak{S}(S)$ with the desired properties. □

Proposition 4.4.34. *The class of completing functors is closed under countable products and coproducts.*

Proof 1. Let $(\mathfrak{T}_i)_{i \in I}$ be a countable family of completing functors. Setting $\mathfrak{T} = \prod_{i \in I} \mathfrak{T}_i$, we obtain a diagram ($j \in I$)

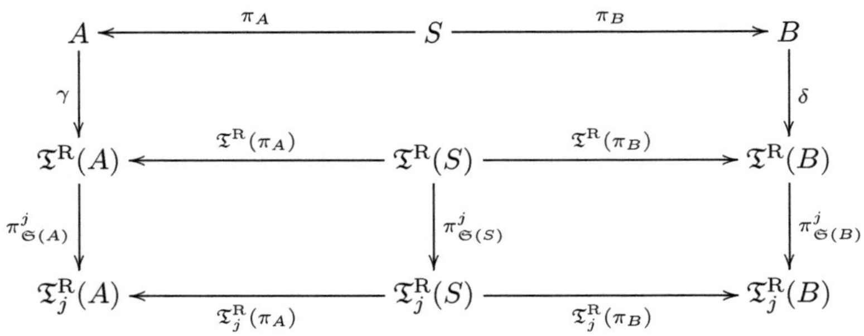

The bottom half commutes by definition of $\mathfrak{T}$. For each $j \in I$, we obtain a function $s_j : S \to \mathfrak{T}^{\mathrm{R}}_j(S)$ such that

$$\mathfrak{T}^{\mathrm{R}}_j(\pi_A) \circ s_j = \pi^j_{\mathfrak{S}(A)} \circ \gamma \circ \pi_A$$

and

$$\mathfrak{T}^{\mathrm{R}}_j(\pi_B) \circ s_j = \pi^j_{\mathfrak{S}(B)} \circ \delta \circ \pi_B$$

hold. The family $(s_j)_{j \in J}$ induces a function $s : S \to \mathfrak{T}^{\mathrm{R}}(S)$ with $\pi^j_{\mathfrak{S}(S)} \circ s = s_j$ for all $j \in I$.

For each j, we obtain

$$\begin{aligned}\pi^j_{\mathfrak{S}(A)} \circ \mathfrak{T}^{\mathrm{R}}(\pi_A) \circ s &= \mathfrak{T}^{\mathrm{R}}_j(\pi_A) \circ \pi^j_{\mathfrak{S}(S)} \circ s \\ &= \mathfrak{T}^{\mathrm{R}}_j(\pi_A) \circ s_j \\ &= \pi^j_{\mathfrak{S}(A)} \circ \gamma \circ \pi_A,\end{aligned}$$

and thus $\mathfrak{T}^{\mathrm{R}}(\pi_A) \circ s = \gamma \circ \pi_A$ by uniqueness of the induced morfism. Similarly, $\mathfrak{T}^{\mathrm{R}}(\pi_B) \circ s = \delta \circ \pi_B$, and thus s is the sought morfism.

2. Turning to the case of countable coproducts, write $\mathfrak{G} = \coprod_{i \in I} \mathfrak{T}_i$ and suppose that (A, γ) and (B, δ) are $\mathfrak{G}^{\mathrm{R}}$-coalgebras. For any pair $\langle a, b \rangle$ with $f(a) = g(b)$ we obtain that $\gamma(a) \in \mathfrak{T}_i^{\mathrm{R}}$ and $\delta(b) \in \mathfrak{T}_i^{\mathrm{R}}$ for some $i \in I$. Since $\mathfrak{T}_i$ is completing, we find $t \in \mathfrak{T}_i^{\mathrm{R}} S$ with

$$\gamma(a) = \mathfrak{T}_i^{\mathrm{R}}(\pi_A)(t) \text{ and } \delta(b) = \mathfrak{T}_i^{\mathrm{R}}(\pi_B)(t),$$

and thus also

$$\gamma(a) = \mathfrak{G}^{\mathrm{R}}(\pi_A)(t) \text{ and } \delta(b) = \mathfrak{G}^{\mathrm{R}}(\pi_B)(t).$$

Consequently, $\mathfrak{G}$ is the coproduct of the given family. □

Example 4.4.35. Consider functor $\mathfrak{Z} := (- \times \mathbb{R}_+)^\infty$ from Example 4.2.3. We conclude from Proposition 4.4.34 in conjunction with Proposition 4.4.33 and Example 4.4.32 that $\mathfrak{Z}$ is completing. ⊣

We will establish that the existence of map s permits us to construct a coalgebra structure σ on S together with suitable morfisms $(S, \sigma) \to (A, \gamma)$ and $(S, \sigma) \to (B, \delta)$. This will be done through a selection argument. In order to explore this construction, we make a brief excursion and show how the functor and a Polish topology which underlies the functor's argument can be modified so that a selection argument applies. After this intermission we will return to the main thread of the discussion.

Excursion: A Selection Argument ...

We use Proposition 1.5.3 for establishing the existence of the dynamics for a mediating coalgebra. It turns out that nonemptiness of the set-valued map is crucial, since we will derive now that a set-valued map associated with our functor is otherwise well behaved. Assume $\mathfrak{T}$ to be **SB**-admissible and let (A, γ) and (B, δ) be $\mathfrak{T}^{\mathrm{R}}$-coalgebras with Standard Borel spaces A and B. Assume that $S \subseteq A \times B$ is product measurable, $\pi_A[S] = A$ and $\pi_B[S] = B$, and that S is equipped with the trace σ-algebra $S \cap \mathcal{A} \otimes \mathcal{A}'$.

Lemma 4.4.36. *We can find admissible topologies $\mathcal{T}_A, \mathcal{T}_B$, and $\mathcal{T}_S$ on A, B, and S as well as admissible topologies $\mathcal{R}_A, \mathcal{R}_B$, and $\mathcal{R}_S$ on $\mathfrak{T}^R(A), \mathfrak{T}^R(B)$, and $\mathfrak{T}^R(S)$ with these properties.*

a. γ is $\mathcal{T}_A$-$\mathcal{R}_A$-continuous; δ is $\mathcal{T}_B$-$\mathcal{R}_B$-continuous.
b. $\mathfrak{T}^R(\pi_A)$ is $\mathcal{R}_S$-$\mathcal{R}_A$-continuous; $\mathfrak{T}^R(\pi_B)$ is $\mathcal{R}_S$-$\mathcal{R}_B$-continuous.
c. Define for $\langle a, b \rangle \in S$

$$\Gamma(a, b) := \{t \in \mathfrak{T}^R(S) \mid \mathfrak{T}^R(\pi_A)(t) = \gamma(a),\ \mathfrak{T}^R(\pi_B)(t) = \delta(b)\};$$

then $\Gamma(a, b)$ is $\mathcal{R}_S$-closed, and $\exists\Gamma(C)$ is a Borel set in S, provided $C \subseteq \mathfrak{T}^R(S)$ is $\mathcal{R}_S$-compact.

Proof 1. Choose admissible topologies $\mathcal{T}_A^0$ and $\mathcal{T}_B^0$ on A and B. Since the product of the Polish spaces $(A, \mathcal{T}_A^0) \times (B, \mathcal{T}_B^0)$ is a Polish space again, and S is a Borel set, we can find an admissible topology $\mathcal{T}_S^0$ on S. $\mathfrak{T}^{\mathrm{R}}(A)$ depends only on the σ-algebra on A (and not on the topology).

2. There exists an admissible topology $\mathcal{R}_A$ on $\mathfrak{T}^{\mathrm{R}}(A)$. Since $\gamma : A \to \mathfrak{T}^{\mathrm{R}}(A)$ is a Borel map, we may refine the topology $\mathcal{T}_A^0$ to an admissible topology $\mathcal{T}_A$ in such a way that γ is $\mathcal{T}_A$-$\mathcal{R}_A$-continuous. Similarly, we find admissible topologies $\mathcal{T}_B$ and $\mathcal{R}_B$ on B and $\mathfrak{T}^{\mathrm{R}}(B)$ such that δ is $\mathcal{T}_B$-$\mathcal{R}_B$-continuous. Applying Proposition 1.4.19 twice, we find an admissible topology $\mathcal{R}_S$ on $\mathfrak{T}^{\mathrm{R}}(S)$ such that $\mathfrak{T}^{\mathrm{R}}(\pi_A)$ is $\mathcal{R}_S$-$\mathcal{R}_A$-continuous, and $\mathfrak{T}^{\mathrm{R}}(\pi_B)$ is $\mathcal{R}_S$-$\mathcal{R}_B$-continuous as well.

3. $\mathcal{R}_S$-closedness of $\Gamma(a, b)$ follows from the continuity of the respective maps. Let $C \subseteq \mathfrak{T}^{\mathrm{R}}(S)$ be a $\mathcal{R}_S$-compact subset. We will show that $\exists\Gamma(C)$ is $\mathcal{T}_S$-closed. In fact, let $\left(\langle x_n, y_n\rangle\right)_{n\in\mathbb{N}}$ be a sequence in $\exists\Gamma(C)$ such that $\langle x_n, y_n\rangle \to \langle x, y\rangle \in S$ with respect to topology $\mathcal{T}_S$. We want to show that some $\langle x, y\rangle \in \exists\Gamma(C)$. We can find for each $n \in \mathbb{N}$ some $t_n \in \Gamma(x_n, y_n) \cap C$. Since C is compact, we find a converging subsequence of $(t_n)_{n\in\mathbb{N}}$ with limit $t \in C$. By passing to a subsequence, if necessary, it is no loss of generality to assume that $t_n \to t$. From the continuity of $\mathfrak{T}^{\mathrm{R}}(\pi_A)$ and $\mathfrak{T}^{\mathrm{R}}(\pi_B)$ we obtain $\mathfrak{T}^{\mathrm{R}}(\pi_A)(t) = \gamma(x)$ and $\mathfrak{T}^{\mathrm{R}}(\pi_B)(t) = \delta(y)$; hence $t \in \Gamma(x, y)$. Consequently, $\langle x, y\rangle \in \exists\Gamma(C)$. Thus $\exists\Gamma(C) \neq \emptyset$ is closed, hence Borel. □

. . . And Happily Back Again

Proposition 4.4.37. *Using the notation from Definition 4.4.31, the following conditions are equivalent:*

a. Functor $\mathfrak{T}$ is completing.
b. For any $\langle a, b\rangle \in S$ there exists $t \in \mathfrak{T}^R(S)$ such that

$$\gamma(a) = \mathfrak{T}^R(\pi_A)(t) \text{ and } \delta(b) = \mathfrak{T}^R(\pi_B)(t).$$

c. There exists a coalgebra structure $\sigma : S \to \mathfrak{T}^R(S)$ such that

$$(A, \gamma) \xrightarrow{\pi_A} (S, \sigma) \xleftarrow{\pi_B} (B, \delta)$$

is a span of morfisms.

Proof 1. The implications $a \Rightarrow b$ as well as $c \Rightarrow a$ are trivial; thus we are left to establish the implication $b \Rightarrow c$.

2. The set $S = \{\langle a, b\rangle \mid f(a) = g(b)\}$ is measurable in $A \times B$, and the projections $\pi_A : S \to A$ and $\pi_B : S \to B$ are onto by assumption. Hence we may apply Lemma 4.4.36; all topological notions will be with respect to the topologies obtained from Lemma 4.4.36.

3. Fix $\langle a, b\rangle \in S$. The set

$$\Gamma(a, b) = \{t \in \mathfrak{T}^{\mathrm{R}}(S) \mid \mathfrak{T}^{\mathrm{R}}(\pi_A)(t) = \gamma(a), \mathfrak{T}^{\mathrm{R}}(\pi_B)(t) = \delta(b)\}$$

is closed by Lemma 4.4.36 and nonempty by assumption. Since the weak inverse

$$\exists\Gamma(C) := \{\langle a,b\rangle \mid \Gamma(a,b) \cap C \neq \emptyset\}$$

is measurable for each compact $C \subseteq \mathfrak{T}^{\mathrm{R}}(S)$, we obtain by Proposition 1.5.3 a measurable function $\sigma : S \to \mathfrak{T}^{\mathrm{R}}(S)$ with $\sigma(a,b) \in \Gamma(a,b)$ for all $\langle a,b\rangle \in S$.

4. By definition of Γ, the latter condition is equivalent to the commutativity of

$$\begin{array}{ccccc}
A & \xleftarrow{\pi_A} & S & \xrightarrow{\pi_B} & B \\
\downarrow{\scriptstyle\gamma} & & \downarrow{\scriptstyle\sigma} & & \downarrow{\scriptstyle\delta} \\
\mathfrak{T}^{\mathrm{R}}(A) & \xleftarrow[\mathfrak{T}^{\mathrm{R}}(\pi_A)]{} & \mathfrak{T}^{\mathrm{R}}(S) & \xrightarrow[\mathfrak{T}^{\mathrm{R}}(\pi_B)]{} & \mathfrak{T}^{\mathrm{R}}(B)
\end{array}$$

But this means that (A,γ) and (B,δ) are bisimilar. □

This yields the following as an immediate consequence.

Corollary 4.4.38. *If the right $\mathfrak{T}$-coalgebras* F *and* G *are logically equivalent, and if $\mathfrak{T}$ is completing, then* F *and* G *are bisimilar.* □

Basing Models on Standard Borel Spaces

Assume that the models $\mathcal{M} = (A,\gamma,V)$ and $\mathcal{N} = (B,\delta,W)$ are both based on Standard Borel spaces and are behaviorally equivalent. Thus there exists a span

$$\mathcal{M} \xrightarrow{f} \mathcal{Q} \xleftarrow{g} \mathcal{N}$$

of surjective model morfisms. Write $\mathcal{N} = (B,\delta,W)$. Discarding valuations momentarily, we obtain a span of surjective coalgebra morfisms, and thus, if $\mathfrak{T}^{\mathrm{R}}$ is completing, a commutative diagram

$$\begin{array}{ccccc}
 & & (S,\epsilon) & & \\
 & \swarrow{\scriptstyle\pi_A} & & {\scriptstyle\pi_B}\searrow & \\
(A,\gamma) & & & & (B,\delta) \\
 & {\scriptstyle f}\searrow & & \swarrow{\scriptstyle g} & \\
 & & (B,\delta) & &
\end{array}$$

of surjective coalgebra morfisms results. The underlying set S is given as above through $\{\langle a,b\rangle \mid f(a) = g(b)\}$. Observe that we have $\mathfrak{B}(\pi_A) \circ V = \mathfrak{B}(\pi_B) \circ W$. Indeed, for $x \in \mathsf{V}$ and $\langle a,b\rangle \in S$ we have

$$\langle a,b\rangle \in \mathfrak{B}(\pi_A) \circ V(x) \iff a \in V(x) \iff f(a) \in W(x),$$

since f is a model morfism. By symmetry, we also have $\langle a, b\rangle \in \mathfrak{B}(\pi_B) \circ W(x)$ iff $g(b) \in W'(x)$. Hence, $\mathfrak{B}(\pi_A) \circ V = \mathfrak{B}(\pi_B) \circ W$ by definition of S. Thus, $U = \mathfrak{B}(\pi_A) \circ V$ turns

$$\mathcal{M} \xrightarrow{\pi_A} (S, \epsilon, U) \xleftarrow{\pi_B} \mathcal{N}$$

into a diagram of surjective model morfisms; that is, $\mathcal{M}$ and $\mathcal{N}$ are bisimilar. Thus, we have proved:

Theorem 4.4.39. *If* $\mathfrak{T} : \mathbf{SB} \to \mathbf{SB}$ *is completing, then behavioral equivalence implies bisimilarity for all models based on Standard Borel spaces.* □

We will now use the Standard Borel covers introduced earlier to reduce the general case to the case just proven.

The General Case

For every **SB**-admissible functor $\mathfrak{T} : \mathbf{ANL} \to \mathbf{ANL}$ we can define a restriction of $\mathfrak{T}$ to **SB**. We say $\mathfrak{T}$ is completing iff this restriction has this property.

Now suppose $\mathfrak{T}$ is **SB**-admissible and is completing. Whenever we have a cospan

$$\mathcal{M} \xrightarrow{f} \mathcal{Q} \xleftarrow{g} \mathcal{N}$$

of surjective model morfisms, we can form covers by models based on Standard Borel spaces according to Proposition 4.4.30. This enables us to apply Theorem 4.4.39, which leads us to the following commutative diagram of surjective model morfisms:

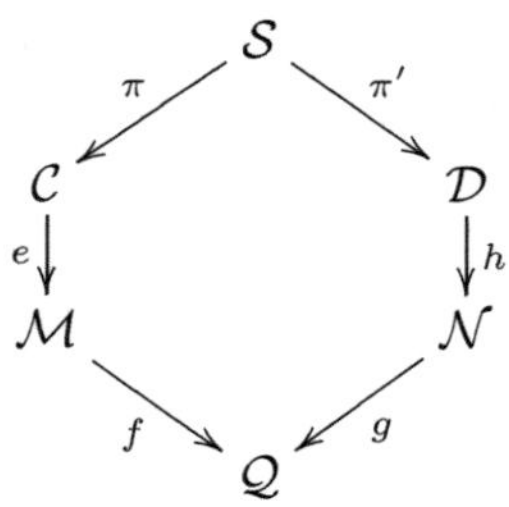

Inspecting the upper part of the above diagram, we see that we have shown:

Proposition 4.4.40. *If* $\mathfrak{T}$ *is* **SB**-*admissible and is completing, then behavioral equivalent models are bisimilar. Moreover, the mediating model can be chosen to be based on a Standard Borel space.* □

It is established in Proposition 1.6.26 that given a stochastic relation A which is based on an analytic space there exists a stochastic relation P based on a Polish space and a surjective morfism $f : \mathsf{P} \to \mathsf{A}$. This was shown in [19] to be a consequence of the existence of semi-pullbacks for certain types of stochastic relations and was established through a measure extension argument involving the Hahn–Banach Theorem as well as a selection theorem. It is remarkable that a similar statement holds as well for the case of right coalgebras (without having to resort to such heavy artillery as the Hahn–Banach Theorem).

Concluding, we present the following Theorem collecting the main results of this section:

Theorem 4.4.41. *Let $\mathfrak{T} : \mathbf{ANL} \to \mathbf{ANL}$ be* **SB**-*admissible and completing, and assume that Λ is a separating set of predicate lifting for $\mathfrak{T}^R$. Then the following statements are equivalent for models $\mathcal{M}$ and $\mathcal{N}$:*

a. $\mathcal{M}$ and $\mathcal{N}$ are logically equivalent;
b. $\mathcal{M}$ and $\mathcal{N}$ are behaviorally equivalent;
c. $\mathcal{M}$ and $\mathcal{N}$ are bisimilar. □

Example 4.4.42. Let $\mathfrak{T} := \coprod_{n\in\mathbb{N}} -^n$; then $\mathfrak{T}$ is completing by Proposition 4.4.33 and Proposition 4.4.34. Define for rational q, $k \in \mathbb{N}$ and the Borel set $G \subseteq A$

$$\lambda_{q,k,A}(G) := \{ (0, \dots 0, \overbrace{\mu}^{k}, 0, \dots 0) \mid \mu \in \mathfrak{S}A, \mu(G) \geq q \}$$

with 0 as the zero measure. Then $\lambda_{q,k} : \mathfrak{B} \to \mathfrak{B} \cdot \mathfrak{T}^{\mathrm{R}}$ is a natural transformation; put $\Lambda := \{\lambda_{q,k} \mid k \in \mathbb{N}, q \in \mathbb{Q} \cap [0, 1]\}$.

Define the formulas of logic $\mathfrak{L}^{\sharp}_{\maltese}$ through

$$\phi ::= \top \mid \phi_1 \wedge \phi_2 \mid \phi_1 \vee \phi_2 \mid \langle\lambda\rangle\phi \mid x \mid \mu\, x.\phi$$

with x taken from the set V of all variables and the familiar μ-operator; we take $\langle A, B\rangle \mapsto A \cup B$ as a binary bridge operator. The μ-operator is interpreted through the infinitesimal operator $\iota^{(4)} : (R_n)_{n\in\mathbb{N}} \mapsto \bigcup_{n\in\mathbb{N}} R_n$ from Example 4.4.18. This logic models the finite, stubborn replication of a basic process; it is stubborn because the μ-operator permits formulating fixed points, i.e., decisions, once met, will be adhered to infinitely.

Let $\mathcal{M}$ be a model. Then it is not difficult to see that $[\![\mu\, x.\phi]\!]_{\mathcal{M}}$ is the smallest fixed point for $R \mapsto [\![\phi]\!]_{\mathcal{M}[x\backslash R]}$; see the argumentation for the more complicated logic μ**CSL** in Section 2.4.3, in particular Proposition 2.4.6. One shows as in Example 4.4.23 that Λ is separating. Thus logical equivalent models for $\mathfrak{L}^{\sharp}_{\maltese}$ are bisimilar, as one would intuitively expect. ⊣

4.4.5 A Local Construction

Given two models $\mathcal{M}$ and $\mathcal{N}$, we call *the state a of $\mathcal{M}$ and b of $\mathcal{N}$ bisimilar* iff we can find a bisimulation $\mathcal{M} \xleftarrow{f} \mathcal{Q} \xrightarrow{g} \mathcal{N}$ with $a = f(q), b = g(q)$ for some state q on $\mathcal{Q}$. While we discussed so far bisimulations globally, i.e., for entire models, the bisimilarity of states requires the construction of a span such that the corresponding morfisms produce the given states from a state in the mediating model. If two states are bisimilar, then they satisfy exactly the same formulas; this follows from Lemma 4.4.20. We will discuss a remarkable construction due to Ch. Schubert [76] for reversing this implication; technically, it shows that the existence of a global bisimulation can be used for constructing a local one.

We say that a Borel measurable map $f : A \to B$ between the analytic spaces A and B is *of class S* iff it has these properties: f is injective, the Borel sets on A form the initial σ-algebra with respect to f and B, and $f[A] \subseteq B$ is a Borel set. The condition on initiality of f entails

$$\mathcal{B}(A) = \{f^{-1}[C] \mid C \in \mathcal{B}(B)\}$$

by Lemma 1.3.4. Consider as an example an injection $i_A : A \to A + B$ into a coproduct; then it is easy to see that i_A is of class S; in fact, it is easily established that these injections are exactly the Borel maps of class S.

Since we will make use of these maps, we will study their properties briefly. First, the subprobability functor preserves this property.

Lemma 4.4.43. *If $f : A \to B$ is of class S, so is $\mathfrak{S}(f) : \mathfrak{S}(A) \to \mathfrak{S}(B)$.*

Proof 1. If $\mu \neq \mu'$, then there exists $C \in \mathcal{B}(A)$ with $\mu(C) \neq \mu'(C)$. But $C = f^{-1}[D]$ for some Borel set $D \in \mathcal{B}(B)$. Thus

$$\mathfrak{S}(f)(\mu)(D) = \mu\big(f^{-1}[D]\big) \neq \mu'\big(f^{-1}[D]\big) = \mathfrak{S}(f)(\mu')(D),$$

so that $\mathfrak{S}(f)$ is injective.

2. In order to establish initiality of $\mathfrak{S}(f)$, consider

$$\mathcal{D} := \{H \in \mathcal{B}(\mathfrak{S}(A)) \mid H = \mathfrak{S}(f)^{-1}[G] \text{ for some } G \in \mathcal{B}(\mathfrak{S}(B))\}.$$

Then $\mathcal{D}$ is a σ-algebra which contains for $H \in \mathcal{B}(A)$ the set

$$\Gamma_{A,H,q} := \{\mu \in \mathfrak{S}(A) \mid \mu(H) \geq q\}.$$

This is so since given $H \in \mathcal{B}(A)$, there exists $G \in \mathcal{B}(B)$ with $H = f^{-1}[G]$, so that $\mu(H) \geq q$ iff $\mathfrak{S}(f)(\mu)(G) \geq q$; consequently,

$$\Gamma_{A,f^{-1}[G],q} = \mathfrak{S}(f)^{-1}[\Gamma_{B,G,q}].$$

But this entails

$$\mathcal{B}(\mathfrak{S}(A)) = \sigma\big(\{\Gamma_{A,H,q} \mid H \in \mathcal{B}(A), q \in [0,1]\}\big) \subseteq \mathcal{D},$$

showing that in fact each member of $\mathcal{B}(\mathfrak{S}(A))$ is the inverse image of an element of $\mathfrak{S}(B)$.

3. Let $f[A] = G \in \mathcal{B}(B)$; then

$$\mathfrak{S}(f)[\mathfrak{S}(A)] = \{\nu \in \mathfrak{S}(B) \mid \nu \blacktriangleright G\}$$

(the relation *lives on* $\blacktriangleright$ is defined on page 181). In fact, $\mathfrak{S}(f)(\mu)(B \setminus G) = \mu\big(f^{-1}[B \setminus G]\big) = 0$. Now let $\nu \in \mathfrak{S}(B)$ with $\nu \blacktriangleright G$ be given, and restrict ν to ν_0 on the Borel sets of G. Because $\mathfrak{B}(f) : \mathcal{B}(G) \to \mathcal{B}(A)$ is an isomorphism of Boolean σ-algebras, $\mu(\mathfrak{B}(f)(H)) := \nu_0(H)$ defines a measure μ on $\mathcal{B}(A)$ such that $\mathfrak{S}(f)(\mu)(D) = \nu(D)$ for all $D \in \mathcal{B}(B)$. □

Let us have a look at how pullbacks and the subprobability functor $\mathfrak{S}$ deal with measurable maps of class S.

Lemma 4.4.44. *Assume that for the analytic spaces B, X, and Y and the Borel map $f : X \to Y$ a pullback*

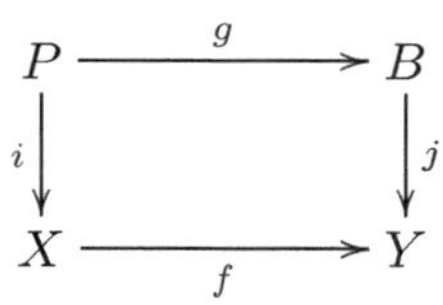

is given, where j is of class S. Then

a. i is of class S,
b. $\mathfrak{S}$ preserves the diagram.

Proof 1. Let for $h, h' : Q \to P$ for some analytic space Q with $i \circ h = i \circ h'$. Because the diagram is a pullback, and since $f \circ h \circ h = f \circ i \circ h'$, we may conclude that $h = h'$, so that i is injective. In order to establish that i is initial, assume that $i \circ k$ is Borel measurable for some map $k : Q \to P$. Because $i \circ k$ is measurable, we conclude that $f \circ i \circ k$ is measurable, hence also $j \circ g \circ k$. By initiality of j, we conclude that $g \circ k$ is measurable; so we obtain from the diagram being a pullback a unique measurable map $t : Q \to P$ with $i \circ k = i \circ t$, which implies $k = t$; hence k is measurable. From the characterization of initial σ-algebras in Lemma 1.3.4 we conclude that i is initial. Finally, we claim that

$$i[P] = f^{-1}[j[B]]$$

holds. Suppose there exists $p \in P$ such that $i(p) \in f^{-1}[Y \setminus j[B]]$, so that $j(b) \in Y \setminus j[B]$ with $b := g(p) \in B$. This is a contradiction. On the other hand, because the diagram is a pullback diagram, we find for any given $x_0 \in f^{-1}[j[B]]$ an element $p_0 \in P$ with $x_0 = i(p_0)$: Put

$$Q := \{\langle x, b\rangle \mid x \in f^{-1}[j[B]], f(x) = j(b)\},$$

and let $\pi_1 : Q \to X, \pi_2 : Q \to B$ be the projections. Q is analytic, and the projections are measurable. Since $f \circ \pi_1 = j \circ \pi_2$, we find $t : Q \to P$ with $\pi_1 = i \circ t$ and $\pi_2 = g \circ t$; consequently, $p_0 := t(x_0, b)$ for some $\langle x_0, b \rangle \in Q$ will do. Thus $x_0 \in i[P]$. This shows that i is of class S.

2. Now assume that we have an analytic space Q and measurable maps $a : Q \to \mathfrak{S}(X)$, $b : Q \to \mathfrak{S}(B)$ with $\mathfrak{S}(f) \circ a = \mathfrak{S}(j) \circ b$. Then $a(q) \blacktriangleright i[P]$ for any $q \in Q$. This is so since for $H \cap f^{-1}[j[B]] = \emptyset$

$$\begin{aligned} a(q)\big(H \cup f^{-1}[j[B]]\big) &= a(q)(H) + a(q)\big(f^{-1}[j[B]]\big) \\ &= a(q)(H) + b(q)\big(j^{-1}[j[B]]\big) \\ &= a(q)(H) + b(q)(B) \\ &= a(q)(H) + a(q)(X); \end{aligned}$$

thus $a(q)(H) = 0$. Similarly, one shows that $b(q) \blacktriangleright j[B]$. Let $G \in \mathcal{B}(P)$ be a Borel set; then $i[G] \in \mathcal{B}(X)$. This is so since there exists $F \in \mathcal{B}(X)$ with $G = i^{-1}[F]$ by initiality of i, where we may assume that $F \subseteq i[P]$ (because $\mathcal{B}(P)$ and $\mathcal{B}(i[P])$ are isomorphic as σ-algebras on account of the properties of i). Consequently, $i[G] = F \in \mathcal{B}(X)$. Now define for $q \in Q$ and for $G \in \mathcal{B}(P)$

$$t(p)(G) := a(q)\big(i[G]\big).$$

Because i is injective, $t(p)$ defines a measure on $\mathcal{B}(P)$: if $(G_n)_{n\in\mathbb{N}}$ is a disjoint sequence of sets, $\big(i[G_n]\big)_{n\in\mathbb{N}}$ is disjoint, and $i\left[\bigcup_{n\in\mathbb{N}} G_n\right] = \bigcup_{n\in\mathbb{N}} i[G_n]$. If $H \in \mathcal{B}(X)$ with $H \subseteq i[P]$, then

$$t(q)(i^{-1}[H]) = a(q)\big(i^{-1}[i[H]]\big) = a(q)(H),$$

so that $\mathfrak{S}(i) \circ t = a$. In order to establish $\mathfrak{S}(j) \circ t = b$, we take $E \in \mathcal{B}(B)$; then there exists by initiality of j a set $D \in \mathcal{B}(Y)$ with $E = j^{-1}[D]$. Because $b(q) \blacktriangleright j[B]$ we may and do assume that $D \subseteq j[D]$. Then

$$\begin{aligned} f^{-1}[D] &= i\left[i^{-1}\left[f^{-1}[D]\right]\right] \\ &= i\left[(f \circ i)^{-1}[D]\right] \\ &= i\left[(j \circ g)^{-1}[D]\right] \\ &= i\left[g^{-1}\left[j^{-1}[D]\right]\right], \end{aligned}$$

so that

$$\begin{aligned} t(q)\big(g^{-1}\left[E\right]\big) &= t(q)\big(g^{-1}\left[j^{-1}\left[D\right]\right]\big) \\ &= a(q)\big(i\left[g^{-1}\left[j^{-1}\left[D\right]\right]\right]\big) \\ &= \big(\mathfrak{G}\left(f\right)\circ a\big)(q)(D) \\ &= \big(\mathfrak{G}\left(j\right)\circ b\big)(q)(D) \\ &= b(q)\big(j^{-1}\left[D\right]\big) \\ &= b(q)(E). \end{aligned}$$

It is evident that t is the only possible choice. □

Definition 4.4.45. *Call an endofunctor $\mathfrak{T}$:* **ANL** $\rightarrow$ **ANL** strongly completing *iff $\mathfrak{T}$ is* **SB***-admissible and completing such that $\mathfrak{T}$ preserves pullbacks of the form*

$$\begin{array}{ccc} P & \xrightarrow{\;g\;} & B \\ {\scriptstyle i}\downarrow & & \downarrow{\scriptstyle j} \\ X & \xrightarrow[\;f\;]{} & Y \end{array}$$

with B, X, Y analytic, $f : X \rightarrow Y$ arbitrary, and j of class S.

From Lemma 4.4.44 we obtain

Corollary 4.4.46. *The identity is strongly completing. The product $\prod_{j\in J}\mathfrak{T}_j$ and the coproduct $\coprod_{j\in J}\mathfrak{T}_j$ of a countable family $(\mathfrak{T}_j)_{j\in J}$ of strongly completing functors is strongly completing. If $\mathfrak{T}$ is strongly completing, then so is $\mathfrak{T}^R$.*

Proof (Sketch) The product and the coproduct are strongly completing because the pullback is computed in both cases componentwise. This is established through a direct computation. The reader is referred to the discussion in [57, Sect. 5.3]. The other assertions are obvious. □

Fix a strongly completing functor $\mathfrak{T}$, and assume that Λ is a separating set of predicate lifting for $\mathfrak{T}^{\mathrm{R}}$. Let $\mathcal{M}$ and $\mathcal{N}$ be arbitrary models, and define $\mathcal{Q} := \mathcal{M} + \mathcal{N}$; then $\mathcal{Q}$ is a model by Lemma 4.4.16. The equivalence relation $\mathsf{r}_{\mathcal{Q}}$ is smooth; thus $\mathcal{Q}_1 := \mathcal{Q}/\mathsf{r}_{\mathcal{Q}}$ is a model, and

$$\mathcal{Q} \xrightarrow{\eta_{\mathsf{r}_{\mathcal{Q}}}} \mathcal{Q}_1 \xleftarrow{\eta_{\mathsf{r}_{\mathcal{Q}}}} \mathcal{Q}$$

are surjective morfisms. The injections $i_{\mathcal{M}} : \mathcal{M} \rightarrow \mathcal{Q}$ and $i_{\mathcal{N}} : \mathcal{N} \rightarrow \mathcal{Q}$ are morfisms as well, the underlying maps being of class S. Now let x and y be logically equivalent states in $\mathcal{M}$ and $\mathcal{N}$. Then $\eta_{\mathsf{r}_{\mathcal{Q}}}\big(i_{\mathcal{M}}(x)\big) = \eta_{\mathsf{r}_{\mathcal{Q}}}\big(i_{\mathcal{N}}(y)\big)$ by construction. By Theorem 4.4.41 there exists a bisimulation $\mathcal{Q}_2$ and morfisms p_1, p_2 such that this diagram commutes:

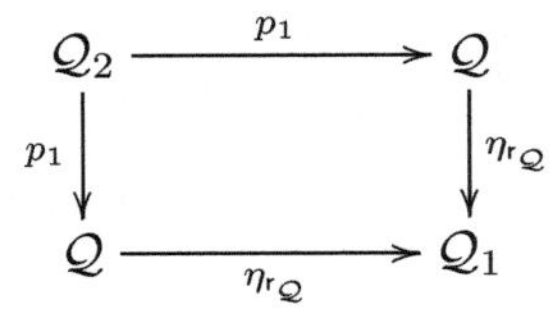

Now form pullbacks, yielding the lower left and the upper right diagrams, yielding models $\mathcal{Q}_3$ and $\mathcal{Q}_4$ with the associated morfisms.

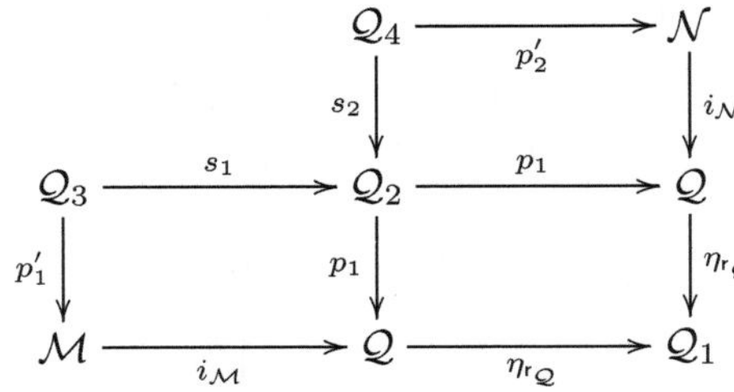

We know from Lemma 4.4.44 that the map underlying s_2 is of class S; so forming the pullback for the upper left diagram yields by that lemma a model $\mathcal{Q}_5$ and suitable morfisms

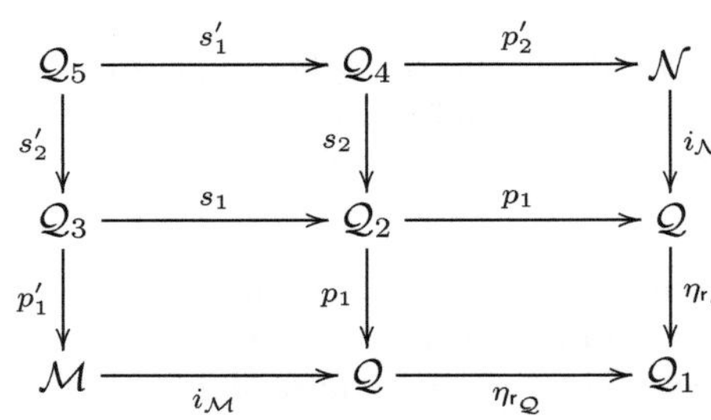

The resulting diagram is a pullback diagram in **ANL** as well; consequently (see part 1 of the proof for Lemma 4.4.44) there exists a state q_1 in $\mathcal{Q}_5$ with $(p_1' \circ s_2')(q_1) = x$, $(p_2' \circ s_1')(q_1) = y$.

We have established

Theorem 4.4.47. *Let $\mathfrak{T} : \mathbf{ANL} \to \mathbf{ANL}$ be strongly completing, and assume that Λ is a separating set of predicate lifting for $\mathfrak{T}^R$. Then logically equivalent states are bisimilar.* $\square$

Markov transition systems are the right coalgebras for the functor $\mathfrak{T} := -^{\mathsf{Act}}$ with Act as a countable set of actions; see Example 4.4.3. Consider the Hennessy-Milner logic $\mathfrak{L}(\mathsf{Act}, \mathbb{Q} \cap [0,1]) = \mathfrak{L}^\sharp(\Lambda, \emptyset, \emptyset, \emptyset)$ with separating set

$$\Lambda := \{\lambda_{q,a} \mid q \in \mathbb{Q} \cap [0,1], a \in \mathsf{Act}\},$$

where

$$\lambda_{q,a,A}(B) := \{m \in \mathfrak{T}^{\mathrm{R}}(A) \mid \pi_a(m)(B) \geq q\},$$

π_a denoting the a-th projection. Thus the logic's formulas are given, as expected, through the grammar

$$\phi ::= \top \mid \phi_1 \wedge \phi_2 \mid \langle a \rangle_q \phi$$

($a \in \mathsf{Act}, q \in \mathbb{Q} \cap [0,1]$). We infer from Proposition 4.4.33 and from Corollary 4.4.46 this local version of Theorem 2.3.11:

Proposition 4.4.48. *Two states in a Markov transition system over analytic spaces are bisimilar iff they accept exactly the same formulas in Hennessy-Milner logic.* □

4.5 Bibliographic Notes

Moss [61] investigates a logic that is defined in terms of the coalgebra involved; similarly, Pattinson [66] uses accessibility properties of the functor for which the coalgebras are defined for a definition of the logic's syntax; see also [54, 56]. Schröder [74] discusses accessibility properties of coalgebraic logics; this approach is extended in [75].

The paper [66] investigates behavioral equivalence locally (for two states) and takes for granted that the functor preserves weak pullbacks; the crucial properties are established essentially through Worrell's elegant proof technique of terminal sequence induction [85]. Since the subprobability functor that looms in the background does not preserve weak pullbacks, Worrell's approach does not work in the present scenario. Thus a direct approach through stochastic relations and their congruences has been chosen.

It is assumed throughout that the coalgebras under consideration are based on analytic spaces. This has several reasons. First, the measure theory necessary for the present development is usually only available in analytic spaces. For example, the existence of a semi-pullback on which the existence of a bisimulation hinges can only be established for stochastic relations based on analytic spaces. A more subtle point that requires an underlying analytic space rather than a general measurable space is the characterization of the factor space for those equivalence relations that are obtained from the logic. The central tools all depend on include Souslin's Separation Theorem 1.4.29, which is not available in general measurable spaces. Thus, although results for stochastic coalgebras without topological assumptions are quite encouraging (e.g. [81, 62]), it would be difficult to omit these assumptions in the present context; see also the discussion in Section 2.6.1. The present approach includes the approach using finite probability spaces that has been investigated, e.g., in [78]. Finite spaces are an important special case of Polish and of analytic spaces, so this case is covered as well.

The logic involved is rather simple, permitting a concentration on the crucial points. It grew out of logics that have been studied in the literature at least since Larsen and Skou's work [58] on probabilistic testing, and since the important paper on bisimilarity of Markov transition systems [17] by Desharnais, Edalat, and Panangaden. Subsequent work includes [3, 20] on different aspects of model checking and on algebraic properties of the corresponding models. The results in this chapter are taken from [29, 21, 76].

It is worthwhile noting that order does not play a rôle in these discussions: in the work on coalgebraic logic based on the power set functor, predicate liftings are usually assumed to be monotonic; this property is crucial for some proofs. This is emphasized by the connection between coalgebras and Stone spaces, as investigated, e.g., in [10, 53]. The absence of the necessity to base arguments on an order structure in the stochastic case discussed here indicates that there are structurally other powers at work. A comparison between the Eilenberg-Moore algebras for the power set monad (sup-complete partial orders [57, Exercise VI.2.1]) and the subprobability monad (positive convex structures [18, 26, 25]; see also [51]) suggests that convexity or positive convexity probably plays the rôle order commands in the set-theoretic scenario.

Logics: The Grammars

This Appendix lists the logics which are investigated in this treatise. Some logics which are discussed merely for illustrating a particular point are omitted.

Basic Modal Language: $\mathfrak{M}_{\mathfrak{b}}(\tau, P)$

The *basic modal language* $\mathfrak{M}_{\mathfrak{b}}(\tau, P)$ is given by the syntax

$$\phi ::= p \mid \top \mid \phi_1 \wedge \phi_2 \mid \neg\phi \mid \triangle(\phi_1, \ldots \phi_{\mathfrak{ar}(\triangle)}),$$

where τ is a set of modal operators, $\triangle \in \tau$ is a modal operator of arity $\mathfrak{ar}(\triangle)$, and $p \in P$ with P a set of propositional letters: defined in Section 2.3 on page 76.

The Extended Modal Language: $\mathfrak{M}_{\mathfrak{s}}(\tau, P)$

The extended modal language $\mathfrak{M}_{\mathfrak{s}}(\tau, P)$ is defined through the syntax

$$\phi ::= p \mid \top \mid \phi_1 \wedge \phi_2 \mid \neg\phi \mid \triangle_q(\phi_1, \ldots \phi_{\mathfrak{ar}(\triangle)}),$$

where $q \in \mathbb{Q} \cap [0, 1]$ is a rational number, τ is a set of modal operators, $\triangle \in \tau$ is a modal operator of arity $\mathfrak{ar}(\triangle)$, and $p \in P$ with P a set of propositional letters: defined in Section 2.3 on page 77.

E.-E. Doberkat, *Stochastic Coalgebraic Logic*, Monographs in Theoretical Computer Science. An EATCS Series, DOI 10.1007/978-3-642-02995-0,

Hennessy-Milner Logic: $\mathfrak{L}(\mathsf{Act}, \mathbb{Q} \cap [0,1])$

The formulas for $\mathfrak{L}(\mathsf{Act}, \mathbb{Q} \cap [0,1])$ are given through

$$\top \mid \phi_1 \wedge \phi_2 \mid \langle a \rangle_r \phi,$$

$a \in \mathsf{Act}$ is an action, $r \in [0,1] \cap \mathbb{Q}$: Section 2.3.1 on page 79.
Variant $\mathfrak{L}(\mathsf{Act}, [0,1])$: r may take arbitrary real values in the unit interval $[0,1]$; the variant defined in Section 2.6 on page 100, see also page 133. The discussion in Section 3.4 is based on it.

Continuous Time Stochastic Logic With Fixed-Point Operators: μCSL(AP, SV, PV)

This logic is usually abbreviated as μ**CSL**.

- *State formulas*:

$$\phi ::= \top \mid a \mid Z \mid \neg\phi \mid \phi \wedge \phi' \mid \mathcal{S}_{\ltimes p}(\phi) \mid \mathcal{P}_{\ltimes p}(\psi).$$

  $a \in \mathsf{AP}$ is an atomic proposition, $Z \in \mathsf{SV}$ is a state variable, ψ is a path formula, $\ltimes \in \{<, \leq, >, \geq\}$ is a relational operator, and $p \in [0,1]$ is a rational number.
- *Path formulas*:

$$\psi ::= P \mid \neg\psi \mid \psi \wedge \psi' \mid \mathcal{X}^I \psi \mid \phi\, \mathcal{U}^I \phi' \mid \mu P.\psi.$$

  $P \in \mathsf{PV}$ is a path variable, ϕ, ϕ' are state formulas, $I \subseteq \mathbb{R}_+$ is a closed interval with rational bounds.

  These operators are available in μ**CSL**:

1. $\mathcal{S}_{\ltimes p}(\phi)$: steady state probability.
2. $\mathcal{P}_{\ltimes p}(\psi)$: path quantifier.
3. $\mathcal{X}^I \phi$: next operator; I is an interval.
4. $\phi_1 \mathcal{U}^I \phi_2$: until-operator; I is an interval.

The syntax von μ**CSL** is defined in Section 2.4.1 on page 84.

Coalgebraic Logic (Left Case):$\mathfrak{L}^\flat(\Lambda, \Theta, \Gamma)$

Λ and Θ are at most countable sets of predicate liftings resp. $\mathfrak{T}$-predicate liftings (Definition 4.2.2). $\mathfrak{L}^\flat = \mathfrak{L}^\flat(\Lambda, \Theta, \Gamma)$ is defined through

- *State formulas*: $\phi ::= \top \mid \phi_1 \wedge \phi_2 \mid \langle\lambda\rangle\phi \mid \langle\vartheta\rangle\psi$ with $\lambda \in \Lambda$ a predicate lifting and $\vartheta \in \Theta$ a $\widetilde{\mathfrak{T}}$-predicate lifting.
- $\mathfrak{T}$-*formulas*: $\psi ::= \widetilde{\top} \mid \psi_1 \wedge \psi_2 \mid b(\phi_1, \ldots, \phi_{\mathfrak{ar}(b)})$ with $b \in \Gamma$ a bridge operator with arity $\mathfrak{ar}(b)$ (Definition 4.2.6).

$\mathfrak{L}^\flat$ is defined in Section 4.3.2 on page 167.

The logic $\mathfrak{L}^\flat(\emptyset, \Theta, \Gamma)$ is discussed as an application of left coalgebras to modal logic in Section 4.3.4, it is given on page 182.

Coalgebraic Logic (Right Case): $\mathfrak{L}^\sharp(\Lambda, \mathsf{Br}, \mathsf{Inf}, \mathsf{V})$

Λ, Br and Inf are countable sets of $\mathfrak{T}^{\mathrm{R}}$-predicate liftings, bridge operators and infinitesimal operators (Definition 4.4.17), resp., V is a countable set of variables.

The formulas of the *(full) logic* $\mathfrak{L}^\sharp = \mathfrak{L}^\sharp(\Lambda, \mathsf{Br}, \mathsf{Inf}, \mathsf{V})$ are given through

$$\phi ::= \top \mid \phi_1 \wedge \phi_2 \mid \langle\lambda\rangle\phi \mid \beta(\phi_1, \ldots, \phi_{\partial(\beta)}) \mid x \mid \iota x.\phi$$

with $\lambda \in \Lambda$, $\beta \in \mathsf{Br}$, $x \in \mathsf{V}$ and $\iota \in \mathsf{Inf}$. The *basic* or *kernel logic* is the special case $\mathfrak{L}^\sharp_b = \mathfrak{L}^\sharp(\Lambda, \emptyset, \emptyset, \emptyset)$, given by the grammar

$$\phi ::= \top \mid \phi_1 \wedge \phi_2 \mid \langle\lambda\rangle\phi$$

for $\lambda \in \Lambda$. These logics are defined in Section 4.4.2 on page 193.

References

[1] M. Abramowitz and I. A. Stegun. *Handbook of Mathematical Functions*. Dover Publications, New York, 1965.

[2] W. Arveson. *An Invitation to C*-Algebra*. Graduate Texts in Mathematics. Springer-Verlag, New York, 1976.

[3] C. Baier, B. Haverkort, H. Hermanns, and J.-P. Katoen. Model-checking algorithms for continuous time Markov chains. *IEEE Trans. Softw. Eng.*, 29(6):524 – 541, June 2003.

[4] M. Barr and C. Wells. *Category Theory for Computing Science*. Les Publications CRM, Montreal, 1999.

[5] H. Bauer. *Wahrscheinlichkeitstheorie und Grundzüge der Maßtheorie*. Walter de Gruyter, Berlin, 1968.

[6] P. Billingsley. *Convergence of Probability Measures*. John Wiley & Sons, New York, 1st edition, 1968.

[7] P. Billingsley. *Convergence of Probability Measures*. John Wiley & Sons, New York, 2nd edition, 1999.

[8] P. Blackburn, M. de Rijke, and Y. Venema. *Modal Logic*. Number 53 in Cambridge Tracts in Theoretical Computer Science. Cambridge University Press, Cambridge, UK, 2001.

[9] P. Blackburn, J. van Benthem, and F. Wolter, editors. *Handbook of Modal Logic*, volume 3 of *Studies in Logic and Practical Reasoning*. Elsevier, Amsterdam, 2007.

[10] M. M. Bonsangue and A. Kurz. Duality of logics of transition systems. In V. Sassone, editor, *Proc. FOSSACS'05*, volume 3441 of *LNCS*, pages 455–469, 2005.

[11] F. Borceux. *Handbook of Categorical Algebra 1: Basic Category Theory*, volume 50 of *Encyclopedia of Mathematics and Its Applications*. Cambridge University Press, Cambridge, UK, 1994.

[12] F. Borceux. *Handbook of Categorical Algebra 2: Categories and Structures*, volume 51 of *Encyclopedia of Mathematics and Its Applications*. Cambridge University Press, Cambridge, UK, 1994.

[13] S. Burris and H. P. Sankappanavar. *A Course in Universal Algebra.* Springer-Verlag (The Millennium Edition), 1981.

[14] C. Castaing and M. Valadier. *Convex Analysis and Measurable Multifunctions.* Number 580 in Lect. Notes Math. Springer-Verlag, Berlin, Heidelberg, New York, 1977.

[15] C. Cîrstea and D. Pattinson. Modular construction on modal logics. In P. Gardner and N. Yoshida, editors, *Proc. CONCUR'04*, number 3170 in Lect. Notes Comp. Sci., pages 258 – 275, 2004.

[16] E. M. Clarke, O. Grumberg, and D. A. Peled. *Model Checking.* The MIT Press, Cambridge, MA, 1999.

[17] J. Desharnais, A. Edalat, and P. Panangaden. Bisimulation of labelled Markov-processes. *Information and Computation*, 179(2):163 – 193, 2002.

[18] E.-E. Doberkat. Eilenberg-Moore algebras for stochastic relations. *Information and Computation*, 204:1756 – 1781, 2006.

[19] E.-E. Doberkat. Semi-pullbacks for stochastic relations over analytic spaces. *Math. Struct. Comp. Sci.*, 15:647 – 670, 2005.

[20] E.-E. Doberkat. *Stochastic Relations. Foundations for Markov Transition Systems.* Chapman & Hall/CRC Press, Boca Raton, New York, 2007.

[21] E.-E. Doberkat. Stochastic coalgebraic logic: Bisimilarity and behavioral equivalence. *Ann. Pure Appl. Logic*, 155:46 – 68, 2008.

[22] E.-E. Doberkat. Stochastic relations: congruences, bisimulations and the Hennessy-Milner theorem. *SIAM J. Computing*, 35(3):590 – 626, 2006.

[23] E.-E. Doberkat. The converse of a probabilistic relation. *J. Logic and Algebraic Progr.*, 62(1):133 – 154, 2004.

[24] E.-E. Doberkat. Factoring stochastic relations. *Information Processing Letters*, 90(4):161 – 166, May 2004.

[25] E.-E. Doberkat. Kleisli morphisms and randomized congruences for the Giry monad. *J. Pure Appl. Alg.*, 211:638–664, 2007.

[26] E.-E. Doberkat. Erratum and Addendum: Eilenberg-Moore algebras for stochastic relations. *Information and Computation*, 206:1476 – 1484, 2008.

[27] E.-E. Doberkat. Semi-pullbacks and bisimulations in categories of stochastic relations. In *Proc. ICALP'03*, volume 2719 of *Lect. Notes Comp. Sci.*, pages 996 – 1007, Berlin, 2003. Springer-Verlag.

[28] E.-E. Doberkat. Lattice properties of congruences for stochastic relations. Technical Report 178, Chair for Software Technology, Technische Universität Dortmund, March 2009.

[29] E.-E. Doberkat and Ch. Schubert. Coalgebraic logic for stochastic right coalgebras. *Ann. Pure Appl. Logic*, 159:268–284, 2009.

[30] L. Dubins and D. Freedman. Measurable sets of measures. *Pac. J. Math.*, 14:1211 – 1222, 1964.

[31] A. Edalat. Semi-pullbacks and bisimulation in categories of Markov processes. *Math. Struct. Comp. Science*, 9(5):523 – 543, 1999.

[32] G. A. Edgar. *Integral, Probability, and Fractal Measures.* Springer-Verlag, New York, 1998.
[33] R. Engelking. *General Topology*, volume 6 of *Sigma Series in Pure Mathematics.* Heldermann-Verlag, Berlin, revised and completed edition edition, 1989.
[34] M. Fisz. *Wahrscheinlichkeitsrechnung und Mathematische Statistik.* VEB Deutscher Verlag der Wissenschaften, Berlin, 1971.
[35] M. Fitting. Modal proof theory. In P. Blackburn, J. van Benthem, and F. Wolter, editors, *Handbook of Modal Logic*, volume 3 of *Studies in Logic and Practical Reasoning*, pages 85–138. Elsevier, Amsterdam, 2007.
[36] D. H. Fremlin. *Measure Theory*, volume 1 – 4. Torres Fremlin, Colchester, 2000 – 2003.
[37] E. Gamma, R. Helm, R. Johnson, and J. Vlissides. *Design Patterns — Elements of Reusable Object-Oriented Software.* Professional Computing Series. Addison-Wesley, Reading, Mass., 1994.
[38] M. Giry. A categorical approach to probability theory. In *Categorical Aspects of Topology and Analysis*, number 915 in Lect. Notes Math., pages 68 – 85, Berlin, 1981. Springer-Verlag.
[39] R. Goldblatt. Deduction systems for coalgebras over measurable spaces. *Journal of Logic and Computation*, (in print), 2008.
[40] R. Goldblatt. *Mathematics of Modality.* Number 43 in CSLI Lecture Notes. CSLI Publications, Stanford, 1993.
[41] I. P. Goulden and D. M. Jackson. *Combinatorial Enumeration.* Wiley-Interscience Series in Discrete Mathematics. John Wiley & Sons, New York, 1983.
[42] G. Grätzer. *Universal Algebra.* The University Series in Higher Mathematics. Van Nostrand, Princeton, N.J., 1968.
[43] P. R. Halmos. *Measure Theory.* Van Nostrand Reinhold, New York, 1950.
[44] M. Hennessy and R. Milner. On observing nondeterminism and concurrency. In *Proc. ICALP'80*, number 85 in Lect. Notes Comp. Sci., pages 395 – 409, Berlin, 1980. Springer-Verlag.
[45] E. Hewitt and K. R. Stromberg. *Real and Abstract Analysis.* Springer-Verlag, Berlin, Heidelberg, New York, 1965.
[46] C. J. Himmelberg and F. Van Vleck. Some selection theorems for measurable functions. *Can. J. Math.*, 21:394 – 399, 1969.
[47] K. Hinderer. *Foundations of Non-stationary Dynamic Programming with Discrete Time Parameter.* Number 33 in Lect. Notes Op. Res. Math. Syst. Springer-Verlag, Berlin, 1970.
[48] C. Jones and G. Plotkin. A probabilistic powerdomain of evaluations. In *Proc. LICS'89*, pages 186 – 195, IEEE Computer Society Press, 1989.
[49] A. S. Kechris. *Classical Descriptive Set Theory.* Graduate Texts in Mathematics. Springer-Verlag, Berlin, Heidelberg, New York, 1994.
[50] J. Keener and J. Sneyd. *Mathematical Physiology.* Interdisciplinary Applied Mathematics. Springer-Verlag, 2004.

[51] K. Keimel. The monad of probability measures over compact ordered spaces and its Eilenberg-Moore algebras. Preprint. TU Darmstadt, July 2008.

[52] D. E. Knuth. *The Art of Computer Programming. Vol. I, Fundamental Algorithms.* Addison-Wesley, Reading, Mass., 2 edition, 1973.

[53] C. Kupke, A. Kurz, and Y. Venema. Stone coalgebras. *Theor. Comp. Sci.*, 327(1 - 2):109 – 134, 2004.

[54] C. Kupke, A. Kurz, and Y. Venema. Completeness of the finitary Moss logic. In C. Areces and R. Goldblatt, editors, *Adv. Mod. Logic AiML'08*, volume 193 – 218. College Publications, 2008.

[55] K. Kuratowski. *Topology*, volume I. PWN – Polish Scientific Publishers and Academic Press, Warsaw and New York, 1966.

[56] A. Kurz. Specifying coalgebras with modal logic. *Theor. Comp. Sci.*, 260:119 – 138, 2001.

[57] S. Mac Lane. *Categories for the Working Mathematician.* Graduate Texts in Mathematics. Springer-Verlag, Berlin, 1997.

[58] K. G. Larsen and A. Skou. Bisimulation through probabilistic testing. *Information and Computation*, 94:1 – 28, 1991.

[59] G. W. Mackey. Borel structure in groups and their duals. *Trans. Am. Math. Soc.*, 85:134 – 165, 1957.

[60] E. Moggi. Notions of computation and monads. *Information and Computation*, 93:55 – 92, 1991.

[61] L. M. Moss. Coalgebraic logic. *Ann. Pure Appl. Logic*, 96:277 – 317, 1999.

[62] L. M. Moss and I. D. Viglizzo. Harsanyi type spaces and final coalgebras constructed from satisfied theories. *ENTCS*, pages 279 – 295, 2004.

[63] L. M. Moss and I. D. Viglizzo. Final coalgebras for functors on measurable spaces. *Inf. Comput.*, 204:610–636, 2006.

[64] P. Panangaden. Probabilistic relations. In C. Baier, M. Huth, M. Kwiatkowska, and M. Ryan, editors, *Proc. PROBMIV*, pages 59 – 74, 1998.

[65] K. R. Parthasarathy. *Probability Measures on Metric Spaces.* Academic Press, New York, 1967.

[66] D. Pattinson. Expressive logics for coalgebras via terminal sequence induction. *Notre Dame J. Formal Logic*, 45(1):19 – 33, 2004.

[67] D. Pumplün. Positively convex modules and ordered normed linear spaces. *J. Convex Analysis*, 10(1):109 – 127, 2003.

[68] D. Pumplün. *Elemente der Kategorientheorie.* Spektrum Akademischer Verlag, Heidelberg, 1999.

[69] B. v. Querenburg. *Mengentheoretische Topologie.* Springer -Lehrbuch. Springer-Verlag, Berlin, 3rd edition, 2001.

[70] U. Rieder. Bayesian dynamic programming. *Adv. Appl. Prob.*, 7:330 – 348, 1975.

[71] W. Rudin. *Real and Complex Analysis.* Tata McGraw-Hill, 2nd edition, 1974.

[72] J. J. M. M. Rutten. Universal coalgebra: a theory of systems. *Theor. Comp. Sci.*, 249(1):3 – 80, 2000. Special issue on modern algebra and its applications.
[73] S. Schneider and S. Thomas. Countable Borel equivalence relations. Lecture Notes, Appalachian Set Theory Seminar, Athens, OH, November 2007.
[74] L. Schröder. Expressivity of coalgebraic modal logic: the limits and beyond. *Theor. Comp. Sci.*, 390:230–247, 2008.
[75] L. Schröder and D. Pattinson. Modular algorithms for heterogeneous modal logics. In *Proc. ICALP*, number 4596 in Lect. Notes Comp. Sci., pages 459–471, 2007.
[76] Ch. Schubert. Coalgebraic modal logic over analytic spaces. Technical Report 170, Chair for Software Technology, Technical University of Dortmund, Dezember 2007.
[77] A. N. Shiryaev. *Probability*, volume 95 of *Graduate Texts in Mathematics*. Springer-Verlag, Berlin, Heidelberg, New York, second edition, 1996.
[78] A. Sokolova. *Coalgebraic Analysis of Probabilistic Systems*. PhD thesis, Department of Computer Science, University of Eindhoven, 2005.
[79] S. M. Srivastava. *A Course on Borel Sets*. Graduate Texts in Mathematics. Springer-Verlag, Berlin, 1998.
[80] Y. Sun. Economics and nonstandard analysis. In P. A. Loeb and M. Wolff, editors, *Nonstandard Analysis for the Working Mathematician*, Mathematics and Its Applications, pages 259 – 305. Kluwer Academic Publishers, Dordrecht, 2000.
[81] I. D. Viglizzo. Final sequences and final coalgebras for measurable spaces. In *Proc. CALCO*, number 3629 in Lect. Notes Comp. Sci., pages 395 – 407. Springer-Verlag, 2005.
[82] D. H. Wagner. A survey of measurable selection theorems. *SIAM J. Control Optim.*, 15(5):859 – 903, August 1977.
[83] H. S. Wilf. *generatingfunctionology*. Academic Press, Boston, 1990.
[84] D. J. Wilkinson. *Stochastic Modelling for Systems Biology*. Mathematical and Computational Biology Series. Chapman & Hall/CRC–Press, Boca Raton, New York, 2006.
[85] J. Worrell. On the final sequence of a finitary set functor. *Theor. Comp. Sci.*, 338(1–3):184–199, 2005.
[86] C. Zhou. *Complete Deductive Systems for Probabilistic Logic with Application to Harsany Type spaces*. PhD thesis, Department of Mathematics, University of Indiana, 2007.

List of Symbols

Index

MIX
Papier aus verantwortungsvollen Quellen
Paper from responsible sources
FSC® C105338

If you have any concerns about our products,
you can contact us on
ProductSafety@springernature.com

In case Publisher is established outside the EU,
the EU authorized representative is:
Springer Nature Customer Service Center GmbH
Europaplatz 3, 69115 Heidelberg, Germany

Printed by Libri Plureos GmbH
in Hamburg, Germany